GREEN PROFITS

The Manager's Handbook for ISO 14001 and Pollution Prevention

Nicholas P. Cheremisinoff, Ph.D.
Avrom Bendavid-Val

Boston Oxford Auckland Johannesburg Melbourne New Delhi

 A member of the Reed Elsevier Group

 Recognizing the importance of preserving what has been written, Butterworth-Heinemann prints its books on acid-free paper whenever possible.

Library of Congress Cataloging-in-Publication Data

Cheremisinoff, Nicholas P.
Green Profits: the manager's handbook for ISO 14001 and pollution prevention /
Nicholas P. Cheremisinoff, Avrom Bendavid-Val.
p. cm.
Includes indexes.
ISBN 0-7506-7401-6 (alk. paper)
1. ISO 14000 Series Standards. 2. Manufacturing industries—Environmental aspects. I. Bendavid-Val, Avrom. II. Title.

TS155.7.C454 2001
658.4;08—dc21

2001025366

British Library of Congress Cataloging-in-Publication Data
A catalogue record for this book is available form the British Library.

The publisher offers special discounts on bulk orders of this book.
For information, please contact:
Manager of Special Sales
Butterworth-Heinemann
225 Wildwood Avenue
Woburn, MA 01801-2041
Tel: (781) 904-2500
Fax: (781) 904-2620

For information on all Butterworth-Heinemann publications available, contact our World Wide Web home page at: http://www.bhusa.com

10 9 8 7 6 5 4 3 2 1

CONTENTS

PREFACE

Environmental management systems, or EMSs, are planned and organized means that an enterprise can use to manage its interactions with the environment — in particular, those interactions that contribute to resource consumption, environmental degradation, and human health risk. An EMS is a structured program of continual environmental improvement that follows procedures drawn from established business management practices. EMS concepts are straightforward, and EMS principles can be easily applied if supported by top management.

The best known EMS worldwide is ISO 14001. ISO 14001 is an international standard for EMSs that embodies essential elements of other EMSs, such as Britain's BS7750 and the European Union's EMAS, and clearly is becoming universally recognized among industry and the public as *the* standard for an EMS.

A proper EMS is a program of continual improvement in environmental performance that follows a sequence of steps based on established sound business-management practice. Accordingly, it allows an enterprise to understand and track its environmental performance, and provides a framework for identifying and carrying out improvements that may be desirable for financial or other corporate reasons, or that may be required to meet regulatory requirements. Under the best conditions, an EMS is built on an existing quality-management system. A typical sequence of planning and management steps that structure an EMS is:

1) Establish an environmental policy for the enterprise
2) Review the environmental consequences of enterprise operations
3) Establish environmental performance goals
4) Develop and carry out an action plan to achieve the goals
5) Monitor performance
6) Report the results
7) Review the system and the outcomes and strive for continual improvement

As this sequence of steps suggests, perhaps the most important characteristic of an EMS, and one that can yield significant long-term benefits, is that it represents a *systematic strategic management approach* through which an enterprise deals with environmental issues. The opposite of strategic management is reactive decision-making — when something breaks, fix it; when markets fail, scramble for new ones; when the public clamors for greater corporate responsibility, figure out what to do. Winners in the marketplace today are increasingly those with good strategic planning and management systems.

Environmental regulators — certainly in Western countries — are slowly but definitely moving from media-based to systems-based compliance enforcement approaches that reward enterprises for engaging in strategic environmental management through an EMS. For recent evidence of this see the U.S. Environmental Protection Agency's new National Environmental Performance Track program (www.epa.gov/perftrac/program/program.html).

The public also is demanding that corporations deal with environmental, quality, health and safety, and social-justice issues in a systematic, integrated, and strategic way, under the overall heading of "corporate social accountability" or similar terms.

Already, corporations at the leading edge are exploring ways to integrate their quality, environmental, and occupational health and safety (OHS) management systems.

System-based strategic management is *good* management, and a sound EMS complements and reinforces a systematic strategic approach to managing an enterprise. An EMS helps management identify and address environmental priorities in terms of their ecological and health implications, as well as in terms of the strategic requirements of the enterprise, rather than just checking and enforcing compliance with a collection of unrelated and unprioritized standards. This makes an EMS an instrument for promoting enterprise sustainability and long-term prosperity.

A key feature of ISO 14001 is its specification of the elements of a system that can be independently audited and certified as conforming to an internationally accepted EMS standard. The question of certification underlies much of the discussion about environmental management systems, and the real or hoped-for benefits of certification have provided the principal motivation for implementing ISO 14001 in many enterprises. But the benefits associated with certification represent only a small portion of overall benefits gained by implementing an EMS. Enterprises need not focus on certification when they begin the process of implementing an ISO 14001-based EMS. An enterprise can use ISO 14001 either as guidance for implementing a very simple EMS as the first step on a long road that may ultimately involve seeking certification, or as a set of precise specifications for setting up an EMS that can be certified relatively soon.

The greatest direct benefits to an enterprise of implementing an EMS usually come from the savings generated by pollution prevention (P2) practices and technologies. About 50 percent of the pollution generated in a typical "uncontrolled" plant can be prevented with minimal investment by adopting simple and inexpensive process improvements. In industrial countries, increased discharge fees and waste-disposal charges provide much of the incentive for cost-effective pollution reduction. A major consequence of implementing an EMS is the identification of waste-minimization and cleaner production opportunities for the enterprise. The process of introducing the EMS can be a catalyst for worker support for environmental-performance improvements (including the simple changes that make up "good housekeeping"), and also for making the best use of existing P2 and control equipment.

Pollution prevention diverts attention away from pollution controls and post-treatment practices (that is, treatment of pollution after it has been generated), and focuses instead on replacing technologies that generate pollution and undesirable by-products with those that do not, or that generate them at greatly reduced levels. Ultimately, P2 through waste minimization and cleaner production is more cost-effective and environmentally sound than traditional pollution control and post-treatment methods. P2 techniques apply to any manufacturing process or business operation, and range from relatively easy operational changes and good housekeeping practices to more extensive changes, such as finding substitutes for toxic substances, investing in clean technologies, and using state-of-the-art materials-recovery equipment. Pollution prevention can improve plant efficiency, enhance the quality and quantity of natural resources for production, and make it possible to invest more financial resources into enterprise competitiveness. To illustrate:

- The EPA, through its voluntary *Green Lights* program, introduced the concept of using low life-cycle-cost lighting systems rather than the lowest first-cost lighting systems. Pollution prevention activities under the energy-efficiency program include reducing wattage per square foot; changing from incandescent to fluorescent bulbs; changing ballast materials; using sodium- and mercury-vapor lamps; and installing motion detectors to detect when an area is occupied, and activate lights only then. Based on a survey of 200 participating companies, on average each enterprise was able (through decreased electricity demand) to reduce their emissions of carbon dioxide (CO_2) by 826 metric tons (kkg) per year; to reduce their emissions of sulfur dioxide (SO_2) by 6.5 kkg per year; and to reduce their emissions of nitrogen oxides (NO_x) by 2.7 kkg per year. This is equivalent to burning about 2,100 barrels of oil per year. But the participants did much more than reduce air-pollution emissions. Over the life of the program, participants have saved more than 380 million kilowatt-hours annually. This is enough electricity to run 42,000 American households for a full year. The cost savings for the representative group of companies is impressive. The average total cost for such P2 activities per company was $245,550; the average annual savings per company was $113,431. In other words, on average, a one-time investment of $245,550 returned $113,341 for every year thereafter. Payback periods ranged from less than one year to more than four years *(source — EPA doc. # 742/96/002)*.

- Motorola Inc.'s Government Systems and Technology Group, in an effort to eliminate ozone-depleting chemicals (ODCs), developed a soldering process for circuit-board manufacturing that is so clean that the chemical fluxes normally used to remove oxides prior to soldering was no longer necessary. The fluxes leave corrosive residues, which must be removed with chemical rinses. Freon 113 and trichloroethane (TCA), both ODCs, were commonly used as part of these chemical-rinse activities. Old-style soldering machines use up to 8,000 pounds of cleaner per month (i.e., 48 tons of cleaners per year). The newer technology has eliminated the need for a rinsing stage and, therefore, has eliminated the use of Freon 113 and TCA cleaners — and their associated air emissions. Each machine that employs this new soldering process saves between $50,000 and $245,000 per year in chemical use alone. Conventional wave solder machines can be retrofitted for $40,000 to $100,000, thereby providing very attractive payback periods *(source – EPA doc. # 742/96/002)*.

- A recent study of the Polish chemical industry identified 43 pollution prevention projects among manufacturers of fertilizers, pesticides, synthetic rubber, plastics, dyes and pigments, coke chemicals, inorganic chemicals, and pharmaceuticals. P2 activities ranged from simple housekeeping practices, to materials substitutions, to technology changes. The table below summarizes the results of the study. Total investment for all 43 projects amounted to $1,439,075. *In just the first year,* these projects in combination yielded a return of $7,184,490 — or nearly 500 percent! The payback periods for these P2 investments ranged from less than 1 month to more than 2 years, and the projects will continue to generate savings in both pollution and money in the years to come. What is important to note from the table is that more than 60 percent of the cost savings were achieved by relatively low levels of investments (less than $50,000 per activity). Just as the first example illustrates, P2 is not necessarily a high capital investment. Instead, it can be a lot of common sense *(source: study performed by the author for the United State's Agency for International Development – USAID).*

Investment Threshold	Total emissions reduction, tons/year	Raw materials savings, tons/year	Cost savings, $/year
Low-cost/no-cost: $10,000 or less	455,578	832,034	$ 1,423,430
Moderate cost: $10,000–50,000	2,263	226,652	$ 2,898,060
Major investment: $50,000 or more	71,108	420,545	$ 2,863,000

The financial benefits of P2 to a company are most often derived from incremental savings over time, and from a number of successful projects. Pollution prevention is every bit a corporate religion that must be adopted throughout an organization — starting with top management and working its way throughout the ranks of a company. The concept of eliminating or reducing polluting wastes (as opposed to controlling them or treating them after the fact) must be practiced on a daily basis, in the same way that an enterprise would approach building quality-management principles and philosophy into its products and services. Enterprises must also carefully monitor the financial performance of P2 projects, and take corrective actions as needed. For this reason, P2 practices provide the best chances for success and profitability when they are implemented through an overall environmental management system.

We have argued that the greatest direct benefits of an EMS will be in the bottom-line benefits of P2 practices and technologies, and we have argued that P2 practices and technologies have their greatest power in the context of an EMS. That is why we — one of us an engineer, the other an economist and planner — decided that we needed to write a single volume that would provide managers with the basic understanding, approaches, tools, and techniques to pursue either EMS or P2 in their enterprises and, even more important, to pursue *both*, integrally. A review of the literature reveals that though the integral nature of P2 and EMS is widely recognized and understood, the two subjects are never covered together, at least not in any depth. But they need to be covered together, to a reasonable depth, if the needs and convenience of managers and practitioners concerned with enterprise environmental performance are to be served well. They also need to be covered together to help enterprises capture the full financial benefits EMS and P2 can deliver in combination.

This volume is structured in two parts. Part I is about how an EMS works, particularly an EMS based on the ISO 14001 standard. It's also about how an enterprise can establish an EMS that will serve it well, in light of its particular circumstances and requirements. Part II is about P2 principles, practices, and the tools for implementing P2 activities — auditing, energy and material balances, and methods of calculating costs and returns of individual P2 investment and of optimizing P2 investments overall. Both parts are organized into four chapters.

In the EMS presentation in Part I, the chapter titles are:

Chapter 1) EMS: Principles and Concepts
Chapter 2) EMS: Applied Models
Chapter 3) EMS: Tools and Techniques
Chapter 4) EMS: First Steps

This represents a progression that begins with a sweeping description of EMSs in general, and ISO 14001 in particular, and then goes to a detailed element-by-element discussion of EMS and ISO 14001. It then examines a catalog of practical implementation tools, and concludes by discussing how an enterprise can take the first steps toward establishing its own EMS.

In the P2 presentation in Part II, the chapter titles are:

Chapter 5) Pollution Prevention: Principles and Concepts
Chapter 6) Industry-Specific Pollution Prevention Practices
Chapter 7) The Pollution Prevention Audit
Chapter 8) Financial Planning Tools

This represents a progression that begins with a sweeping description of P2; then goes to detailed real-life examples of P2 practices; then to a step-by-step guide to conducting a P2 audit to identify real P2 opportunities in an enterprise; and finally to a catalog of tools for assessing potential P2 investments from a bottom-line point of view.

Part I provides what managers need to embark on EMS implementation in their enterprises, including a thorough understanding of the requirements of the international ISO 14001 EMS standard. Readers who were hoping for a step-by-step handbook that shows exactly what to do or, in effect, does it for you by providing boilerplate documentation and procedural prescriptions to achieve ISO 14001 certification need to know that, despite any claims to the contrary, there simply can be no such thing. Every enterprise is unique; the idea of a one-size-fits-all template for establishing an EMS is one that will occur only to a person who thinks of an EMS as a set of fixed requirements rather than as an environmental management system that needs to be tailored to the circumstances and preferences of each enterprise.

Though Part I provides a thorough and practical understanding of the elements of EMSs, and of the tools and techniques needed to implement an EMS, and help with getting the implementation process started, managers will still have to bring their own creativity, commitment, effort, and knowledge of their enterprise and its operations to bear on the process of establishing an EMS. This is true whether they are establishing their own unique model of EMS, establishing one that conforms fully to the requirements of ISO 14001 and will be able soon to pass a certification audit, or establishing one that will conform only partially to the requirements of ISO 14001 while aiming to gradually expand to full conformance and certification. Whatever the case, managers of enterprises will have to invent their own "applied model" — that is, invent their own versions of some of the tools and techniques provided here — and develop their own unique environmental policy statements, environmental procedures, monitoring systems, and the like. Part I provides the essentials of how to implement an

EMS; it certainly gives managers what they need to get the process started and it equips them to hire an EMS implementation consultant and supervise him or her knowledgeably. The aim of Part I is not to provide a substitute for EMS capabilities that an enterprise lacks. The aim of Part I is to empower those involved with establishing and operating an enterprise's EMS.

In Part II, the reader will find useful tables and matrices offering proven P2 practices and technologies for specific industry sectors. No less important, Part II provides a concise approach, through the auditing method and the use of project financing tools, to implementing P2 practices step-by-step. We place heavy emphasis on simple but effective techniques for determining the economic viability of P2 projects, and on implementing P2 audits that can identify cost savings as well as pollution-reduction measures.

The reader will find ample industry examples and case studies in Part II that strengthen his or her understanding of P2 concepts, methods, and techniques. We describe techniques for financial analysis of P2 projects, and ways to calculate and demonstrate the importance of a P2 investment on a life-cycle or total-cost basis in terms of revenues, expenses, and profits, and give practical examples taken from industry. Our goal in this section is to equip readers not only with the ability to identify true P2 opportunities in their enterprises, but also with the ability to sell the P2 investment to top management in terms that are meaningful to them.

Our discussions in Part II apply to different levels of P2 practices or investments. Some industry sectors require high-tech solutions, and the P2 investments are substantial. The electric utility industry is one example. If we look at the investments required to convert older coal-fired plants to natural gas, investments in P2 technologies are staggering, with a single low-NO_x gas turbine costing tens of millions of dollars. These levels of investments may or may not carry with them immediate returns. On the other hand, an old coal-fired plant can present a range of inexpensive to moderately expensive P2 opportunities, which could allow for short-term investment strategies that enable savings to accrue over time. These savings could then defray costs of the larger-scale investments. For P2 programs to offer enterprises the opportunity to optimize their investments and savings while achieving environmental goals, they have to examine several levels of investment. That's why Part II of this book focuses on identifying and optimizing P2 investment opportunities.

Part II is followed by an appendix titled "Additional Resources," which lists print publications and Web sites first for ISO 14001-related resources, and then for those related to P2. These additional sources of information are integral to getting the most out of this book, and for this reason we reference them often throughout its chapters. Two additional resources are provided to assist the reader. First, at the beginning of the book the reader will find an extensive list of abbreviations that are referred to throughout the volume. Second, the authors have provided, in addition to a subject index, an index to the many tables that we use throughout the volume. The index to tables can be found immediately following the appendix.

Nicholas P. Cheremisinoff
Avrom Bendavid-Val
Washington, DC

Acknowledgements

Special thanks to Michael Forster and the others at Butterworth-Heinemann who helped bring this book to light; and to Todd Bernhardt, of Enterworks, Inc. for bringing his amazing editing talents to bear on the manuscript. Finally, a special thanks to Tatyana Davletshina for final proof readings of the manuscript.

We'd like also to express our gratitude to the following individuals and organizations for their many and diverse contributions to this work:

- Peter Bittner, Chemonics International Inc., Washington, D.C.
- Angela Crooks, U.S. Agency for International Development, Washington, D.C.
- David Gibson, Chemonics International Inc., Washington, D.C.
- Alex Keith, Chemonics International, Inc., Washington, D.C.
- Kevin Kelly, Chemonics International Inc., Washington, D.C.
- Henry Koner, Chemonics International, Inc., Washington, D.C.
- Dr. Svatislov Kurulenko, Ministry of Environmental Protection & Nuclear Safety, Ukraine
- Sergey V. Makarov, Mendeleev University of Chemical Technology, Moscow
- Dan Marsh, Chemonics International Inc., Washington, D.C.
- Jennifer McGuinn, Chemonics International Inc., Washington, D.C.
- William Moore, Edelman Communications International, Washington, D.C.
- Ashraf Rizk, Chemonics International Inc., Washington, D.C.
- John Shideler, Futurepast: Inc., Washington, D.C.
- Thurston Teele, Chemonics International Inc., Washington, D.C.
- Robert Wilson, IQuES, LLC., Detroit
- CAST SA, Bucharest
- CityProf Consulting, Krakow
- Energo-Sistem, Skopje
- POVVIK-EP, Sofia
- Russian Engineering Academy, Volga District, Samara
- The many participants in EMS/ISO 14001 implementation and internal auditor courses conducted by Chemonics International Inc. in Russia, Eastern Europe, Asia, and the Middle East.

Abbreviations

A

AATCC	American Association of Textile Chemists and Colorists
ABS	acrylonitrile butadiene styrene
acfm	actual cubic feet per minute
ACM	asbestos-containing materials
ACRS	accelerated cost recovery system
ACS	American Chemical Society
ADP	air-dried pulp
AHE	acute hazards event
AIRS	aerometric information retrieval system
AMD	acid mine drainage
AN	ammonium nitrate
API	American Petroleum Institute
ASN	ammonium sulfate nitrate

B

BAT	best available technology
BATNEEC	best available technology not entailing excessive cost
BB	butane/butylene
B/C	benefit-to-cost ratio
BIFs	boilers and industrial furnaces
BOD	biochemical oxygen demand
BOF	basic oxygen furnace
BPT	best practicable technology
BS	black smoke or British smokeshade method
BS	British Standard

C

CAA	Clean Air Act
CAAA	Clean Air Act Amendments
CAN	calcium ammonium nitrate
CCD	continuous countercurrent decanting
CERCLA	Comprehensive Environmental Response, Compensation, and Liability Act
CFCs	chlorofluorocarbons
cfm	cubic feet per minute
CFR	Code of Federal Regulations
CI	color index
COD	chemical oxygen demand
COG	coke oven gas
CP	cleaner production

CPI	chemical process industry
cpm	cycles per minute
CSM	continuous stack monitoring
CTC	carbon tetrachloride
CTMP	chemithermomechanical pulping
CTSA	Cleaner Technology Substitute Assessment
CWA	Clean Water Act
D	
DAF	dissolved air flotation
DAP	diammonium phosphate
DDT	dichlorodiphenyltrichloroethane
DEA	diethanolamine
DIPA	di-isopropanolamine
DMT	dimethyl terephthalate
DO	dissolved oxygen
E	
EA	environmental assessment
EAF	electric arc furnace
ECF	elemental chlorine-free (bleaching)
EIA	environmental impact assessment
EIS	environmental impact statement
EMAS	Eco-Management and Auditing Scheme
EMS	environmental management system
emf	electromotive force
EMS	environmental management systems
EPA	Environmental Protection Agency
EPCRA	Emergency Planning and Community Right-to-Know Act
EPT	Environmental Policy and Technology
ERNS	Emergency Response Notification System
ESP	electrostatic precipitator
EU	European Union
F	
FBC	fluidized bed combustion
FCC	fluidized catalytic cracking
FCCUs	fluidized-bed catalytic cracking units
FGD	flue gas desulfurization
FGR	flue gas recirculation
FGT	flue gas treatment
G	
GHG	greenhouse gas
GJ	gigajoule
GLPTS	Great Lakes Persistent Toxic Substances

GMP	Good Management Practices
gpm	gallons per minute
gr	grain
GW	gigawatt
GWP	global warming potential
H	
HAPs	hazardous air pollutants
HCFCs	hydrochlorofluorocarbons
HCN	hydrogen cyanide
HCs	hydrocarbons
HDPE	high density polyethylene
HEPA	high efficiency particulate air filter
HFC	hydrofluorocarbon
HSDB	Hazardous Substances Data Bank
HSWA	Hazardous and Solid Waste Amendments
I	
IAF	International Accreditation Forum
IARC	International Agency for Cancer Research
ID	identification
IER	Initial Environmental Review
IFC	International Finance Corporation
IPCC	Intergovernmental Panel on Climate Change
IQ	intelligence quotient
IRIS	integrated risk information system
ISO	International Organization for Standardization
L	
LAB	linear alkyl benzene
LCA	life cycle analysis
LCC	life cycle costing or life cycle checklist
LDAR	leak detection and repair
LDPE	low density polyethylene
LDRs	land disposal restrictions
LEA	low excess air
LEPCs	local emergency planning committees
LLDPE	linear low density polyethylene
LPG	liquefied petroleum gas
M	
MACT	maximum allowable control technology
MAP	monoammonium phosphate
MCLGs	maximum contaminant level goals
MCLs	maximum contaminant levels
MEK	methyl ethyl ketone

MIBK	methyl isobutyl ketone
MLAs	multilateral agreements
MMT	methylcyclopentadienyl magnesium tricarbonyl
MOS	metal oxide semiconductor
MSDSs	Material Safety Data Sheets
MTBE	methyltertbutylether
N	
NAAQSs	National Ambient Air Quality Standards
NCP	National Contingency Plan
NESHAPs	National Emission Standards for Hazardous Air Pollutants
NGO	non-governmental organizations
NGVs	natural gas vehicles
NIS	Newly Independent States of the former Soviet Union
NPDES	National Pollutant Discharge Elimination System
NPK	nitrogen-phosphorus-potassium
NPL	National Priority List
NRC	National Response Center
NSCR	nonselective catalytic reduction
NSPSs	New Source Performance Standards
O	
ODP	ozone depleting potential
ODSs	ozone depleting substances
OFA	overfire air
OSHA	Occupational Safety and Health Act
OHS	occupational health and safety
OTA	Office of Technology Assessment
OTC	over-the-counter (medicines)
P	
PAH	polynuclear aromatic hydrocarbons
P2M	pollution prevention matrix
P2	pollution prevention
P3	pollution prevention practices
P3AW	Pollution Prevention Project Analysis Worksheet
PAHs	polynuclear aromatic hydrocarbons
PBR	polybutadiene rubber
PCB	polychlorinated biphenyls
PCE	perchloroethylene
PFA	pulverized fly ash
PICs	products of incomplete combustion
PM	particulate matter
PM_{10}	particulate matter smaller than 10 microns in size
PMN	premanufacture notice

POM	prescription only medicines
POTW	publicly owned treatment works
PP	propane/polypropylene
ppm	parts per million
ppb	parts per billion
ppmv	parts per million by volume
PUP	per unit of product
PVC	polyvinyl chloride
PVNB	present value of net benefits
PWB	printed wiring board
Q	
QA	quality assurance
QC	quality control
R	
R&D	research and development
RCRA	Resource Conservation and Recovery Act
ROI	return on investment
rpm	revolutions per minute
S	
SARA	Superfund Amendments and Reauthorization Act
SBR	styrene butadiene rubber
scfm	standard cubic feet per minute
SCF	supercritical cleaning fluid
SCR	selective catalytic reduction
SCW	supercritical fluid
SDWA	Safe Drinking Water Act
SERCs	State Emergency Response Commissions
SIC	Standard Industrial Code
SIP	state implementation plan
SMT	surface mount technology
SNCR	selective noncatalytic reduction
SS	suspended solids
SSP	single phosphate
T	
TAME	tertiary amyl methyl ether
TCA	total cost accounting or total cost analysis
TCE	1, 1, 1 -trichloroethane
TCF	total chlorine-free
TCLP	toxic characteristic leachate procedure
TDI	toluenediisocyanate
TEWI	total equivalent waming impact
tpy	tons per year

TQEM	Total Quality Environmental Management
TQM	Total Quality Management
TRI	Toxic Release Inventory
TRS	total reduced sulfur
TSCA	Toxic Substances Control Act
TSD	transport, storage and disposal
TSP	total suspended particulates
TSS	total suspended solids
U	
UIC	underground injection control
UNEP	United Nations Environmental Programme
USEPA	United States Environmental Protection Agency
USTs	underground storage tanks
UV	ultraviolet
V	
VAT	value added taxes
VCM	vinyl chloride monomer
VOCs	volatile organic compounds
W	
WBO	World Bank Organization
WEC	World Environmental Center
WHO	World Health Organization

Road Map to Part I: ENVIRONMENTAL MANAGEMENT SYSTEMS

Part I of this book is designed to provide what the reader needs to put an EMS in place. It includes information for convincing top management and others that establishing an EMS is in the best interests of the enterprise, and shows how an EMS fits in with and reinforces existing management systems. It provides the means for assessing what kind of EMS would be best for a particular enterprise, and shows how to take the first steps toward establishing one. It offers tools and techniques for fully implementing an EMS, whether it's a minimal EMS or one that will pass an ISO 14001 certification audit. And, most important, it aims to provide a real understanding of what an EMS is, how it works, how it makes or saves money for an enterprise, and how it can be adapted to the unique and changing needs of a particular enterprise.

Our purpose in the four chapters of Part I is to serve anyone with an interest in the subject of environmental management systems. We make the point (perhaps too often) that an EMS is not about ISO 14001 certification; rather, it's about running an enterprise better, more efficiently, more competitively, and more sustainably, and about making and saving money. Running an enterprise is a creative endeavor; accordingly, Part I is built on the assumption that readers want to understand what an EMS is all about and then apply themselves creatively to building one that operates efficiently and delivers huge benefits for their enterprises. Consequently, Part I does not dwell so much on the details of what different EMS procedures enterprises should include as it does on the knowledge and tools managers need to figure this out for themselves in the context of their enterprise's operations.

In some cases, top management may be interested for the time being only in experimenting lightly with a simple and limited EMS, perhaps to see if it offers a better way to ensure the enterprise's compliance with environmental regulations or to minimize the risk of environmental liabilities. At the other extreme, top management may be interested in achieving ISO 14001 certification in a few months. The material in Part I has been developed to serve those extremes, as well as everything between, while keeping in mind students and others who just want to be well-informed on the subjects of EMS and ISO 14001.

Chapter 1 starts with a short overview of the basic principles and concepts of EMSs, what ISO 14001 is all about, and the core elements of ISO 14001. As with all chapters in this book, the last section of Chapter 1 contains a brief review of the main points. At the end of the chapter are some questions for individual consideration or classroom discussion (other chapters in Part I have questions *and* exercises, or just exercises at the end). The reader who picks up this book only because he or she is wondering exactly what an EMS is or perhaps what ISO 14001 is, and no more, will have their curiosity completely satisfied by Chapter 1. If the chapter stimulates a deeper interest, or if for other reasons the reader wants to go beyond basic principles and concepts, he or she should move on to Chapter 2.

Chapter 2 fully explains:

- Each element of an EMS and what role it plays in the overall management system
- Each element's minimum requirements
- How each element relates to other EMS elements
- How each element can be adapted to the needs of different enterprises
- Exactly what ISO 14001 requires for each element
- What an enterprise needs to have in place to pass an ISO 14001 certification audit

Chapter 2 follows the outline of ISO 14001, which is organized as a series of clauses and subclauses that each represent a distinct element of the EMS. Except for the introductory and review sections, the sections of Chapter 2 correspond to the elements of ISO 14001. Though the material stands on its own, enterprises concerned with ISO 14001 certification will benefit most by reading Chapter 2 with a copy of the ISO 14001 standard at hand. The chapter is called "EMS: Applied Models" because it contains information that enables the reader to create an EMS model uniquely appropriate for the circumstances of his or her enterprise.

After reading a section of Chapter 2, some readers will want immediately to see the tools and techniques for implementing that EMS element. Those readers can jump to the counterpart section of Chapter 3 before going on to the next section of Chapter 2. For example, after reading the section of Chapter 2 on environmental policy, learning what an environmental policy is, and what ISO 14001 requires for the environmental policy of an enterprise, a reader may want to explore the processes and tools for establishing an environmental policy before moving on to an explanation of the next EMS element. The counterpart section of Chapter 3 will provide that information, and will be easy to find because the counterpart sections in the two chapters have exactly the same names. In this case, both are called "Environmental Policy (Clause 4.2)."

Chapter 3 presents tools and techniques for establishing and maintaining an EMS, and is organized the same as Chapter 2 — by ISO 14001 clauses and subclauses. If Chapter 2 is the overall "what and why" of each element of an EMS, Chapter 3 is the "how and what does it contain?" chapter. Chapter 3 offers lists, worksheets, outlines, tips, and so on — practical help for creating an EMS and maintaining it, whether it's an ISO 14001 or some other type of EMS. Readers who already have a full understanding of the details of an EMS, have already taken the first basic steps toward setting up an EMS in their enterprises, and want only to access concrete tools for establishing, maintaining, or upgrading their EMSs, can limit their reading to Chapter 3. However, we recommend reading Chapter 2 as well, because the spirit of the presentation in Chapter 3 flows from the presentation in Chapter 2, and there is some material in Chapter 2 that is equally relevant to Chapter 3 but is not repeated there. Material in counterpart sections of Chapter 2 is referenced frequently in Chapter 3, and knowledge of the former will help in the understanding of the latter.

Chapter 4 discusses what first needs to be done in an enterprise on the road to establishing an EMS — convince top management of the value of an EMS, figure out what sort of EMS is most appropriate and plan how to put it in place, and possibly experiment with a minimal EMS that can serve later as the foundation of a full-scale EMS based on the ISO 14001 standard. Readers who feel they fully understand what an EMS is and just want some help getting started can begin with Chapter 4, and next will want to read Chapter 3.

We have assembled the four chapters of Part I to address a sequence of four questions:

- What's an EMS in general, and specifically, what is ISO 14001?
- What are the individual elements of an EMS in general, and what are the ISO 14001 requirements for each of them?
- What are the tools and techniques for creating each EMS element in an enterprise?
- How do I get started?

This sequence of questions imagines that after the first three chapters a reader would conclude that his or her enterprise *should* and *could* establish an EMS, and at that point would want to know how to get started. But, as we have seen, the material in Part I can be approached from many different points of entry, depending on the background and needs of each reader and enterprise.

Throughout the chapters of Part I, we make frequent reference to the material on pollution prevention (P2) in Part II of this book. This is because P2 improvements in the production process of an enterprise are the ultimate focus of an EMS, and the bottom-line benefits of an EMS most commonly come from the P2 improvements it generates. The material in Part I is mostly of a planning and management nature, while the material in Part II is mostly of an engineering and financial nature. The presentations are different in approach, because the nature of the material and the backgrounds of the principal authors of each part are different, and yet the subjects fit together and reinforce each other in important ways. An EMS yields big payoffs only through P2 practices, and P2 practices and audits will have their greatest benefits for an enterprise in the context of an EMS. Sound engineering and financial calculations applied through a sound strategic-planning and management system is exactly what an enterprise needs to thrive in an increasingly competitive and environmentally demanding marketplace.

Chapter 1
EMS: PRINCIPLES AND CONCEPTS

THE BASICS OF AN EMS: WHAT IT IS, AND WHY DO IT

An environmental management system, or EMS, is an approach ... a tool ... a set of procedures ... a planned and organized way of doing things ... a *system.* It is any planning and implementation system that an enterprise employs to manage the way it interacts with the natural environment.

An EMS is built around the way an enterprise operates. It focuses on an enterprise's production processes and general management system — *not* on its emissions, effluents, and solid waste, as environmental regulations do. An EMS enables an enterprise to address major and costly aspects of its operations proactively, strategically, and comprehensively, as any good manager would want to do. Without an EMS, an enterprise can only *react* to environmental disasters ... to environmental regulations ... to threats of fines and lawsuits ... to being undercut by more progressive and efficient competitors.

An EMS is integrated into the overall management system of an enterprise. Like an overall management system, it represents a process of continual analysis, planning, and implementation; it requires that top management commit and organize such resources as people, money, and equipment to achieve enterprise objectives; and it requires that resources be committed to support the management system itself. Not too surprisingly, productive EMSs are found only in enterprises that are fairly well managed in general. In some cases, these enterprises adopted EMSs because they already had good management systems in place. In other cases, installing and maintaining an EMS led to better overall management because it showed the way to improve control over the enterprise's operations.

> ***Without an EMS, an enterprise can only react to environmental disasters ... to environmental regulations ... to threats of fines and lawsuits ... to being undercut by more progressive and efficient competitors.***

There are lots of types of EMSs around. Some are industry specific, with guidelines often issued by industry associations. Examples of this kind of EMS include the Forest Stewardship Council's *SmartWood* EMS for forest property and forest products; the World Travel and Tourism Council's *Green Globe 21* for the travel and tourism industry; and the U.S. Government's *Code of Environmental Management Principles* for federal agencies.

Many EMSs are uniquely designed for a particular facility: these range from a simple "plan-act-review-revise" model to the high-tech FEMMS, or Facility

Environmental Management and Monitoring System, of the Tobyhanna Army Depot (for more information, visit www.femms.com). And some EMSs are "global," meaning that they are meant to be very broadly applicable, at least across manufacturing enterprises. Two well-known examples of global EMSs are the British Standards Institution's BS 7750, which served as the point of departure for developing ISO 14001, and the European Union's EMAS, or Eco-Management and Auditing Scheme, which permits ISO 14001 to serve as its core EMS component.

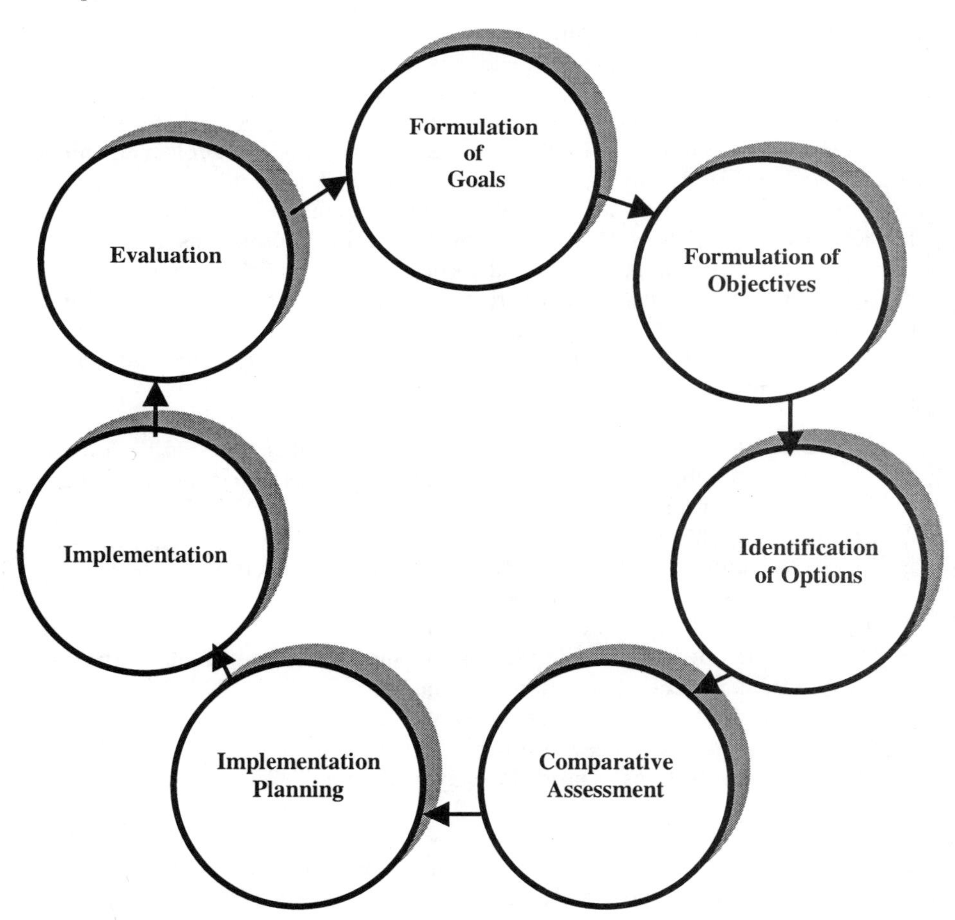

Figure 1. *A planning and implementation process.* (Adapted from: Bendavid-Val, Avrom, *Regional and Local Economic Analysis for Practitioners*, Praeger Publishers, New York, 1983 edition, p. 200.)

In its broadest outlines, an EMS is like any other system of managing of planning and implementing for continual improvement. The same basic steps apply to managing an enterprise, managing a production line, managing your commute to work, or even managing economic development. In other words, these basic steps are not new.

Figure 1, taken from a 1983 book on local economic development planning, supposedly illustrates the basic steps and the circular process used by local officials in planning for economic development. But Figure 1 really illustrates a local economic development *management* system: it represents a continuous planning *and* implementation process, grounded in the idea of continual improvement of the local economy.

Though the words are different, and the planning and implementation cycle is illustrated with fewer steps, the four-step EMS model mentioned earlier ("plan-act-review-revise") is essentially the same as the local economic development management system in Figure 1. "Plan" in the four-step model is broken down into five constituent steps in Figure 1: goals, objectives, options, comparative assessment, and planning. "Act" is called "implementation;" "review" translates into "evaluation;" and "revise" — which involves making corrections that feed into a further round of planning, acting, and so on — is illustrated in Figure 1 by the arrow showing the results of evaluation feeding into the next round of goal formulation.

Really, how could it be otherwise? Sound planning and implementing, in one form or another, necessarily involves:

1) Establishing an overall policy (broad goals, aims, mission, values) to guide everything that follows (this can be considered part of the planning activity or a step that comes before it)
2) Assessing the current situation
3) Determining exactly what you want to achieve (setting explicit goals, objectives, targets, performance standards)
4) Examining different ways of achieving it
5) Working out in detail what seems like the best course of action (type of program, project, plan, action plan, initiative)
6) Carrying out the plan (implementation)
7) Monitoring how things are going
8) Making corrections as needed to stay on course

If the planning and implementation is for a single purpose — say, a one-time initiative to reduce waste from a particular production line — then "making corrections as needed to stay on course" would be the last step. But if the planning and implementation is for continual improvement — say, for an enterprise's product quality, or efficiency or environmental performance — then the process is necessarily continuous; the process is a *management* system. In that case, each planning and implementation cycle leads into the next. The final step in the cycle involves a review of performance results and how well the management system itself is working. This information is fed into the next cycle of planning and implementation. In enterprise terms, the process can be summarized as: a) management direction, b) planning, c) implementing, d) monitoring and correcting, and e) reviewing and revising.

This brings us to ISO 14001. The basic steps in the ISO 14001 EMS are a) environmental policy, b) planning, c) implementation and operation, d) checking and corrective action, and e) management review. One could quibble with the choice of words and the particular way the steps are broken out, but what the authors of the ISO 14001 EMS did is quite remarkable: They took the familiar basic elements of any continuous planning and implementation system — of any management system — and adapted them to the needs of continually improving environmental performance in enterprises worldwide. They made ISO 14001 so broad that it is applicable to almost any organization of almost any size almost anywhere in the world ... yet it has 52 very specific requirements for procedures, actions, resources, and documentation. They built ISO 14001 on the proven general model for a sound enterprise management system so it would fit in easily with existing management systems and reinforce them, and so it would introduce principles of good management where they are lacking. In this way committees of the ISO formulated an international standard for an EMS.

> ***The basic steps in the ISO 14001 EMS are:***
> ***a) Environmental policy***
> ***b) Planning***
> ***c) Implementation and operation***
> ***d) Checking and corrective action***
> ***e) Management review***

Here we are interested in a general EMS for enterprises. We will use ISO 14001 as the reference model because it contains the core elements of any proper EMS; because it is an international EMS standard; because though it has its detractors and will surely evolve over the years, it is being accepted more and more widely throughout the world as *the* EMS standard; and because it is meant to be enterprise friendly.

The aim in this and the next three chapters of this book is to show how enterprises can implement an EMS to reap the benefits of reduced manufacturing costs, greater efficiency, higher product quality, and improved control over production processes. These basic benefits accrue from putting in place and maintaining any sound EMS, with or without certification — not just ISO 14001. Nevertheless, in the course of these chapters readers will learn both the essentials of any EMS and the specifics of an EMS that certifiably conforms to the requirements of the ISO 14001 standard.

WHAT THE ISO 14001 STANDARD IS ALL ABOUT

The International Organization for Standardization, based in Geneva, is a federation of national standards bodies representing about 130 countries. Its mission is to promote international standardization that will facilitate international exchange of goods and services, and to promote international cooperation in the spheres of intellectual, scientific, technological, and economic activity. The ISO was established in 1947 as a non-governmental organization, and since then has dealt with thousands of international standardization issues. All of us come into contact with the results of its work daily — the ISO has developed the international standards for such things as film-speed codes, the format of banking cards, freight containers, symbols for automobile controls, and screw threads. There is a reason why the International Organization for Standardization goes by the acronym "ISO," but we will leave it for readers to find elsewhere.

After a successful experience developing and encouraging international standards for quality management (the ISO 9000 series), in late 1996 the ISO issued the first few standards in the environmental management series — the ISO 14000 series. ISO 14001 is the ISO standard for EMSs, and this is the ISO 14000 standard that we are concerned with here. Other ISO 14000 standards, including some still under development, deal with such subjects as auditing EMSs, environmental considerations in life cycles of products, and environmental labeling.

One of the last steps in the ISO 14001 EMS cycle is a self-audit. When the self-audit confirms that your enterprise has "implemented" (has established and maintains) all the requirements of ISO 14001, you can do any or all of the following:

a) Be content and enjoy the benefits of a good EMS
b) Declare to the world that your enterprise conforms to the ISO 14001 standard ("conformance" to ISO 14001 is voluntary; "compliance" with environmental regulations is not voluntary)
c) Call that big, important client that demanded that you establish an EMS if you want to remain their supplier, and invite them to audit to establish to their satisfaction that your enterprise's EMS conforms to ISO 14001 (this is referred to as a "second-party" audit)
d) Engage an accredited certification company ("registrar") to field a team of professional ISO 14001 auditors to conduct a "third party" audit that, if successful, will result in your enterprise receiving formal certification of conformance to ISO 14001; your firm will be, as they say, "ISO 14000 certified."

The requirements of the ISO 14001 EMS standard are spelled out in a booklet published by the ISO called *ISO 14001: Environmental Management Systems — Specification with Guidance for Use*. You'll find that this is a 21-page booklet of which the first nine pages are introductory material, including definitions of terms. The last nine pages contain appendices with guidance on using the standard, and the middle three pages contain the actual requirements of the standard.

That's right — the 52 explicit requirements of ISO 14001 are detailed in just three pages, and even these three pages have a lot of white space. When you decide to purchase a copy of ISO 14001 (www.iso.ch), you would do well also to purchase a copy of *ISO 14004: Environmental Management Systems — General Guidelines on Principles, Systems, and Supporting Techniques*. This is a companion booklet to ISO 14001 that is meant, together with Appendix A of ISO 14001, to help users understand what is intended by some of the more cryptic passages in the standard, provide practical tips on implementation approaches, tools, and techniques for the standard, and give insights into what ISO 14001 auditors may be looking for during a certification audit. Though checking the implementation suggestions in ISO 14004 is a good idea, nothing in ISO 14004 is a requirement. It contains only what the ISO hopes will be helpful ideas — not "auditable" ideas — for implementing ISO 14001.

The requirements of the ISO 14001 EMS standard are spelled out in a booklet published by the ISO called **ISO 14001: Environmental Management Systems—Specification with Guidance for Use** *(www.iso.ch).*

The basic elements of ISO 14001 are summarized in Figure 2. They all appear in Chapter 4 of the standard (the part that is three pages long), grouped under "Clauses" 4.2 through 4.6 representing the five basic steps of the EMS. What happened to Clause 4.1? Clause 4.1 is titled, "General Requirements." It is three lines long and contains the first of the 52 explicit requirements of the ISO 14001 standard: the requirement that the enterprise's EMS conform to the other 51 requirements. More on this in Chapter 2.

Clause 4.2 of ISO 14001 requires that "top management" establish an overall environmental policy for the enterprise, and requires that the policy include certain commitments on the part of top management. This policy statement puts in writing the overall environmental aims and values that set the stage and establish the conditions for the rest of the EMS. Why insist on this initial step? Because many of the authors of the ISO 14001 standard came from business backgrounds, they knew that if top management is not prepared to make a serious written commitment to an EMS as the very first step in the process, it will be a waste of time to go any farther.

Clause 4.3, "Planning," includes assessing the current situation, both to identify aspects of the enterprise's operations that significantly affect the environment, and to inventory the enterprise's compliance with environmental regulations. Then, in light of the findings of the assessment, the EMS team works its way through a process of establishing environmental objectives and targets for the enterprise. Finally, programs ("action plans" might be a better term) are devised for achieving the objectives and targets. Implicit in this sequence of planning steps is a number of additional activities, including ranking the environmental impacts of the enterprise's operations in terms of significance (allowing priorities for addressing them to be set), and examining alternative approaches for achieving targets.

Clause 4.4, "Implementation and Operation," spells out the requirements for the capabilities and supporting systems that the enterprise must have to carry out its environmental programs and control its environmental performance. There are seven of these capabilities and requirements, as shown in Figure 2.

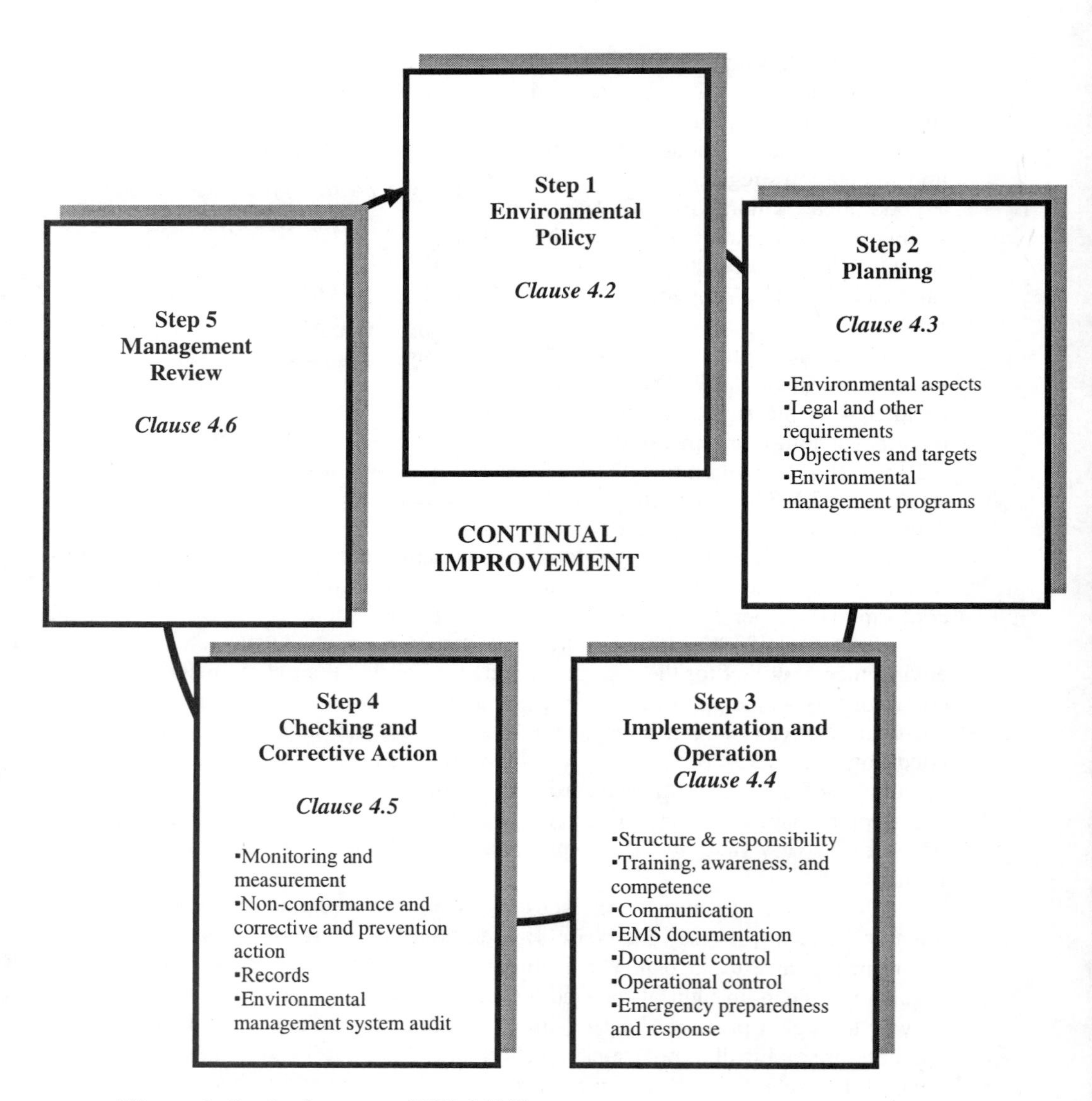

Figure 2. *Basic elements of ISO 14001.*

Clause 4.5, "Checking and Corrective Action," contains requirements for procedures to track progress toward environmental performance targets and, if necessary, for taking corrective action. It requires that proper records be kept, so there is a documented history of performance under the EMS that can be reviewed and analyzed as needed. Finally, it requires a periodic EMS audit, commonly referred to as an "internal audit," as the basis for evaluating and improving the

EMS itself. There are two areas of concern under this clause: the environmental performance of the enterprise, and effectiveness of the EMS as an instrument for improving environmental performance of the enterprise.

Clause 4.6, "Management Review," requires that top management periodically review the overall operation of the EMS. This clause also concerns itself with both the environmental performance of the enterprise under the EMS, and the performance of the EMS as a management system. Obviously, the EMS audit required under Clause 4.5 provides much of the information for the management review. Clause 4.6 of ISO 14001 requires top management to carefully consider every element of the EMS and consider changes that would improve it in light of its performance and changing conditions.

Many view the ISO 14001 EMS cycle as annual, consistent with the general business cycle. Under this view, the five basic steps represented in ISO 14001 Clauses 4.2 through 4.6 would be carried out more or less in sequence, over the course of a year. The management review and top management's decisions regarding changes in the EMS would mark the end of one annual EMS cycle, and the beginning of the next. But there is nothing that requires any EMS, including ISO 14001, to operate on an annual cycle. Depending on what's suitable for the size and complexity of a particular enterprise and the nature of its operations, the EMS cycle could range from six months to three years. In fact, as we work our way through the details of an ISO 14001-based EMS in the coming chapters of this book, it should become increasingly clear that the best form of EMS does not express itself as a sequence of discrete steps. Rather, the most sophisticated and fully developed form of EMS comprises a sequential *pattern* of steps that are all moving forward to one degree or another at the same time, with each step informing the next in the sequence more or less at all times. Environmental policy is frequently reconsidered more than once a year; constant monitoring means constant evaluating and corrective action; constant corrective action means constantly reconsidering the environmental objectives and targets of the enterprise and the programs devised to achieve them. Even an internal audit can be conducted several times a year; and if it's conducted only once a year, it does not have to be conducted all at one time.

ISO 14001 is an international standard for environmental management systems; it is* not *a standard for environmental performance.

One point about ISO 14001 that is very important to understand: ISO 14001 is an international standard for environmental management systems. It is *not* a standard for environmental performance. Conformance to the ISO 14001 standard is voluntary, and the ISO 14001 accreditation and certification mechanisms are (or should be) operated completely by nongovernmental organizations.

Compliance with legal environmental performance standards on the other hand, is a concern of governmental agencies, and is considered by the ISO 14001 standard to be a matter apart from voluntary conformance to ISO 14001. But not completely apart, because if an enterprise EMS truly conforms to the requirements and spirit of the ISO 14001 standard, its environmental performance and level of regulatory compliance (not to mention its business performance) will necessarily improve.

Still, the separation of responsibilities for ISO 14001 conformance on the one hand, and for regulatory compliance on the other, results in the awkward fact that an enterprise can be "ISO 14000 certified" but not in compliance with environmental regulations.

ESSENTIAL ELEMENTS OF THE ISO 14001 EMS

One could argue that it's not quite right to declare "essential" just a few of the many elements of ISO 14001. They are all essential because they work together as part of a well thought-out system. For example, how could you consider "training, awareness, and competence" or "non-conformance and corrective and preventive action" anything less than essential? Besides, an EMS cannot be certified to conform to the ISO 14001 standard unless it incorporates *all* elements of the standard.

The essential elements of an EMS are:

- **_Top management commitment and involvement_**
- **_The enterprise's environmental policy_**
- **_Environmental management programs (action plans, projects, initiatives)._**

But here we want to get to the core essence of an EMS — to the minimum that an enterprise would have to have in place to experience at least some of the bottom-line benefits possible. Though ISO 14001 purists would be made uncomfortable by it, we can still use the international standard as our reference EMS for this purpose. The essential elements of an EMS, and those we will spend the most time on in Chapter 2, are:

- Top management commitment and involvement
- The enterprise's environmental policy
- Environmental management programs (action plans, projects, initiatives)

Top Management Commitment and Involvement

The first words in the first step (Clause 4.2) of the ISO 14001 EMS cycle are "Top management...". The first words in the last step (Clause 4.6) of the ISO 14001 EMS cycle are "The organization's top management...". Because many of the people who helped formulate ISO 14001 came from industry, they knew that an EMS begins and ends with the top management of the enterprise.

For an EMS to be effective, top management has to be committed to improving the environmental performance of the enterprise, and to making money, people, and equipment available to establish and maintain the EMS. But it's not enough for top management to be committed; their commitment has to be well-known to everyone in the enterprise. Top management has to participate and be seen to participate actively in the operation of the EMS. At a minimum, top management should actively participate by promulgating the enterprise's environmental policy and by conducting a widely reported periodic management review of the EMS. It is best if top management also has day-to-day involvement — say, in carrying out specific environmental initiatives under the EMS. By declaration and example, management must make it clear that a continual effort to reduce the impact of enterprise operations on the environment is a prominent part of the company's culture, and that success in this effort will be rewarded.

The CEO of a medium-sized company once complained to me that he had spent a fortune on implementing ISO 14001 in his enterprise. He said that becoming "ISO 14000 certified" seemed "like an awfully high price for some feel-good stuff, and a little clean-up on the shop floor." He was seriously disappointed that business benefits promised by the ISO 14001 consultant had not materialized, and he was considering "just letting the whole thing fizzle out and not bothering with recertification." The consultant had implemented ISO 14001 in the enterprise in just two months by basing all of the necessary procedures, as well as the enterprise's environmental policy and the report of the first management review meeting, on boilerplate materials he had developed working with other clients.

Had the CEO made clear his commitment to improving his company's environmental performance? "Absolutely. I sent out an all-hands memo saying that this enterprise is committed to rapid ISO 14001 certification, that an experienced consultant had been engaged to write the necessary environmental procedures, and that everyone was expected to cooperate with the consultant and follow those procedures." Not too much subtlety there. Middle management and workers understood what top management was committed to — some sort of environmental certification. As usual, management was going to accomplish its latest kick by making everyone else's life more difficult. "Okay," they must have thought, "we'll do what we have to do until this passes."

There really are two related principles that this CEO's viewpoint and actions call attention to. First, an EMS may begin and end with the top management of the enterprise, but in-between are the employees, and without their commitment, the EMS won't work. Second, if top management is not fully and practically committed to continually improving environmental performance — and is not *seen* by employees to be personally committed — the EMS won't work. A real commitment of top management is essential because only top management can make available the resources required for a successful EMS at the enterprise, and because only top management can create rewards for workers that contribute to effective EMS operation. That sort of commitment also is essential because employees take their cues from management, are adept at sensing management's real intent and values, and will conduct themselves accordingly.

The Enterprise's Environmental Policy

According to the ISO 14001 EMS standard, an enterprise's environmental policy has to be formulated and disseminated by top management. It has to include three written commitments:

1) A commitment to comply with environmental regulations
2) A commitment to continual improvement
3) A commitment to preventing pollution

If top management commitment, environmental policy, and environmental programs are the heart of ISO 14001, then the enterprise's commitment to regulatory compliance, continual improvement, and pollution prevention are its soul. The 52 explicit requirements of ISO 14001 represent the letter of the standard; these three overarching commitments set the tone for the way the enterprise meets the requirements, and thus represent the spirit of the standard.

> ***A productive EMS requires three commitments by top management that are expressed in visible actions:***
> ***1) A commitment to comply with environmental regulations;***
> ***2) A commitment to continual improvement;***
> ***3) A commitment to preventing pollution.***

What these three commitments say in combination is this: Our enterprise will do everything it can, without undermining the economic sustainability of the business, to comply with all applicable environmental regulations as quickly as possible. But we won't stop there. We'll keep on trying to improve our environmental performance even beyond what is required by law. We'll set ourselves the unattainable goal of zero negative impact on the environment — we know we'll never get there, but it's the trying that counts. And the way we will progress in that direction is by finding ever more creative means of reducing pollution at the source — that is, by *preventing* pollution, not just by cleaning it up after we've created it. Cleaning up pollution *costs* money; preventing pollution *saves* money, because pollution is nothing more than waste from the production process. In short, our enterprise is committed to endlessly increasing efficiency, quality, sustainability, and prosperity by continually improving its environmental performance.

Environmental Management Programs

Clause 4.3 of ISO 14001 is titled "Planning," but the real subject of the clause is environmental management programs. ISO 14001 requires that these programs be designed and carried out to achieve the enterprise's environmental objectives and targets. It requires that when environmental objectives and targets are being formulated, careful consideration be given to the most significant environmental impacts of the enterprise's operations, and to areas where it doesn't comply with regulations or will soon need to comply with new regulations.

In its barest outlines, here is the ISO 14001 planning process that leads to one or more environmental management programs in the enterprise: First, the EMS Committee, working with management and other employees, identifies the aspects of the enterprise's operations that significantly affect the environment, and also checks the status of the enterprise's compliance with environmental regulations. Top management establishes environmental objectives and targets on the basis of this information.

Objectives are broader than targets. An objective might be to "reduce or eliminate toxic waste generated in our production process." Its corresponding target might be to "reduce the volume of toxic waste from finishing operations by 25 percent in 12 months." After establishing environmental objectives and targets for the enterprise, the EMS committee — again working closely with management and other employees — examines different possible approaches and devises one or more environmental management programs to achieve the environmental targets.

Then steps have to be taken to carry out the environmental management programs. ISO 14001 does not say anything explicit about the process an enterprise must follow to carry out environmental management programs, because there is no meaningful way to describe program implementation generically, without referring to a specific program. Instead, the ISO 14001 standard spells out the conditions required for an enterprise to be able to properly carry out programs it has developed under the EMS. Employees have to be trained, and given responsibilities and authority related to the environmental management programs and to achieving environmental targets and objectives. New environmental procedures and work instructions may have to be created and documented. A system for measuring and assessing the results of environmental management programs has to be devised and implemented so that progress can be monitored. And a procedure for checking progress and making corrections to the environmental management programs has to be put in place.

Finally, top management has to review the results of implementing the programs and then consider revisions to the programs, targets, and overall EMS. In a sense, Clauses 4.3 through 4.6 of ISO 14001 — that is, everything except the clause on environmental policy — can be thought of as elaborating on the subject of environmental management programs.

ONE MORE TIME

Figure 3 pulls together and summarizes many of the ideas discussed in this chapter. An EMS is a system for continually improving the environmental performance, and thereby the business performance and sustainability, of an enterprise. It is a tool that enables an enterprise to address its environmental performance based on the way it operates, rather than just checking the status of its compliance with a collection of unrelated and unprioritized environmental standards. It is a means for top management to proactively, strategically, and comprehensively deal with the way the enterprise interacts with the larger environment, in accordance with basic principles of good management. And it is a tool that enables an enterprise to make good environmental performance a source, rather than a consumer, of its revenues. In concept, an EMS is actually a rather simple tool. It's nothing more than the steps of a classical planning and implementation model, made into a closed circle of continual improvement. The ISO 14001 EMS standard is built around the five-step planning and implementation model shown in Figure 3. Using this model, an enterprise systematically examines its operations, establishes environmental objectives and targets that represent improved environmental performance, and carries out environmental management programs, or projects, to reach its targets.

Thus, the EMS generates environmental management programs that, if well thought-out and well carried-out, result in improved environmental performance and regulatory compliance on the part of the enterprise. If the environmental management programs have been well thought-out and well carried-out, they will also result in reduced costs of production, reduced risk of environmental liability, and reduced regulatory penalties.

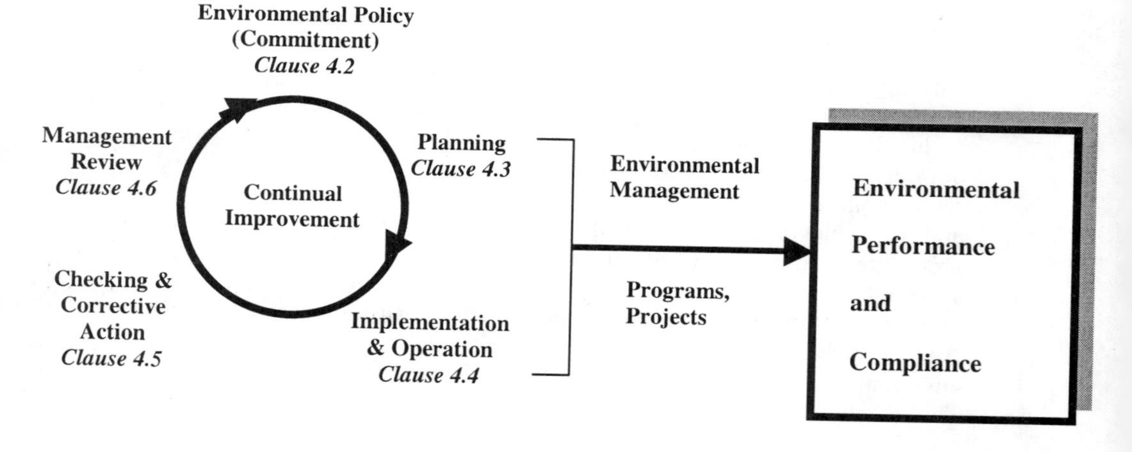

Figure 3. *Schematic of ISO 14001 EMS.*

QUESTIONS FOR THINKING AND DISCUSSING

1) Does your enterprise have a deliberate organized program for managing the way its operations interact with the environment? If not, why not? If yes, what motivated top management to put it in place?

2) Does your enterprise have a process for monitoring and regularly evaluating its process flows and materials balances — that is, the relationship between what comes into the plant, and what leaves it as product and processing waste?

3) Does your enterprise have a process for monitoring and regularly evaluating energy consumption (energy balance)?

4) Does your enterprise have a process for tracking applicable environmental regulations and regularly evaluating its compliance with them?

5) Would top management of your enterprise easily commit itself to complying with applicable environmental regulations as quickly as possible; to continually improving environmental performance even beyond regulatory requirements; and to finding ways to reduce the pollution it generates, rather than creating it and then cleaning it up? If not, why not?

6) Has your enterprise ever received complaints from the public about environmental problems? Does your enterprise have a program for communicating with the public about its environmental performance?

7) Would you generally describe the approach of your enterprise to environmental matters as reactive, proactive, or a little bit of both? Please explain.

8) Does the overall management system of your enterprise more or less follow the pattern of one of the models mentioned in this chapter? Please explain.

9) Has your enterprise ever undertaken an environmental management program, or project, to achieve a particular environmental performance target? Describe and analyze the experience.

10) What are the five biggest obstacles you would face in promoting adoption of an EMS in your enterprise, and how would you deal with each of them?

Chapter 2
EMS: APPLIED MODELS

ISO 14001 AND IMPLEMENTATION OVERVIEW

In the United States and some other places, organizations authorized to conduct ISO 14001 registration audits are known as *accredited*; individuals authorized as ISO 14001 auditors are *certified*; and enterprises verified as conforming to the ISO 14001 EMS standard are *registered to ISO 14001*, or just *registered*. This terminology is good and clear, but across the world, and even in casual conversation in the United States, people tend to refer to enterprises as being *certified* rather than *registered* to ISO 14001. In this book we'll use the more common term, and hope that no readers will be confused by it.

Table 1. The ISO 14001 Framework.

Clause	Sub-clause	Title (topic)
4.2		ENVIRONMENTAL POLICY
4.3		PLANNING
	4.3.1	Environmental Aspects
	4.3.2	Legal and Other Requirements
	4.3.3	Objectives and Targets
	4.3.4	Environmental Management Programs
4.4		IMPLEMENTATION AND OPERATION
	4.4.1	Structure and Responsibility
	4.4.2	Training, Awareness, and Competence
	4.4.3	Communication
	4.4.4	Environmental Management System Documentation
	4.4.5	Document Control
	4.4.6	Operational Control
	4.4.7	Emergency Preparedness and Response
4.5		CHECKING AND CORRECTIVE ACTION
	4.5.1	Monitoring and Measurement
	4.5.2	Nonconformance and Corrective and Preventive Action
	4.5.3	Records
	4.5.4	Environmental Management System Audit
4.6		MANAGEMENT REVIEW

It is important to remember that the ISO 14001 EMS standard must be followed to the letter only by enterprises seeking certification for the first time in the near term or wanting to maintain certification once they have received it. Enterprises that, for the time being, just want the benefits of a productive EMS and don't want to spend unnecessary amounts of money, and enterprises that want to move slowly toward certification, learning as they go, are free to select from the elements of ISO 14001 and modify their selections as they see fit.

In fact, even enterprises that follow ISO 14001 to the letter can adapt the standard to their particular situations in a wide range of ways. In this chapter, we explore the elements of the ISO 14001 framework so we can help readers tailor and apply EMS models to the needs and circumstances of their own enterprises.

The ISO 14001 EMS standard is organized as shown in Table 1. This framework is built around a sequence of five basic planning and implementation steps: the ISO 14001 clauses. The three middle steps — the things that the workers have to do in between top management's environmental policymaking on one end and its reviewing on the other — each have a number of components: the subclauses.

Although the logic of the ISO 14001 framework is one of sequential elements, the logic of an operating EMS is one of simultaneous continuing activities.

Ultimately, of course, all elements of an EMS need to operate more or less simultaneously, because the enterprise operates and interacts with the environment all the time. Though the logic of the ISO 14001 framework is one of sequential elements, the logic of an operating EMS is one of simultaneous, continuing activities.

However, a model based on a set of roughly sequential steps is needed to establish an EMS for the first time in an enterprise. The enterprise, after all, has to begin at the beginning and wants to finish with an operating EMS in place. If the "initial installation" model is based on the ISO 14001 framework, the initial steps might look like this:

1) Top management makes and announces a commitment to implement an EMS and constantly improve the environmental performance of the enterprise. It appoints an EMS Officer (the ISO 14001 standard uses the title "Management Representative" for this position) and pulls together an EMS Committee made up of people from different departments and levels across the enterprise. Top management uses use the term "committee" to signal that the responsibilities of this body cut across all units of the enterprise and that its responsibilities will continue into the future — unlike a "task force," which is disbanded when the job is done.

2) The EMS Management Representative and Committee work with top management to develop and promulgate an environmental policy for the enterprise.

3) The EMS Committee and the enterprise's employees, with the approval of top management, identify:
    - The environmental aspects of the enterprise's operations, and their impacts

- Applicable environmental regulations, and areas where the enterprise does not comply

4) The EMS Committee and other employees, with the review and approval of top management:
    - Assess the significance of the environmental aspects and impacts
    - Determine strategic environmental objectives and "first round" targets for significant environmental aspects
    - Devise environmental management programs to achieve the targets

5) Top management:
    - Assigns EMS responsibilities and authorities to staff
    - Allocates financial and other enterprise resources to the EMS and its environmental management programs
    - Approves a program of EMS-related training proposed by the EMS Committee
    - Approves programs of internal and external communications procedures proposed by the EMS Committee

6) The EMS Committee and other employees develop EMS documentation and EMS document control procedures for the enterprise.

7) The EMS Committee and other employees, with the review and approval of top management, develop procedures to minimize environmental impacts from the operations of the enterprise, and prepare emergency plans.

Top management and the EMS Committee need to be careful to use consultants only as teachers, advisors, and expert resources; they should not use consultants as hired hands to do the work required. If the enterprise intends to create a sound EMS that delivers real benefits, only employees can do that work.

8) The EMS Committee and other employees establish an environmental monitoring program; develop procedures for analyzing monitoring data and undertaking the indicated corrective and preventive actions; and develop procedures for maintaining environmental records.

9) The EMS Committee conducts an audit of EMS operations and their consequences when considered in terms of the environmental and business performance of the enterprise.

10) The EMS Committee reports audit and other findings to top management, which considers this and other information in the course of its review of the operations and

performance of the EMS, leading to improve-ments in its design and implementation.

11) If top management is seeking ISO 14001 certification, the enterprise hires an accredited certification firm, or registrar, to conduct a certification audit.

Many enterprises choose to employ a consultant or consulting firm to help implement ISO 14001. It's important that the enterprise have staff who are reasonably well educated in ISO 14001 before calling in a consultant. A consultant can contribute lessons of experience with many different firms, which can save a great deal of implementation time, labor, money, and mistakes. But a consultant cannot implement an EMS *for* an enterprise; in fact, the very concept of implementation by an outsider runs counter to the spirit of ISO 14001 and the basic principles of EMSs in general. Top management and the EMS Committee need to be careful to use consultants only as teachers, advisors, and expert resources; they should not use consultants as hired hands to do the work required. If the enterprise intends to create a sound EMS that delivers real benefits, only employees can do that work.

Although the introductory clause of ISO 14001 — Clause 4.1, "General Requirements" — says no more than that "the organization shall establish and maintain an environmental management system" that satisfies all the requirements contained in Clauses 4.2 through 4.6, it does, subtly, set the tone for the entire standard. It does this by the first of 52 uses of the word "shall" — a strong word that unambiguously signals an absolute requirement. The message: ISO 14001 is not cosmetic; there are concrete requirements that must be met to conform to the standard and achieve the benefits of conformance.

Clause 4.1 also sets the tone by using the phrase "establish and maintain" for the first of many times in ISO 14001. "Maintain" means kept in action and up-to-date through regular review and, if necessary, revision and redocumentation. The message: an EMS is not a one-time thing. An enterprise cannot simply install an EMS, get itself "ISO 14000 certified," and move on to other initiatives. There's really no point in establishing an EMS if it is not going to be productive. To be productive, an EMS must be constantly working and constantly supported and improved, so that it is constantly producing benefits. This is possible only if it is integrated permanently into the regular management and operational routines of the enterprise.

The ISO drafters aimed to produce an international standard on which the broadest range of applied EMS models could be based. So, the general approach of the standard is not to state *how* something should be done, and usually not even entirely *what* should be done. Most often, ISO 14001 simply says that top management, or "the organization," is required to do something (like write environmental policy) that has certain specified characteristics (the things the policy must contain), or develop a procedure for doing something that accomplishes a specific purpose (for example, properly maintaining EMS documents so they can be readily located). The implication is that the specified characteristics or purposes are the minimum for ISO 14001 conformance, but an enterprise can choose to go well beyond the minimum.

Tools and techniques for implementing each element of ISO 14001 in turn are provided in the corresponding sections of Chapter 3 of this book.

ENVIRONMENTAL POLICY (CLAUSE 4.2)

Top Management Takes the Lead

Everything in an enterprise, including an EMS, starts with top management — the person or group of people with executive responsibility for the enterprise. Top management uses published policy to convey the mission, operational approach, and values of the enterprise to employees and others ... but especially to employees. Policies issued by top management set the ultimate standards for evaluating the performance of employees and the enterprise as a whole.

The first step is for top management to issue a statement of environmental policy, along with a clear demonstration of its determination to commit the necessary resources, and to reward performance that supports the policy.

In short, in the case of an EMS, top management's policy is that the enterprise will pursue continual improvement in its environmental performance by strategically planning the management of its interactions with the environment. This is the same approach that good managers use for other aspects of enterprise operations, like marketing or new product development.

Top management has to issue environmental policy because only top management governs the entire sweep of enterprise activities and products that interact with the environment; because policy issued by top management is understood by everyone in the enterprise as the unquestioned law of the land; and because to be meaningful the policy must be issued by those who control the resources — including rewards to employees — for making it happen. So if top management wants to establish an EMS in the enterprise for the purpose of improving business and environmental performance, the first step is to issue a statement of environmental policy, along with a clear demonstration of top management's determination to commit the necessary resources, and to reward performance that supports the policy.

The fact that the environmental policy has to be issued by top management does not mean that only top management should be involved in formulating the policy — in fact, just the opposite is true. All departments and levels of the enterprise — in other words, the people closest to day-to-day operations and with the greatest knowledge of how the enterprise interacts with the environment — should contribute to developing the policy. In fact, the need to develop environmental policy can help focus attention on environmental issues and raise the visibility of environmental concerns throughout the enterprise. In this way, development of environmental policy can be the first step in an enterprise's environmental awareness program.

Scope of the Policy and the EMS

What should an environmental policy include as the foundation for a productive EMS? First, the policy needs to establish the area of focus, or scope, of the EMS. Obviously, the concern is ultimately with all operations, facilities, equipment, inputs, products, and services of the enterprise. But in many enterprises it may be best to start on a small and assuredly manageable scale — for example, concentrating on one facility or on a discrete operation, such as the painting or metal-finishing operation within a larger production process. In other words, top management needs to consider whether or not it wants to start the process of establishing an EMS by limiting the initial scope of the EMS.

> ***In many enterprises it may be best to start EMS implementation on a small and assuredly manageable scale — for example, concentrating on one facility or on a discrete operation.***

With a limited initial EMS scope, small-scale mistakes can more easily and quickly be corrected; successful initial implementation will sustain enthusiasm by avoiding frustrations; and small-scale implementation efforts can provide important lessons for later EMS implementation throughout the enterprise. If top management takes this approach — restricting the initial EMS scope to learn and build toward enterprise-wide implementation — it must clearly state that in the environmental policy. In any case, the first essential element of an enterprise's environmental policy is a statement of the scope of the policy and the associated EMS.

Commitment to Comply with Environmental Laws

Environmental laws and regulations represent the minimum environmental standards or practices acceptable to society. Put another way, they reflect the maximum environmental price society is willing to pay for the benefits provided by various enterprises. If top management at an enterprise is not prepared to go along with the environmental standards of the society that provides its inputs, purchases its outputs, and is home to its facilities and employees, why bother with an EMS? An environmental policy would be meaningless unless it included a commitment to comply with environmental laws and regulations.

But what does a commitment to comply with environmental laws and regulations really mean? Top management can decide that the first most important objective under its EMS is to bring the enterprise into full compliance with environmental law, and to keep it in compliance as the regulatory framework evolves. Another approach would be to assign highest priority to eliminating the environmental impacts of enterprise operations that pose the greatest dangers to human and ecological health. In this case, a commitment to comply with environmental laws does not mean that regulatory compliance takes precedence over every other consideration at all times.

Top management has to consider the need for possible tradeoffs among these sorts of concerns, and through the environmental policy provide guidance to employees regarding environmental priorities. This is one way that the environmental policy of an enterprise serves as a framework for determining the environmental objectives of its EMS.

Commitment to Continual Improvement

A third essential element of an enterprise's environmental policy is a commitment to continual improvement. This commitment means that even if the enterprise is in compliance with environmental regulations, it will continue to improve its environmental performance — just as in quality management, an enterprise continually works to improve the quality of its products.

There is no such thing as "good enough"; perfection, though unattainable, is the goal. If it were possible to have the wealth of goods and services, and the richness of life that these give us, with *no* damage to the natural environment, is there anyone who would not want it? There is no reason for undertaking an EMS, a tool for systematically and strategically improving environmental and business performance, if top management's purpose is only to achieve a certain amount of improvement and stop there. That can be done with a one-time fix. A statement of commitment to continual improvement in the environmental policy of the enterprise is really a statement of commitment to a meaningful EMS.

Besides, from a purely economic point of view, pollution is waste. So long as there is waste, there is opportunity for greater efficiency, improved enterprise sustainability, and improved business performance. It is true that, in the short run, skewed pricing structures can sometimes make it cheaper to waste than not waste, but that is gradually changing. Only the poor or irresponsible manager — the manager without vision and without a strategic approach, or without regard for the long-term effects of that approach — would ignore the trends and fail to get his or her enterprise on a course of long-term sustainability.

At the very least, the environmental policy statement associated with an EMS should include:
a) a statement of the scope of the policy and the associated EMS;
b) a commitment to comply with environmental laws and regulations;
c) a commitment to continual improvement in environmental performance;
d) a commitment to preventing pollution.

Commitment to Pollution Prevention

Finally, at least as far as a general EMS is concerned, the environmental policy of an enterprise must include a commitment to *preventing* pollution. In the United States, the common term for the body of knowledge, approaches, techniques, practices, and technologies aimed at minimizing the creation of pollution is "pollution prevention," or "P2." Terms often used as near synonyms for P2 are "waste minimization" and "clean production." Together these terms sum up the idea nicely — the enterprise that practices P2 prevents pollution, minimizes waste, and engages in relatively clean production. Terms referring to after-the-fact approaches include "pollution control," "pollution treatment," and "end-of-pipe measures."

The interesting thing about end-of-pipe pollution abatement measures is that after creating pollution — which wastes and degrades natural resources — through this approach an enterprise consumes even more resources to deal with the problem. In other words, the enterprise first wastes money by buying inputs and then discarding part of them as waste in the form of pollution; it then compounds that waste of money by spending even more money to control the impacts of that waste through an end-of-pipe approach, such as smokestack scrubbers or on-site wastewater treatment facilities.

P2 measures include such things as process redesigns for greater efficiency of materials use; post-process heat capture and reuse; conversion to high-efficiency burners; conversion to non-toxic inputs; internal recycling; less packaging; automated process controls; routine energy-efficiency measures, such as building insulation and double-glazed windows; improved monitoring and maintenance routines; optimizing delivery routes; and much much more (please see Chapter 6 in Part II of this book for dozens of examples of P2 measures in many different industry sectors).

Notice that most of these P2 measures save an enterprise money that otherwise would be spent on inputs that end up as waste; on waste disposal, pollution control, or cleanup; or, possibly, on special permits, fees, or fines. P2 measures range from "no-cost" housekeeping improvements to multimillion-dollar high-tech investments that dramatically reduce use of raw materials, process water, or energy needs. P2 investments typically yield rates of return far in excess of competing uses for capital. No matter where P2 measures reduce pollution — on site, upstream (e.g., as a result of improved electric energy efficiency) or downstream (e.g., as a result of improved fleet efficiency) — the economic benefits tend to accrue in the same place: the enterprise's bottom line.

Because end-of-pipe measures increase consumption of resources per-unit of useful output, a meaningful commitment by top management to continually improve environmental performance of the enterprise cannot but encompass a commitment to pollution prevention. A systematic way of identifying pollution prevention opportunities is to conduct a P2 audit. P2 audits come in many varieties and make use of many different levels of technical expertise, and may be aimed at identifying pollution prevention measures of many different scales. Many enterprises incorporate P2 audits into their EMSs, and that is the approach that this book encourages and supports.

It is easy to see that P2 is fundamental to the concept of an EMS, of a strategic approach to managing interactions with the environment and improving enterprise sustainability. If an enterprise carries out a pollution prevention audit, then implements the priority P2 measures suggested by the audit, then repeats the audit and follows up in continuing periodic cycles, it will most certainly comply with all regulations soon enough and will reach continual improvement in environmental performance thereafter. Though some EMS devotees might disagree, the simplest form of EMS could be thought of roughly as successive rounds of carefully designed P2 audits, with follow-up. A common pattern is for an enterprise first to carry out a P2 audit, then undertake some P2 investments, and then — convinced of and able to demonstrate to, say, stockholders the business benefits of improving environmental performance — move on to implement an EMS.

Because end-of-pipe measures increase consumption of resources per-unit of useful output, a meaningful commitment by top management to continually improve environmental performance of the enterprise cannot but encompass a commitment to pollution prevention.

The environmental policy that serves as the foundation for the EMS of an enterprise can encompass more than the statement of scope and three commitments described previously. Indeed, the environmental policy required under the ISO 14001 standard requires more, as we will see in a moment. The policy can lay out a larger list of core values and higher purposes of the enterprise; can detail specific objectives and priorities; can include assignment of EMS responsibilities and commitment of resources; and more.

It is a good idea for top management to issue a statement that designates the individual with primary responsibility for the EMS and establish the EMS Committee, and specify at least an initial commitment of resources (such as a basic EMS-related training program) either as part of the environmental policy or as a supplemental document issued with it. This provides a clear demonstration of top management's commitment to doing what is needed to carry out what it has publicly said is its own policy.

The environmental policy could — in fact, *should* — be reviewed by top management as a periodic agenda item for its regular meetings. It's especially important that the policy be reviewed frequently if the enterprise is taking an incremental approach to EMS implementation. Top management needs to review progress, consider lessons learned, and determine how the EMS can be improved and extended to additional parts of the enterprise, so as to be sure that it covers the entire enterprise, and does so efficiently and productively, without unneeded delay.

Requirements of ISO 14001

Clause 4.2 of the ISO 14001 EMS standard requires the enterprise to have an environmental policy that:

a) Is issued by top management
b) Is appropriate to the nature, scale, and environmental impacts of the enterprise's operations and products
c) Includes a commitment to comply with applicable environmental laws and regulations
d) Includes a commitment to comply with other environmental standards and practices to which the enterprise has obligated itself
e) Includes a commitment to continual improvement
f) Includes a commitment to pollution prevention
g) Provides the framework for establishing and reviewing the enterprise's environmental objectives and targets
h) Is in writing
i) Is communicated to all employees
j) Is implemented
k) Is reviewed regularly and kept up-to-date
l) Is available to the public

Item b), the requirement that the environmental policy be appropriate to the enterprise, can be thought of as referring to a few things. First and most obvious is the idea that a steel mill would require an environmental policy statement of a different order altogether from that of a vendor of computer games, because of their different natures, scales, and environmental impacts. Second, the requirement can be taken to mean that the policy must contain an explicit statement of *scope* — that is, it should make reference to the specific operations or facilities to which the policy and the EMS apply. Without a statement of scope, there is no way to consider the appropriateness of the policy. Third, this requirement means that top management can't declare that its environmental concerns end at the property line of the enterprise. The products that it sends out into the world may also have environmental impacts because they contain toxic chemicals, emit exhaust gasses, consume energy, etc., and so they must remain a concern of the enterprise.

The requirements of the environmental policy clause of ISO 14001 mean that top management can't declare that its environmental concerns end at the property line of the enterprise.

Keep in mind that for an enterprise to qualify for or maintain its ISO 14001 certification, the "organization" in which the EMS is implemented must have its own functions and administration — in other words, it must be substantially independently run. However, an enterprise can choose not to pursue certification or to pursue certification only after lengthy experience with its EMS. There is no

The requirements of the ISO 14001 standard need be met only if and when the enterprise is ready to call for a certification auditor or wants to maintain its certification.

reason it cannot begin the process of EMS implementation in a very limited way, as we discussed earlier. The EMS could initially be limited in terms of scope, or of other parameters, such as the extensiveness of documentation or initial employee training. The idea would be, of course, to start small, learn from experience, and gradually expand to cover the entire enterprise.

Item c), the requirement for a commitment to comply with applicable environmental laws and regulations, does not *require* compliance — only a *commitment* to compliance. However, this cannot be an empty commitment. There are all sorts of reasons, not the least bit sinister or cynical, why an enterprise might not be in compliance with regulations. But when the certification audit is performed, the auditor will look for confirmation that specific actions have been taken by the enterprise to bring it into compliance as quickly as possible. The auditor will look for concrete evidence of the commitment.

Item d), the requirement for a commitment to comply with other environmental standards and practices to which the enterprise has obligated itself, refers to environmental performance standards or practices to which the enterprise has voluntarily agreed. Examples include trade association codes of practice, agreements with citizen groups, self-declared environmental practices, special agreements with local authorities, and agreements with labor organizations.

Item f), the requirement for a commitment to pollution prevention, does not prohibit the use of end-of-pipe pollution control measures. There may be situations where P2 is not feasible because the materials, processes, or technology for achieving it are not available. Or there may be situations where achieving cleaner production requires a larger investment than the enterprise can undertake at the present time. Or there may be situations where the P2 solution to a pollution problem will take a considerable amount of time. In these sorts of situations, the responsible thing for an enterprise might be to employ an end-of-pipe measure to minimize environmental damage in the short run while it seeks P2 solutions for the long run. Again, a certification auditor will be satisfied with concrete and convincing evidence of the enterprise's *commitment* to P2.

Item g), the requirement that the environmental policy provide the framework for establishing and reviewing the enterprise's environmental objectives and targets is easily understood in principle. But how can the policy to be a meaningful "framework" in practice? Sure, in the broadest sense, the commitments to regulatory compliance, continual improvement, pollution prevention, and so on form a "framework for establishing and reviewing." But surely the crafters of the standard meant something more than that.

The guidelines in ISO 14004 tell us that environmental objectives and targets should be based on *knowledge* about the environmental aspects and significant environmental impacts of the operations and products of the enterprise. This knowledge, then, should be reflected in the environmental policy, which in turn provides a "framework for revising and reviewing" objectives and targets.

So it seems that we have another reason why the policy must be reviewed and reconsidered fairly frequently. As environmental aspects and impacts of the enterprise are better understood through operation of the EMS, and as the EMS alters the environmental performance of the enterprise, the information gained must help shape the constantly evolving environmental policy. Through an iterative process, the environmental policy of the enterprise becomes ever more refined and explicit as a framework for establishing and reviewing its environmental objectives and targets, while the environmental objectives and targets become more directly and explicitly linked to the environmental policy.

Item k), the requirement that the environmental policy be reviewed regularly and kept up to date, is necessary because the environmental policy of an enterprise can very quickly be overtaken by events. External conditions — such as the findings of environmental health risk research, changes in the ISO 14001 standard, changes in government regulations, the demands of local citizen groups and authorities, and social values, including concepts of corporate responsibility — can change rapidly. The facilities, equipment, processes, products, and markets of the enterprise may also change. If the EMS is productive, there certainly will be change in the way that the enterprise interacts with the environment. So, to serve meaningfully as the EMS foundation and as a framework for revising and reviewing environmental objectives, the environmental policy of the enterprise must be reviewed, and its particulars reconsidered, frequently.

The ISO 14001 policy clause implicitly establishes the mission, overall operational approach, and values for an EMS, just as top management uses policy statements to convey to employees the mission, operational approach, and values of the enterprise. Hopefully, readers have perceived how much connectedness there is among the internal elements of the environmental policy, and also between the policy and other elements of the EMS. That's why ISO 14001 serves so well as a standard for an EMS. It is highly integrated internally, and integrates easily into — and reinforces — the overall management system of an enterprise. The standard looks simple, but behind this apparent simplicity is a great deal of understanding of sound enterprise management and operations. In that sense, the ISO 14001 EMS standard is a very sophisticated construction.

PLANNING: ENVIRONMENTAL ASPECTS (SUBCLAUSE 4.3.1)

Item j) in our list of ISO 14001 requirements for environmental policy is a requirement that the policy be implemented. Essentially, everything else in the EMS standard deals with procedures for implementing the environmental policy of the enterprise. But because of the special contents of the policy statement, implementing the policy forces the enterprise into a strategic mode of managing its interactions with the environment. In other words, because the environmental policy of the enterprise embodies strategic principles, there really is no way to carry out the policy other than through strategic planning.

Policy to Planning

If you're going to plan, the first thing you need to do is identify the *object* of the planning — the thing on which the planning is focused. With respect to the EMS the question is, "What things must we change to carry out the environmental policy?" The answer is: the environmental aspects of the operations and products of the enterprise. The *purpose* of the planning is to change the environmental aspects of the enterprise — that is, its interactions with the environment — in ways that represent movement toward regulatory compliance, better environmental performance, pollution prevention, and specific objectives that may be in the environmental policy of the enterprise.

For EMS planning the first question is, "What things must we change to carry out the environmental policy?" The answer is: the environmental aspects of the operations and products of the enterprise.

But there is no need to try to change all the environmental aspects of the enterprise — just those that have negative environmental impacts. Most enterprises will have trouble addressing all of those, so the EMS Committee will have to find a way to set priorities among its environmental management programs. In setting these priorities, the EMS Committee will have to consider current and anticipated regulatory compliance issues, as well as other strategic considerations.

Once they have figured out priorities among the enterprise's environmental aspects, the EMS Committee will need to set objectives for altering those aspects. In other words, the committee will set objectives for changing specific ways that the enterprise and its products interact with the environment.

Next, for each objective, they will have to determine concrete performance targets that represent achieving the objective, or an acceptable increment of progress in that direction. Once that is done, the stage will have been set for devising environmental management programs — action plans, projects, initiatives — that achieve the targets and thus implement the environmental policy.

To recap, the sequence of EMS planning steps to implement the enterprise's environmental policy is:

1) Identify the environmental aspects of the operations and products of the enterprise
2) Identify the negative environmental impacts associated with the environmental aspects
3) Assess the significance of the environmental impacts
4) Determine strategic environmental objectives for the significant environmental aspects
5) Determine specific performance targets for achieving the objectives
6) Develop environmental management programs to achieve the targets

These planning steps are what is addressed under the planning clause, Clause 4.3, of ISO 14001. The following clause — Clause 4.4, Implementation and Operation — deals with the capabilities and support systems for carrying out the environmental management programs.

But it's not so simple. How do you set a realistic target without knowing how — or if — the target can be achieved? Again, there must be an iterative process. The effort to devise environmental management programs, or gain experience with the programs, may result in a rethinking of targets and, in turn, a reassessment of objectives and priorities. Perhaps it's reasonably simple after all: You figure out your priorities, objectives, and targets, and then adjust those priorities, objectives, and targets based on additional information that comes to light as you implement your plan. This is the dynamic of refinement that is basic to all realistic planning and implementation processes.

Activities and Their Environmental Aspects

In simple terms, an enterprise can be thought of as acquiring inputs, carrying out production processes, and shipping out products. Yes, enterprises also have administrative activities, acquire and dispose of equipment, purchase and perhaps provide commercial services, set up new facilities, engage in marketing, and much more; and they have a connection with the use and disposal of their products. Though we have been referring to all this as "operations" or the "operations and products" of the enterprise, the ISO 14001 standard uses the phrase "activities, products, and services." What is meant in either case is everything an enterprise does or causes to be done (by contractors, for example), and the use and disposal of the products it creates and sends out into the world.

The first thing the EMS Committee will need to do for this element of the EMS is break down the operations and products of the enterprise into relevant components, or activities, for analysis. In doing this, the range of operations and products to be considered will be limited to what is encompassed by the scope of the environmental policy. In fact, to gain experience with the process of identifying relevant discrete activities and their associated environmental aspects and impacts, the EMS Committee could begin with a scope that is even less than what will ultimately be addressed by the EMS. Surely top management would prefer a small job done well to a large one done poorly.

There is no standard for what constitutes a "relevant activity" for purposes of considering environmental aspects — it depends on the nature of enterprise operations, the size of the enterprise, the experience of the EMS Committee with this sort of analysis, past practice, and the like. In the first instance the EMS Committee might begin simply by reviewing operations and product uses for activities that obviously have environmental connections, and focusing only on those. Obviously the job could be done in a much more sophisticated, scientific, analytical, and comprehensive way as well. Whatever the case, the aim is to generate a list of discrete activities like the one in Table 2 that can be examined for their environmental aspects. Discrete activities associated with the operations and products of the enterprise could have no environmental aspects, or they could each have one or more environmental aspects.

> ***Eventually the enterprise should get to the point where it considers the full range of its activities and their environmental aspects in the EMS: those that are past, existing, or planned; those that represent normal operating conditions, as well as startups, shut-downs, retooling, peak seasons, etc.; and possible emergency situations.***

An *environmental aspect* answers the following question: "What is the nature of the enterprise's interaction with the environment through this activity?" Table 2 provides examples of activities and their environmental aspects. Notice that environmental aspects (and their impacts) include those that routinely take place, as well as those that could take place by accident. Indeed, ultimately the full range of activities and their environmental aspects should be considered — those that are past, existing, or planned; those that represent normal operating conditions, as well as startups, shut-downs, retooling, peak seasons, etc.; and possible emergency situations.

Environmental Impacts and Their Significance

Each environmental aspect can cause no, one, or more than one negative environmental impacts. If an environmental aspect answers the question, "What is the nature of the enterprise's interaction with the environment through this activity?", an *environmental impact* answers the question, "So what?" While the environmental aspect is the interaction, the environmental impact is the *consequence* of the interaction — the consequence for the air, water, soil, and other natural resources, and for the living things that depend on them. Table 2 provides examples of different ways of describing negative environmental impacts.

Once environmental impacts have been identified, they and the environmental aspects that they result from have to be assessed and ranked according to their *environmental significance*. ISO 14001 does not provide a list of criteria for environmental significance, but such a list would obviously include factors like:

- Scale of the environmental impact
- Severity of the impact
- Duration of the impact
- Likely frequency of the impact
- Ecological health risks associated with the impact
- Human health risks associated with the impact

Ranking environmental aspects by environmental significance of their impacts is important because most enterprises will not be able to address all significant aspects with the same degree of aggressiveness; seldom will an enterprise have the resources for that. Ranking aspects by environmental significance is a step in prioritizing, a process completed when strategic environmental objectives are established by the enterprise.

Table 2. Activities, Aspects, Impacts.

Activity or operation	**Environmental aspects**	**Negative environmental impact**
1. Generating process steam	Release of NO_x, SO_x, and CO_2 emissions Fly-ash disposal	Air pollution Climate change effects None: fly ash is sold to paving contractor
2. Receiving and handling hazardous chemical inputs	Potential for accidental spills of toxic chemicals	Potential soil and groundwater contamination
3. Core process operations	Use of energy (electricity) Use of water Leaking of lubrication oils Fugitive particulates	Upstream air pollution Consumption of resource River pollution OHS risk Air pollution
4. Product finishing operations	Release of toxic effluents	River pollution
5. Product packaging	Use of paper products	Forest resource consumption
6. Maintaining vehicle fleet	Potential spills of hazardous chemicals Exhaust emissions Hazardous solid waste	Potential soil and groundwater contamination Air pollution Soil and groundwater contamination
7. Front- and back-office operations	Use of electricity Use of paper Generation of solid waste	Upstream air pollution Forest resource consumption Contributes to need for landfills
8. Maintaining grounds	Spraying and spreading of herbicides and pesticides	Soil and groundwater contamination
9. Constructing new storage facility	Soil erosion	Degradation of the site and nearby areas
10. Discarding used batteries supplied with our product (customer activity)	Placement of hazardous materials in household waste that may or may not be landfilled	Soil and groundwater contamination, human health risk

In due course environmental significance of the environmental aspects will be combined with other criteria representing strategic importance to the enterprise, such as whether or not the environmental impact represents a regulatory noncompliance, the financial and liability implications of the impact and its abatement, community priorities, and so on. Through this process the enterprise eventually determines strategic environmental objectives. So ideally objectives would be set for all significant environmental aspects, but those objectives would reflect the enterprise's strategic priorities among its significant environmental aspects: some objectives will be more aggressive, some less, reflecting a prioritized allocation of enterprise resources. An EMS is a tool for strategic environmental management, and this two step process for establishing environmental objectives is a good illustration of what that means. More on the subject of strategic environmental objectives later, in the section on environmental objectives and targets (Subclause 4.3.3).

It is essential that the EMS Committee develop a clear methodology for determining significance of environmental aspects and impacts, and apply that methodology consistently. Only with a consistent methodology — a governing procedure — can an enterprise determine the relative significance of its different environmental aspects and, ultimately, their relative strategic importance in a way that is useful for planning purposes.

A P2 audit would yield important information about environmental aspects of the enterprise's operations, and might also contribute information for assessing the environmental significance and strategic importance of those aspects.

A P2 audit would yield important information about environmental aspects of the enterprise's operations, and might also contribute information for assessing the environmental significance and strategic importance of those aspects.

Requirements of ISO 14001

Subclause 4.3.1 of the ISO 14001 EMS standard requires the enterprise to:

a) Establish and maintain one or more procedures to identify environmental aspects of its operations or products that it can influence or control
b) Determine which of the aspects have, or could have, significant impacts on the environment
c) Consider these significant environmental aspects when setting environmental objectives
d) Keep the information on aspects and impacts up-to-date

Notice that the ISO 14001 standard simply requires that the enterprise have a procedure and that it be implemented and reviewed regularly ("establish and maintain"). The standard does not contain requirements about what the procedure should be. It's up to the enterprise to decide how to go about it.

Item a) of this subclause also recognizes that it makes no sense for an enterprise to spend time and effort working with aspects of its operations and products that it can do nothing about. For example, an enterprise may be able to do nothing about the inefficient use of energy by a supplier of parts to its production line. ISO 14001 is realistic, yet reminds the user that the procedure for identifying aspects must not be an empty one. Don't include aspects you can't control, it says, but be sure you *do* include aspects you *can* control, or that you can at least influence sufficiently to make a difference.

With regard to item b), determining which aspects have significant impacts on the environment, there is room for a small amount of confusion. The ISO 14001 use of "significant" refers only to factors of environmental importance, such as scale, severity, and duration of the environmental impact. But EMS implementation guidance in ISO 14004 suggests a broader interpretation, one that overlaps with *strategic* factors to be taken into account when setting objectives (these are listed in Subclause 4.3.3, "Objectives and Targets"). In fact, if it wishes, an enterprise can choose to simultaneously assess the environmental significance and strategic importance of its environmental aspects as two parts of a single significance assessment activity. If it does this, it can conduct the significance assessment in connection with either Subclause 4.3.1 or Subclause 4.3.3.

An ISO 14001 certification auditor doesn't care whether the two steps are done separately or together, or under one subclause or the other or both. The only important thing is that the requirements of the standard are met. The certification auditor will look for evidence that there is an active procedure for identifying environmental aspects, that somehow a determination of environmental significance is made, and that significant environmental aspects (significant, because their environmental impacts are significant) are considered, along with certain other strategic factors, in determining the enterprise's environmental objectives.

From the point of view of an ISO 14001 certification auditor it doesn't matter how or under what subclause the enterprise assesses relative significance of environmental impacts, so long as in the final analysis the requirements of the standard are met.

In fact, the question of timing of the two steps leading to environmental objectives for significant environmental aspects should arise only when the EMS is being set up for the first time. After that, some facet of identifying environmental aspects and impacts, and of assessing environmental significance and strategic importance, will always be underway to some degree.

That is part of what is meant in the final requirement under this subclause — the requirement that information on environmental aspects and impacts be kept up to date by the enterprise. There needs to be regular review of the procedure and the information it produces. Such regular review is needed because, as described earlier with regard to environmental policy, the procedure and information concerning the environmental aspects and impacts can quickly be overtaken by events within the enterprise and beyond it. This will certainly be true if the EMS is a productive one.

PLANNING: LEGAL AND OTHER REQUIREMENTS (SUBCLAUSE 4.3.2)

Requirements of ISO 14001

Subclause 4.3.2 of the ISO 14001 EMS standard requires the enterprise to establish and maintain a procedure to:

a) Track environmental laws and regulations that relate to the environmental aspects of its operations and products
b) Track other environmental standards and practices to which the enterprise has obligated itself
c) Ensure that employees have access to the requirements of the laws, regulations, and other standards and practices

Note again that the requirement here is for a procedure, and that though we are told what the procedure must accomplish, we're not told what the procedure must be.

Clause 4.3.2 requires a procedure for tracking the legal and other environmental obligations of the enterprise that are related to *any* of its environmental aspects, not just its *significant* environmental aspects.

> ***"Other environmental standards and practices to which the enterprise has obligated itself" might include requirements from corporate headquarters, codes of practice of trade associations, agreements with citizen groups, self-declared environmental practices, special agreements with local authorities, and agreements with labor organizations.***

Again, it's easy to see that once the procedure for tracking environmental obligations is in place, it needs to constantly function to some degree. It would not do to merely to update the information annually, or even twice a year.

Though ISO 14001 does not say so under this subclause, obviously the wise thing for an enterprise to do would be to track relevant regulations and other environmental obligations *and* the status of its compliance with them. In fact, procedures for tracking regulatory compliance *are* required under Subclause 4.5.1 "Monitoring and Measurement," so a certification auditor will check for it.

Examples, most mentioned earlier, of "other environmental standards and practices to which the enterprise has obligated itself" as required by item b) include requirements from corporate headquarters, codes of practice of trade associations, agreements with citizen groups, self-declared environmental practices, special agreements with local authorities, and agreements with labor organizations.

An interesting question is why a requirement for the enterprise to track its nonregulatory environmental obligations is included in the EMS standard. Not that this isn't a good idea, but is an EMS standard the place for what might seem like a business ethics issue? Perhaps the requirement is, among other things, meant to reinforce the idea that even an enterprise that self-declares its conformance to ISO 14001 has to honor all of the standard's requirements, by the internal logic of the system.

PLANNING: OBJECTIVES AND TARGETS (SUBCLAUSE 4.3.3)

Assessing Strategic Importance

As mentioned earlier, environmental aspects need to be assessed and ranked in terms of environmental significance, and then strategic business considerations have to be taken into account. Environmental significance factors and criteria of strategic importance together underpin the process of establishing strategic environmental objectives for the enterprise. After all, an EMS promotes strategic environmental management, and because most enterprises can't do everything at once, they have to set strategic priorities.

Take as an example two environmental aspects of an enterprise's operations that are equally significant in terms of environmental impact — say, gaseous air pollution emissions and particulate air pollution emissions. The enterprise can't deal with both aggressively, so it would establish the more aggressive environmental objective for the aspect that is strategically more important. One aspect might be strategically more important than the other because it represents a regulatory noncompliance or because there has been an outcry about it from the neighboring community.

Actually, the EMS Committee cannot complete its assessment of environmental significance of the different environmental impacts and aspects until it has spent some time working on objectives, targets, and environmental management programs. This is because additional information on the scale, severity, duration, and frequency of the impact, as well as its ecological and human health implications, is likely to surface during the process of establishing environmental objectives and targets, and designing programs to achieve them.

Indeed, the EMS Committee can't even complete its assessment of the strategic importance of environmental aspects and set objectives unless it sets environmental targets and environmental management programs. For example, one of the important criteria for evaluating the strategic importance of an environmental aspect and its impact may be the ability of the enterprise to remedy the situation, or the cost or amount of time it would take. The EMS Committee may be able to develop this sort of information only while determining appropriate environmental targets for the enterprise or while designing environmental management programs to achieve the targets.

Let's consider the first activity/aspect/impact example in Table 2 — generating process steam/release of NO_x, SO_x, and CO_2 emissions/air pollution and climate change effects. Suppose that research conducted while planning an environmental management program to reduce the air pollution shows that, because the equipment and practices of the enterprise are fairly up-to-date, it is possible to achieve only a very slight reduction in these emissions, at very high cost. If one of the factors for strategic importance for this enterprise was "cost to achieve a 50 percent reduction in the impact," this finding might bump this environmental aspect of the enterprise down on the scale of strategic importance. The EMS Committee might decide to shift the resources of the enterprise to where they could do more good, and revise the original objective established for this aspect to a more modest one.

The list of criteria used to assess the strategic importance of significant environmental aspects depends on the nature of the enterprise, its operations, and its markets, as well as on the things that top management and the EMS Committee want to include in the mix of strategic factors.

There obviously has to be iterative interplay among the processes of determining environmental significance, determining strategic importance, establishing environmental objectives, setting environmental targets, and planning environmental management programs. This learning-and-revising process also extends to *implementing* the environmental management programs.

In addition to environmental significance, other factors for evaluating the strategic importance of environmental aspects and their impacts might include:

- Whether or not the environmental impact represents a current regulatory noncompliance, or is related to forthcoming regulatory standards
- Whether or not the environmental impact represents a current nonconformance to other standards and practices to which the enterprise has obligated itself, or plans to obligate itself
- The cost of reducing the environmental impact
- The employee labor required to reduce the environmental impact
- Technology choices
- The findings and results of environmental impact assessments
- The economic benefits of reducing the environmental impact (e.g., the benefit/cost ratio of a P2 solution)
- Future environmental liability concerns
- Community concerns and priorities
- The concerns and priorities of other stakeholders, such as workers, shareholders, customers, trade organizations
- Results of past EMS audits

- The effect on the ability of the enterprise to do business or carry out its business plans
- The effect on the ability of the enterprise to maintain its operations or carry out planned changes in its operations
- The effect on the public image of the enterprise

Apart from a few minimum requirements listed in ISO 14001 that make good sense for any EMS, it's up to top management and the EMS Committee to formulate the list of criteria for evaluating strategic importance. What goes on this list of criteria depends on the nature of the enterprise, its operations, and its markets, as well as on the things that top management and the EMS Committee want to include in the mix of strategic factors.

Though the EMS Committee may do most of the work, the decision about which criteria to use when ranking significant environmental aspects is a strategic decision for the enterprise, so top management needs to participate in the process and give final approval.

Environmental Objectives

At this point in the planning component of the EMS we have:

a) Divided up the operations of the enterprise and the uses of its products into discrete activities, and examined those that seem to have an important environmental aspect
b) Identified the specific environmental aspects of the activities
c) Made at least a first pass at identifying the environmental impacts associated with each of the environmental aspects, though we may quantify and refine how we record these impacts later
d) Begun the process of assessing environmental significance and strategic importance of the environmental aspects, so we can prioritize them, establish appropriate environmental objectives for them, and focus the limited resources of the enterprise particularly on those that are most significant

This gives us the specific *objects* of the planning process, the "things on which the planning is focused" that was mentioned earlier.

Now it's time to establish what the enterprise wants to achieve with respect to its significant environmental aspects - the *purposes* of the planning. Common planning practice establishes specific aims at two levels of detail: objectives and targets. An environmental *objective* reflects the overall intent of the enterprise regarding an environmental aspect, whereas a *target* generally is an explicit performance criterion, quantified and time-framed, that represents progress toward achieving the objective.

Environmental objectives can be thought of as specific goals, representing the type of environmental performance called for in the environmental policy of the enterprise. Obviously, top management has to approve the environmental objectives for the EMS, and the objectives need to be frequently reviewed and adjusted as necessary. Most environmental objectives will be *operational objectives* (such as those in Table 3) that represent environmental performance aims for the enterprise. But there will also be *management objectives*: objectives that focus on achieving improved capabilities and systems to support good environmental performance, such as:

- Achieving improved management systems
- Achieving certain types and levels of training of workers
- Achieving improved communication with the neighboring community
- Establishing a routine of P2 audits

Most objectives will be operational ones that represent environmental performance aims for the enterprise; but there will also be management objectives that are not as "concrete," that will focus on achieving improved capabilities and systems to support good environmental performance.

Table 3 extends the first five examples of enterprise activities and their environmental aspects from Table 2. It shows that any one environmental aspect can have more than one objective associated with it, and that any one environmental objective can be represented by more than one environmental target. Please review Table 3 carefully and note the many different forms environmental objectives and targets can take.

Some of the examples of environmental objectives in Table 3 are extreme, like capturing *all* fugitive particulates from core process operations, and may have to be revised down as the enterprise gains knowledge of what is possible. In fact, after it gains further knowledge, the EMS Committee may want to refine the initial environmental objectives and targets in many ways, such as making them more or less ambitious, or expressing them in different terms. For example, the first objective in Table 3 — reduce emissions of the three types of pollutants together by 80 percent per unit of production — may later be refined into different percentages for each type of pollutant because each of the pollutants has a different environmental impact per unit of weight.

Environmental objectives and, in most cases, targets are matters apart from the methods used to achieve them. For example, the objective for the environmental aspect of the fourth activity in Table 3 is "Eliminate release of toxic effluents to the river." In principle, depending on the nature of the toxic chemicals, the enterprise might pursue this objective by:

- Outsourcing product finishing operations
- Using nontoxic chemicals in the finishing operations (a P2 solution)
- Using a completely different finishing process that doesn't produce a toxic effluent (a P2 solution);
- Installing a wastewater treatment facility at the plant
- Redirecting the effluent to the municipal wastewater treatment system

- Capturing the toxic chemicals after the finishing process but before they mix with other effluent from the plant (if they can be recycled in a closed loop, this becomes be a P2 solution)

It is easy to see why objectives and targets usually can be finalized realistically only after the measures for trying to achieve them have been selected.

Table 3 provides examples of several different ways that environmental objectives can be expressed. In some cases, the EMS Committee states the objective in percentage terms; in other cases, it states the objective in terms of eliminating the environmental aspect or impact; and in still other cases, it simply states the objective in terms of minimizing or reducing the environmental aspect.

Table 3. Aspects, Objectives, Targets. (*Note*: PUP=Per Unit of Production)

Activity or operation	**Environmental aspects**	**Objectives**	**Targets, 1st year (ending *[date]*)**
1. Generating process steam	Release of NO_x, SO_x, and CO_2 emissions	Reduce emissions of pollutants by 80% PUP	Reduce tons of CO_2 equivalents by 20% PUP
	Fly ash disposal	Not applicable	
2. Receiving and handling hazardous chemical inputs	Potential for accidental spills of toxic chemicals	Minimize potential for spills	Receiving and handling procedures and equipment in place
		Contain all spills	Containment procedures and equipment in place
			All handlers trained in hazardous chemical handling and containment
3. Core process operations	Use of energy (electricity)	Reduce energy consumption by 25% PUP	Reduce KWH of energy consumption by 10% PUP
	Use of water	Reduce m^3 of water used by 50% PUP	Reduce M^3 of water used by 10% PUP
	Leaking of lubrication oils	Leak-free process	No leaks for 3 months
	Fugitive particulates	Capture all fugitive particulates	Capture 80% (by weight) of fugitive particulates
4. Product finishing operations	Release of toxic effluents	Eliminate release of toxic effluents to river	Reduce kg of liquid toxins released by 25% PUP

Table 3. Continued.

Activity or operation	**Environmental aspects**	**Objectives**	**Targets, 1st year (ending *[date]*)**
5. Product packaging	Use of paper products	Reduce amount of packaging to minimum	Reduce packaging an average of 10% by weight per unit
		Utilize at least 50% recycled materials in packaging	Average recycled content of 15% by weight in product packaging

Any of these approaches will do, as long as it is suited to the corresponding environmental aspect and the result is a meaningful environmental objective.

Environmental Targets and Indicators

Environmental targets provide the next level of detail: verifiable indicators of achieving the objectives, or of progress toward achieving the objectives, within specific periods of time. *Operational targets* are expressed in terms of indicators of environmental performance; *management targets* are expressed in terms of completion of specific activities.

After further knowledge is gained, the EMS Committee may want to refine the initial environmental objectives and targets in many ways, such as making them more or less ambitious, or expressing them in different terms.

For simplicity's sake, all the environmental targets in Table 3 are given in terms of a one-year time frame, but in practice neither the starting nor the ending dates of time frames have to be the same for all the environmental targets of the EMS. In fact, when you consider the very different sorts of initiatives that will be needed to address the different environmental objectives of an enterprise, it becomes clear that environmental targets must have widely differing time frames. As a practical matter, those time frames might range between three months and three years.

In some of the Table 3 examples, the target represents full achievement of an environmental objective. In most, the target represents only a first increment of progress, and it will take several such increments before the objective is achieved. In at least one case (reducing packaging by 10% per unit), it remains to be determined later whether or not achieving the target also will be considered fully achieving the objective.

Most of the operational targets in Table 3 are expressed in terms of verifiable units of weight, volume, or energy. They are expressed as changes in percentage when compared to baseline data at the beginning of the period. Of course, when you have baseline data, targets can be expressed as absolute environmental performance measures instead of percentages. In most cases, the performance indicators are expressed as "per unit of production," or PUP, so the change achieved is meaningful in terms of the enterprise's operating practices.

The second example in Table 3 concerns a *potential* environmental aspect and impact. In this case, the objectives and targets deal with events that have not yet taken place in the enterprise, but should be prevented from taking place. Because there can be no baseline for comparison, the targets established are necessarily management targets: having procedures and equipment in place, and workers properly trained.

In the case of a similar environmental aspect — leaking lubrication oils in the third example — leaks have occurred often in the past, and so a three-month period free of leaks in the first year will be taken as evidence that successful new preventive measures have been implemented by the enterprise.

Subclause 4.3.3 implies recognition that an enterprise may be able to deal with objectives and targets only for its most significant environmental aspects. It is also meant both to allow and require an enterprise to prioritize strategically.

Just as they must approve environmental objectives, top management must approve environmental targets. It is difficult to say at what point approval of top management should be sought, because approval of environmental targets is necessarily connected to some degree to the time, labor, and money resources required to achieve them, and the resources required for achieving environmental targets may not be known until environmental management programs have been figured out. Obviously, the more top management is involved with the entire process of determining objectives, targets, and environmental management programs for the EMS, the less of an issue obtaining top management approval will be.

Requirements of ISO 14001

Subclause 4.3.3 of the ISO 14001 EMS standard requires the enterprise to:

a) Establish and maintain environmental objectives and targets

- in written form
- for each relevant operational function and level of the enterprise
- that are consistent with its environmental policy

that reflect its commitment to pollution prevention

b) Consider the following when establishing and reviewing its environmental objectives:

- environmental laws and regulations with which it must comply
- other environmental standards and practices to which it has obligated itself
- its significant environmental aspects

- available technological options for achieving the objectives
- its financial requirements
- its operational requirements
- other business requirements
- the views of trade associations, citizen groups, and other outside interested parties

Item a) is the basic requirement that the enterprise establish and regularly review and update a set of environmental objectives and corresponding targets. The EMS standard does not define the terms "objective" and "target." The annex to ISO 14001 offers the guidance that "objectives should be specific and targets should be measurable wherever practicable"; so, the intent is obviously for increasing specificity, and that is what auditors will look for. Auditors will also look for evidence that the objectives and targets are taken seriously by the enterprise, and are actively pursued as well as regularly reviewed and updated. Progress toward a target with a lengthy time frame can be assessed at any time by an auditor, who can simply note where the enterprise stands in the timetable for the environmental management program designed to achieve the target.

ISO 14001 certification auditors will also look for a relationship between the set of environmental objectives and targets, and specific elements of the enterprise's environmental policy. They will try to consider objectively if the enterprise has set P2-based objectives and targets (instead of end-of-pipe-based objectives and targets), wherever practicable. What are the relevant operational functions and levels of the enterprise for establishing objectives and targets? Obviously, this cannot be answered in generic terms, but it's safe to say that any enterprise that has seriously and systematically examined its "activities, products, and services" for significant environmental aspects and impacts will be deemed by auditors to have addressed its relevant functions and levels.

Item b) provides a minimum list of strategic factors to be taken into account when establishing environmental objectives for the enterprise — what we have called the criteria or factors of strategic importance. This is essentially a requirement that environmental objectives be established on a strategic basis. There is nothing in the standard that prohibits an enterprise from considering additional criteria or factors; and the standard contains no requirements regarding the relative weights that should be given to the significance factors, or even how the individual factors should be interpreted.

But it surely is not by chance that environmental laws and regulations are first among the strategic importance factors listed in ISO 14001. Item b) implies a recognition that, as a practical matter, an enterprise may have to focus most of its available resources on achieving objectives for its most strategically important significant environmental aspects. This subclause it is meant both to allow and require an enterprise to prioritize strategically and in a way that reinforces sustainability of the enterprise, but with special attention to regulatory compliance.

PLANNING: ENVIRONMENTAL MANAGEMENT PROGRAMS (SUBCLAUSE 4.3.4)

Environmental Management Programs

At this point in the planning process, the enterprise — or, more precisely, the EMS Committee and top management — has determined in very concrete terms what should be achieved by way of improving environmental performance. It has established environmental objectives and a target, or sequence of targets, for its significant environmental aspects. Now the EMS Committee needs to plan for action — to figure out the *environmental management programs* that will achieve the enterprise's environmental objectives by meeting the targets.

The EMS Committee should think in terms of a set of environmental management programs for the enterprise, or alternatively, one environmental management program containing a number of separate initiatives.

As mentioned earlier, no doubt the EMS Committee will discover the need to alter at least some targets and objectives in the course of coming to terms with what sorts of environmental management programs are actually possible. Planning and carrying out environmental management programs are the active expressions of strategic environmental management. They should be integrated into existing management routines so environmental management is an integral part of designing products, planning and implementing production processes, planning and carrying out the movement of people and materials, marketing, expansion planning, and so on.

The plan for an environmental management program contains the same basic elements as any action plan. It spells out a sequence of specific activities that progress toward an end state (achieving the target); it provides a schedule for carrying out the activities; and it identifies the individuals responsible for each of the activities. To confirm top management's commitment, the plan for an environmental management program should also indicate the basic resources allocated to it — the time of specific individuals, elapsed time (that is, the total amount of time allocated, rather than the amount of time that one or more individuals put into it), equipment, materials, and money. In this way, when senior managers approve a plan for an environmental management program, they perforce approve allocating the resources to carry it out. Also, a good plan should have a number of scheduled benchmarks that make it easy for implementers and managers (and auditors) to assess progress and compare action with stated intent.

How many environmental management programs should an enterprise have? The short answer is that an enterprise should have at least one environmental management program for each significant environmental aspect.

But an enterprise starting the EMS implementation process very incrementally may have set a limited initial scope for its EMS, or may be concentrating on only one or two environmental aspects at first, so it may have only one or two environmental management programs. Moreover, a single environmental management program can address more than one environmental aspect; a single environmental aspect can have more than one objective associated with it, possibly requiring more than one environmental management program; and forward-thinking managers may well want to have environmental management programs for more than just the environmental aspects for which objectives and targets have been established. They may want environmental management programs that truly serve the needs of strategic management.

The EMS Committee should think in terms of a *set* of environmental management programs for the enterprise. An alternative conceptualization is one environmental management program containing a number separate initiatives. Either way, we are talking about two or more distinct planned-out efforts aimed at implementing the environmental policy and improving the environmental performance of the enterprise. Why at least two? Table 4 shows the three basic types of environmental management programs. Any enterprise that takes its EMS seriously and is trying to implement its environmental policy could not possibly satisfy itself with only one initiative among the three different categories.

Table 4. Three Types of Environmental Management Programs.

Type of Program	Purpose	Short or Long Term	Operations or Management
Monitoring	Track environmental aspects without programs	Long term	Operations
Targeted	Achieve targets	Short and long term	Management and/or operations
Installation	Track environmental aspects of a new development	Short term	Management and/or operations

The EMS Committee, taking into account all strategic assessment factors and available resources, will select from one to several of the most significant environmental aspects of enterprise operations, establish objectives and targets for them, and develop environmental management programs to achieve the objectives. But what about the other environmental aspects, the ones that are not most significant? For these, there should be environmental management programs that monitor their environmental impacts and track changes over time.

Targeted programs are the ones most often talked about. They are designed to achieve specific targets and objectives. The programs can be relatively short or long, or they can be increments of a larger, longer-term program. Targeted programs can be management or operations programs, or they can contain elements of both.

Installation programs are programs designed to manage the environmental aspects of something new in the enterprise, like installation of new equipment, a major process change, a new product line, a new facility, and so on. This type of environmental management program usually includes an overall environmental review of all aspects of the new installation, and may involve interventions in the planning, design, construction, and operating stages. In some cases, it may also encompass the disposal stage at the end of the installation's working life.

Programs to Achieve Environmental Targets

Here we'll look a bit more closely at targeted environmental management programs, the second type in Table 4. In looking for ways to achieve a target, the EMS Committee looks not just at the activity and location in the enterprise where the environmental interaction — that is, the environmental aspect — takes place. They also look "upstream" at what takes place earlier in a process, and "downstream" at what follows the environmental aspect in a process, for possible clues to a least-cost and cost-effective environmental management program to address the aspect. And in the case of any basic solution to a particular problem, there may be alternative approaches and technologies, vendors and service contractors, and other things that must be factored into the equation. This means that most often a good bit of research goes into the planning for an environmental management program.

Table 5 extends the earlier examples, adding a column for summaries of the environmental management programs. The rightmost column shows the major activities and timeframes for each environmental management program. For simplicity, we again represent each program as taking place within a 12-month period or less. But the careful reader will see that some of the programs are really just a first increment of a longer-term program. The second footnote indicates that the first month of all the environmental management programs is dedicated to basic research and refinement, and to detailed planning for the programs.

If we look at the program summary of the first example we see that, even after initial research and refinement, the EMS Committee estimated that another two months are needed to explore alternative solutions to reducing the air pollution created by generating process steam. Coal is the fuel being used. Looking upstream, downstream, and directly at the steam plant, emissions of pollutants could be reduced by producing less hazardous emissions, by using less steam and thus producing a lower volume of emissions overall, or by using a combination of the approaches. The environmental management program could entail fuel substitution (natural gas or a mixed-fuels system); firebox improvements (such as thermal insulation); low-NO_x burners; afterburner efficiency improvements; upgraded seals and valves; automated control instruments; or other energy efficiency innovations in the process that uses the steam.

Achieving the environmental target and then the objective will involve a major investment, so it's no wonder that the EMS Committee wants sufficient time to study the matter thoroughly. Even after that, they estimate another three months to fully plan a long-term program to achieve the objective. Only the first phase will be implemented in the first year, and hopefully achieve the target for the first year. We encourage readers to review each of the other examples in Table 5 and to work out in their imaginations the situation at the enterprise and the considerations and intents of the EMS Committee reflected in the environmental management program summaries.

In reviewing the program summaries, note how in these examples a P2 audit serves the needs of three of the environmental management programs. A P2 audit — even a very basic one — can be an excellent source of ideas for ways to achieve targets, as well as for baseline data for planning and for monitoring environmental management programs.

Table 5. Objectives, Targets, and Programs. (*Note*: PUP=Per Unit of Production)

Objectives	Targets, 1st year (ending *[date]*)	Program summary*
Activity or operation		
1. Generating process steam		
Reduce emissions of pollutants by 80% PUP	Reduce tons of CO_2 equivalents by 20% PUP	Explore alternatives in detail (months 2-4) Develop a 2–3 year program of activities** and investments (months 5–7) Implement 1st phase of the program (months 8–12)
2. Receiving and handling hazardous chemical inputs		
Minimize potential for spills	Receiving and handling procedures and equipment in place	Prepare receiving and handling procedures (months 2–5) Acquire and install new equipment (months 2–9)
Contain all spills	Containment procedures and equipment in place	Prepare containment procedures (months 2–5) Acquire and install new equipment (months 2–9)
	All handlers trained in hazardous chemical handling and containment	Develop handling and containment training program (months 5–8) Complete first 3-month training cycle (months 9–11)
3. Core process operations		
Reduce energy consumption by 25% PUP	Reduce KWH of energy consumption by 10% PUP	Conduct P2 audit (month 2) Develop a 2-3 year program of activities** and investments (month 3) Implement 1st phase of the program (months 4–12)
Reduce m^3 of water used by 50% PUP	Reduce m^3 of water used by 10% PUP	After the P2 audit develop a 2-3 year program of activities** and investments, and implement 1st phase (months 3–12)

Objectives	Targets, 1st year (ending *[date]*)	Program summary*
Leak-free process	No leaks for 3 months	Prepare maintenance procedures and work instructions covering all valves and seals (months 2–4) Develop maintenance training program (months 5–7) Complete first training cycle (month 8)
Capture all fugitive particulates	Capture 80% (by weight) of fugitive particulates	Research vent and baghouse systems (months 2–3) Acquire and install new equipment (months 4–6) Train workers on monitoring particulates and using and servicing equipment (months 7–8)
4. Product finishing operations		
Eliminate release of toxic effluents to river	Reduce kg of liquid toxins released by 25% PUP	After P2 audit develop a 2–3 year program of activities** and investments (months 3–4) Implement 1st phase of the program (months 5–12)
5. Product packaging		
Reduce amount of packaging to minimum	Reduce packaging an average of 10% by weight per unit	Engage packaging design consultant (month 2) Conduct experiments with alternative designs and recycling content (months 3–5)
Utilize at least 50% recycled materials in packaging	Average recycled content of 15% by weight in product packaging	Design 2–3 year program to reduce product packaging (months 6–7) Implement 1st phase of the program (months 8–12)

* *Not shown in the table, the first month of all programs is dedicated to basic research (collecting baseline data; case studies on the experiences of others; and data on alternative techniques, technologies, and materials), revision of objectives and targets as necessary, refining target performance measures in terms of indicators and quantities, detailed planning of the environmental management program (including monitoring procedures), and possibly engaging consultant support.*

** *Training programs, development of procedures and detailed work instructions, continuing research into P2 technologies, experimentation with new techniques, etc.*

Requirements of ISO 14001

Subclause 4.3.4 of the ISO 14001 EMS standard requires the enterprise to:

a) Establish and maintain one or more environmental management programs to achieve its objectives and targets, with the plan for each program specifying, among other things:

- the people responsible for carrying out the program and achieving the objectives and targets at each relevant function and level of the enterprise
- the means and timeframe by which objectives and targets are to be achieved

b) Amend these programs as needed to ensure that it applies principles of environmental management to new developments and to new or modified operations or products

In Item a), the phrase "establish and maintain" means that, in addition to creating one or more environmental programs, the enterprise is required to review those programs periodically and revise them if necessary to account for changed circumstances of any type, including: changes to objectives or targets; changed external conditions, such as the availability of new technologies; new information from, say, a P2 audit; and new knowledge arising from experience with implementing the programs. Beyond that, the standard requires only that individuals with responsibility for an environmental management program and its results be designated, that the program include the means for achieving the target, and that the target itself, along with its objective, indicators of success (implicitly), and timeframe, be incorporated into the program.

The standard does not require that environmental management programs be written, but it's hard to see how they could not be. Otherwise, how could anyone, including a certification auditor, determine if the programs exist, are complete, are being carried out, and are being updated?

Item b) can be interpreted a few different ways, because of the language employed. For the enterprise's EMS to be "audit proof" — but, more important, for it to do what is best for the enterprise — item b) should be interpreted as a two-fold requirement. First, care should be taken to amend an environmental management program appropriately if there is a relevant change in the activity or process that gives rise to the environmental aspect being addressed by the program. This is really a reemphasis of, and elaboration on, the term "and maintain" at the beginning of item a). Second, environmental management programs should be devised for new developments. To use the language of Table 4, "installation" programs should be devised for new equipment, a major process change, a new product line, a new facility, and so on. In this case, ISO 14001 means that the *set* of environmental programs should be amended with new programs. Any amendment or new program ought to be documented, so a certification auditor can see that the requirement is being met.

IMPLEMENTATION AND OPERATION: STRUCTURE AND RESPONSIBILITY (SUBCLAUSE 4.4.1)

The environmental management programs have been planned, and now it's time to carry them out — to implement them. ISO 14001's Clause 4.4,

"Implementation and Operation," contains seven subclauses, covering requirements for the capabilities and support mechanisms that underpin implementation of environmental management programs. These same capabilities and support mechanisms also underpin operation of the EMS overall, so we might think of them together as the EMS infrastructure of the enterprise. This EMS infrastructure also enables sound management by the enterprise of its environmental aspects in the course of normal day-to-day operations. That's why both implementation *and* operation are in the title of this clause.

Authority, Capacity, and Resources

The first essential element of what we call the EMS infrastructure is a framework of people with specific ongoing EMS assignments, and with the authority, capacity, and resources to carry out their assignments. A framework of people means the assignment of EMS roles and responsibilities to workers at different levels and in different departments of the enterprise, with clear lines of reporting and accountability. At the center (or top) of the EMS framework is the single individual assigned responsibility for overall coordination of EMS operations and for reporting on EMS operations to top management. As previously mentioned, in many enterprises, especially larger ones, top management establishes an EMS Committee headed by the overall coordinator, to share the ongoing work of coordinating and maintaining the EMS. The EMS Committee is usually made up of workers with key roles in the overall framework of people with specific responsibilities for operating and maintaining the EMS.

At one end of the spectrum, this "framework" could be no more than one person devoting one-quarter of their time to EMS coordination, and reporting to top management thoroughly but casually, without formal meetings (in fact, in a small enterprise, the EMS coordinator might *be* top management). At the other end of the spectrum, the EMS framework could include, say, a full-time EMS coordinator, dozens of key and second-level workers throughout the enterprise, and EMS Committees at different levels that meet regularly and report through the hierarchy of committees to top management in formal, monthly sessions. It all depends on the size and nature of the enterprise, the scope of its EMS, and its approach to establishing and operating its EMS.

But having even a well designed framework of people with EMS assignments doesn't amount to much if those people don't have the authority to make decisions and command resources appropriate to their responsibilities. And having the appropriate authority won't help if they don't also have the capacity (the skills and knowledge) to exercise the authority effectively. Authority and capacity are useless without adequate resources — time, money, equipment, and help — to which their authority and talents can be applied. The framework of people, roles and responsibilities, and resources must be thought of together as a single critical element of the enterprise infrastructure essential to implementing and maintaining a productive EMS.

Requirements of ISO 14001

Subclause 4.4.1 of the ISO 14001 EMS standard requires the enterprise to:

a) Determine EMS roles, responsibilities, and authorities
b) Describe the EMS roles, responsibilities, and authorities in writing
c) Communicate the EMS roles, responsibilities, and authorities
d) Provide essential resources in the form of people, special skills, equipment, and money to implement and control the EMS
e) Appoint one or more Management Representatives who, regardless of their other responsibilities, have defined roles, responsibilities, and authority for:
    - ensuring that EMS requirements are implemented and maintained in conformance to ISO 14001
    - reporting on the EMS to top management so that they can review its performance and operations and make informed decisions about ways to improve it

Certification auditors will look for written documentation of the EMS structure of roles, responsibilities, and authorities, and they will look for evidence that the information has been communicated. A good auditor will also check with workers on the shop floor to see if the communication has reached everyone.

Item b) requires that the framework of EMS roles and responsibilities be put in writing, whether as job descriptions, in list form, as an organization chart, or in some other format. The important thing is that it be easily understood, because the intent of ISO 14001 here is clearly to facilitate communication about the EMS framework. The actual language of the standard is that the information should be communicated "to facilitate effective environmental management." Everyone should know what their own role is and be able readily to determine the roles, responsibilities, and authorities of other employees with specific EMS assignments, so the EMS will operate smoothly.

Certification auditors will look for written documentation of the EMS structure of roles, responsibilities, and authorities, and they will look for evidence that the information has been communicated. A good auditor will also check with workers on the shop floor to be sure that the communication has reached everyone, not just workers directly involved with the EMS, so it can contribute to good environmental management throughout the enterprise.

Item e) refers to the "Management Representative," the ISO 14001 term for the overall EMS coordinator. The ISO 14001 term for this position is a good one because it emphasizes that this person represents the interests of top management.

This person is responsible, on top management's behalf, for seeing that the requirements of the standard are met, that the EMS operates as efficiently as possible and integrates with other enterprise management systems, and that it is effective. Item e) allows for more than one Management Representative, but as a practical matter this means that, especially in larger enterprises, it may be appropriate to establish an EMS Committee in addition to designating a primary Management Representative. In a sense, all members of the EMS Committee are Management Representatives.

It's not by chance (nothing in ISO 14001 is by chance) that the language of the standard points out that the Management Representative must have the specified EMS responsibilities regardless of other responsibilities that he or she may have. In a large enterprise, the EMS Management Representative may have no responsibilities other than coordinating and reporting on the EMS. In mid-sized enterprises, it is common for the job description of one person to include coordination of both quality and environmental management, and sometimes occupational health and safety (OHS) and other corporate social accountability functions as well. In most cases, the job of EMS Management Representative is given to a middle manager who already has plenty to do in some area of routine operations.

The language of this subclause is meant to emphasize that the EMS responsibilities of the management representative are to be taken as seriously as any other responsibilities this person may have. By implication, if it's too much for one person to do and carry out their other responsibilities at the same time, then things need to be reshuffled a bit: the Management Representative needs more resources — perhaps time, perhaps help — on the EMS side, or on the routine operations side, or both. The EMS Management Representative is the continuous link between top management and everyone else involved with the EMS.

It's interesting that the authors of ISO 14001 have explicitly noted in this subclause, and in a discussion about the Management Representative, that the reason for reporting to top management is to enable top management to review the performance of the EMS and make informed decisions about ways to improve it. Here is another example of an explicit internal linkage within ISO 14001: the Management Representative is responsible for ensuring that top management has the appropriate information for carrying out the management review required by Clause 4.6 of the standard. To satisfy the requirements of Subclause 4.4.1, certification auditors will look for evidence that the Management Representative is feeding top management the appropriate information; to satisfy the requirements of Clause 4.6, they will look for evidence that top management has the appropriate information in the EMS management review and takes that information into account when deciding what EMS improvements to make.

IMPLEMENTATION AND OPERATION: TRAINING, AWARENESS, AND COMPETENCE (SUBCLAUSE 4.4.2)

Awareness, Capability, and Motivation

In this element of the EMS infrastructure we go from the level of the framework of EMS roles and responsibilities to the level of the individual worker in the enterprise. And not just the workers who are part of the formal EMS framework, but *all* employees of the enterprise. For an enterprise to achieve maximum environmental and bottom-line benefit from its EMS, the performance of all workers needs to support EMS operation and reflect the values and goals of the enterprise's environmental policy.

If people are to perform in a certain way they need three things: awareness of what should be done, the capability to do it, and the motivation to do it.

If people are to perform in a certain way they need three things: awareness of what should be done, the capability to do it, and the motivation to do it. Awareness is conveyed through awareness training, and capability is conveyed through technical training, though most training programs include elements of both. Motivation in part is created by awareness and capability, or competence, because people who know how to accomplish something tend to feel empowered by the knowledge, and want to do it. But, though work satisfaction is indeed a powerful motivator, so are direct rewards for the right performance.

First, the training: The enterprise needs to have a training program to convey awareness and capability related to environmental management programs, the operations and aims of the overall EMS, and managing its environmental aspects in the course of normal day-to-day operations. When the EMS is first being installed it will probably be pretty easy to identify the exact training needs, because there will be so much that is new to everyone. But in due course there will have to be some form of periodic assessment that examines specific types of awareness training and technical training needs for different levels and categories of workers. All employees need to be aware of the EMS and how their activities interact with the environment directly and indirectly; and all employees need to be competent in the skills, tools, and techniques for performing their tasks efficiently and in a manner consistent with the enterprise's environmental policy. A training needs assessment periodically carried out by the enterprise should point to the training necessary to be sure that is the case.

Topics that could be included in an EMS training program include:

- Environmental awareness, general
- Environmental awareness, by function in the enterprise
- Environmental ethics
- Environmental policy and standards of the enterprise
- Operations of the enterprise as related to environmental performance

- Methods of environmental measuring and monitoring
- Applicable environmental regulations
- Environment-related work competence and skills, by function, in the enterprise
- Emergency response procedures
- Principles of P2
- P2 auditing
- The EMS
- EMS implementation
- Roles in the EMS
- EMS auditing
- For contractors and suppliers — the enterprise's environmental policy, EMS, and relevant procedures

The scope and scale of the EMS training program will obviously depend on the size and nature of the enterprise, the scope of its EMS, and its approach to establishing and operating its EMS. The EMS training program cannot stand alone. Because of the money and employee time involved, no enterprise can afford separate training programs for environmental management, quality management, health and safety management, and other important matters, in addition to basic competence training for its workers. For this reason, EMS training must be integrated with other types of training. As a part of the enterprise's overall training program, EMS training will probably be structured in continuous cycles that over the course of, say, a year's time cover every employee of the enterprise. Moreover, to abide by the spirit and purposes of an EMS, an enterprise has to try to influence its contractors and suppliers to adhere to environmental training standards for their workers similar to those of the enterprise itself. There have been many cases in which an enterprise has invited its contractors and suppliers to send workers to participate in its own training program, or has conducted training jointly with its contractors and suppliers.

The enterprise needs to have a training program to convey awareness and capability related to environmental management programs, the operations and aims of the overall EMS, and managing its environmental aspects in the course of normal day-to-day operations.

The training program addresses the need for two things: awareness and capability. Motivation, the third thing that enterprise workers need for their performance to support the EMS and good environmental management in general, is a bit more elusive. As mentioned earlier, motivation is partly created by awareness and capability, but direct rewards for the right performance are also needed. Some elements of a motivating reward system might be easy to figure out: awards, bonuses, raises, and special gifts for achieving environmental management targets, for especially creative and cost-effective solutions to environmental problems, for the largest annual departmental cost saving from a P2 innovation, for the best departmental record of regulatory compliance, and so on.

But in part, elements of a traditional employee motivation program tend to operate counter to sound environmental management, because they typically provide incentives for direct cost- or time-saving performance within a limited frame of reference, such as a single department or a single reporting period. For example, in any department of an enterprise the immediate least-cost approach to dealing with waste — let's say, a combination of packaging from inputs, shavings, and other waste generated directly by the production process; used lubricants; and expendable supplies and small, replaced machine parts — would be to throw them all in the dumpster and have them removed to the municipal landfill. This would be especially appealing if "tipping" fees (landfill dumping fees) are relatively low. It might be even more appealing to some department managers rewarded for minimizing costs by traditional calculations if departmental waste is combined with other waste from the enterprise and the waste management costs are accounted for at the enterprise level.

An alternative approach would be to separate and recycle the different types of waste as much as possible while taking measures to minimize waste, though making this entire scheme viable may require cooperation with other departments of the enterprise. How could this approach, obviously better from an environmental management point of view but also requiring extra work and additional cost of operations in the short run, be made appealing to the department manager? The incentive system would have to allow him or her to account for the fact that, in the future, the cost of wasted materials and landfilling will increase; that there will be long-term streams of revenues and savings from the sale and reuse of sorted and recycled materials, and very possibly savings for other departments of the enterprise as well; that in the long run, the overall bottom line of the enterprise will improve; and perhaps even that better environmental performance has its own inherent value. To motivate performance that supports the aims and operation of an EMS, top management may have to completely overhaul the reward system so the consequences of performance are calculated with the entire enterprise (some would argue the entire globe) and a longer time period as the frame of reference.

Requirements of ISO 14001

Subclause 4.4.2 of the ISO 14001 EMS standard requires the enterprise to:

a) Identify its EMS training needs
b) Make it a requirement that all employees whose work could have a significant impact on the environment receive appropriate training
c) Establish and maintain procedures to make its employees aware of:
    - the importance of performing in a manner that supports its environmental policy and procedures, and the requirements of the EMS
    - the significant environmental impacts and potential environmental impacts of their work activities, and the environmental benefits of improving their performance

- their roles and responsibilities in achieving conformance to the policy, procedures, and requirements of the enterprise's EMS, including emergency preparedness and response requirements
- the potential consequences of not following specified operating procedures

d) Ensure that, as a result of appropriate education, training, and/or experience, workers performing tasks that can cause significant environmental impacts are competent at their jobs

Item a) seems a bit vague, but, in fact, is elegant. Training needs are always changing because of employee turnover and changing circumstances inside and outside the enterprise. So, for a certification auditor to confirm at any time that the enterprise has identified its relevant EMS training needs, there would have to have been a training needs assessment carried out in the not-too-distant past. And because recertification audits are carried out periodically, training needs assessments also have to be carried out periodically, at intervals no longer than the time between recertification audits.

Items b) and d) are closely related and similarly interesting in that they talk in terms of the enterprise ensuring that the workers receive training or are competent, rather than in terms of the enterprise doing something concrete, such as providing training or devising a program to make workers competent. The reason for this is that the training and competence can come from a variety of sources — earlier training at a different enterprise, work experience, an outside training program, on-the-job training, etc. The enterprise is only obligated to ensure that this group of workers is appropriately trained and competent. Both requirements use the term "appropriate" and, by that, presumably mean that the training should equip employees of the enterprise to avoid the potential significant negative environmental impacts in their work.

Item c) calls for the enterprise to have current procedures (presumably a combination of formal training, on-the-job training, other educational opportunities, and experience requirements) for making its employees aware of four basic things that, on the one hand, are closely connected with the EMS and, on the other, are somewhat broadly described. This is a reflection of the intent by the authors of ISO 14001 to always impose the minimum burden on enterprises implementing the standard, to leave as much room as possible for each enterprise to interpret the standard in a manner suitable to its particular circumstances. At the same time, the authors make clear both the intent of the standard and the absolutely minimum requirements that cannot be "interpreted" away.

This subclause of ISO 14001 again — in the third point under item c) — contains an explicit internal linkage with another subclause. The "emergency management and response requirements" referred to are those discussed in Subclause 4.4.7, "Emergency Preparedness and Response."

Nowhere in this subclause is there a requirement that anything be in writing, so how can ISO 14001 certification auditors verify conformance? How could an enterprise have a meaningful periodic training needs assessment, and respond to that with a meaningful program of awareness and competence training, and *not* have anything in writing? Anyway, a good certification auditor will look not just for the program, but for the results of the program on the employees of the enterprise. After all, the language of the standard places overwhelming emphasis on the awareness, knowledge, and competence that the employees of the enterprise must have, not specifically on a training program that the enterprise must conduct.

IMPLEMENTATION AND OPERATION: COMMUNICATION (SUBCLAUSE 4.4.3)

Four-Way Communication

This element of the EMS infrastructure concerns four-way communication: from the top down (from senior managers of the enterprise to workers on the shop floor and in the offices) and from the bottom up; from the enterprise out (from the enterprise to the neighboring community and the rest of the world) and from the outside in.

For EMS programs and operations to run smoothly, everyone must not only know their responsibility and be properly trained for it, but they must communicate information and decisions among the different levels and functions of the enterprise efficiently. That is what the down and up communication is all about. In addition, an enterprise that is serious about continually improving its environmental performance needs to let the outside world know about its environmental aspects and its EMS, so outsiders — who bring a fresh and possibly different perspective to things — can register their comments and concerns with the enterprise. That's what the out and in communication is all about.

The down and up, or internal, communication involves means by which top management can tell employees of the enterprise about changes in the environmental policy, new EMS developments, and new procedures related to environmental aspects; can report to employees about the overall performance of the EMS and the environmental performance of the enterprise; and can convey to workers the results of internal and third-party EMS audits and management reviews. The enterprise needs to establish a thought-out and systematic set of communication channels, procedures, and media to facilitate that communication. But it needs also to offer workers a means of bottom-up communication — for providing to management their ideas on improving the EMS, and for expressing their views and concerns on environmental matters.

Internal communication involves means by which top management can convey to workers information about the EMS and environmental performance of the enterprise, and by which workers can provide to top management their ideas for improving the EMS and express their views and concerns on environmental matters.

The out and in, or external communication involves means by which the enterprise disseminates information about its environmental aspects and performance, and about its EMS, to the outside world, as well as means by which it receives and responds to information and queries from parties like community groups, trade associations, labor organizations, public agencies, consumer groups, contractors and suppliers, and the general public. There is a wide range of technical information, opinion, regulations, and inquiries that could be reaching the enterprise and be useful for its EMS and environmental performance, not to mention for its related business performance and public image. There is also a wide range of information that the enterprise could and should make available to the outside world to encourage useful feedback in the other direction, not to mention that it could also be good for business and its public image. Here, too, the enterprise needs to use a thought-out and systematic set of communication channels and media, and procedures to optimize the benefits from using them.

The question of just how much communication is needed with the outside world, and what it is about the enterprise's environmental aspects and performance and its EMS that should be communicated, has been the subject of much discussion in recent years. The main reason for this is the fact that, as we shall see in a moment, ISO 14001 makes an enterprise's communication out and in concerning its environmental aspects essentially voluntary, while the European Union's Eco-Management and Auditing Scheme, EMAS, essentially makes it mandatory. Today there is a growing movement, certainly in the Western economies, to utilize market forces to make environmental reporting by enterprises to the public a positive advantage in doing business. For a good example of this, see the Global Reporting Initiative of the Coalition for Environmentally Responsible Economies (www.ceres.org and www.globalreporting.org).

Requirements of ISO 14001

Subclause 4.4.3 of the ISO 14001 EMS standard requires the enterprise to:

a) Establish and maintain procedures for conducting internal communication on its environmental aspects and EMS among levels and functions of the enterprise
b) Establish and maintain procedures for conducting external communication on its environmental aspects and EMS, including receiving, documenting, and responding to relevant communication from outside parties
c) Make and record a decision regarding how it will handle external communication on its *significant* environmental aspects

Items a) and b) require that the enterprise create and regularly update procedures for internal and external communication. With regard to internal communication, the subclause covers procedures for communication on the environmental aspects and EMS of the enterprise. The subclause's treatment of this issue can be given a wide range of interpretations. With regard to external communications, the subclause is both explicit and vague: It covers three specific aspects of dealing with external communications, but limits applicability of the subclause to "relevant" external communication. These procedures do not have to be in writing, but a certification auditor will look for evidence that, in the cases of both internal and external communication, the enterprise handles communication in a consistent manner. The auditor will also look for evidence of the communication itself — internally in the form of knowledge among employees about environmental aspects and the EMS, and externally in the form of documentation regarding communication in and responses out.

Item c) requires only that the enterprise record a decision — any decision — regarding external communication on its significant environmental aspects. The crafters of ISO 14001, many of them from the business community, were sensitive to the fact that if the standard absolutely requires enterprises to report to the public on their significant environmental aspects this might discourage many of them from adopting the standard. The reason for this is not so much that many enterprises would be embarrassed by their environmental performance, but that they might consider some information of this sort proprietary from a business point of view, and that revealing this information would invite an avalanche of queries, accusations, and probes that could be costly to deal with.

IMPLEMENTATION AND OPERATION: EMS DOCUMENTATION (SUBCLAUSE 4.4.4)

Documenting the EMS

This element of the EMS infrastructure can be imagined as a set of loose-leaf books (loose-leaf for easy updating) that contain, first, a detailed description of the EMS of the enterprise and, then, all the related documentation. As a practical matter, only the description of the EMS and some additional core documents, such as the environmental policy and work plans for environmental management programs, would be kept together in one place. Kept with them would be a directory of all related documents that contains instructions on where and how to access them.

The enterprise should create, assemble, and regularly update EMS documentation only to the extent absolutely necessary to maintain control over its EMS and over operations related to its environmental aspects, and to be able to review and improve its EMS and environmental performance.

The description of the EMS covers all of its policy, planning, implementation and operation, checking and corrective action, and management review elements, and how they relate to each other. This constitutes an authoritative source of information on the EMS — an EMS reference book — for anyone in the enterprise who needs it. Moreover, when pages are retired from the main body of this reference book and placed in an appendix of superseded material they constitute an information base for tracking the results of changes, so the lessons of experience are preserved. The additional EMS-related documentation maintained with the EMS description can be as extensive or as limited as top management, the EMS coordinator, and the EMS Committee think appropriate.

If an enterprise is seeking ISO 14001 certification or recertification, then at a minimum it must maintain the written policy, procedures, records, and other documentation specifically required by the standard. But if certification is not a current concern, then both the EMS description *and* the additional related documentation can be as minimal or as extensive as the enterprise thinks necessary. A relatively small enterprise, especially, can have a very effective EMS without much formal documentation. In general, the extent and complexity of EMS documentation will depend on the size and nature of the enterprise, the scope of its EMS, and its approach to establishing and operating its EMS. The guiding principle should be to create, assemble, and regularly update EMS documentation only to the extent absolutely necessary for the enterprise to maintain control over its EMS and over operations related to its environmental aspects, and to be able to review and improve its EMS and environmental performance.

In principle, if the EMS is properly integrated with the overall management system of the enterprise, its documentation could also be integrated with the overall documentation of policies, procedures, regulatory compliance tracking, and the like. However, most enterprises find that for convenience, especially if ISO 14001 certification audits are a factor, it's best to maintain a set of EMS documentation separately, possibly in addition to integrating it with the system of overall documentation.

It's useful to bear in mind the distinction between two basic kinds of EMS documentation: *records* and *documents*. It's a bit confusing, because the term "documentation" means something in writing, whereas in the EMS context "documents" are a specific type of thing in writing. The easiest way to keep the distinction clear is to remember that *anything that is not a record is a document.*

A record contains the facts of a past event, such as a decision, action, measurement, or finding. The facts may include what happened, who participated, the findings, the result, and other characterizations. Because a record is kept as a source of factual information about a specific event, by definition it cannot be superseded, though a similar record of a more recent similar event may be created. A record is not a source of guidance and usually is not referred to often; its purpose is only to preserve historical information for analysis, planning, and future reference. An example of a record is a description of a training program — who and how many attended, how participants evaluated the training, and so on — but not the training materials, which would be documents. Other EMS records include EMS audit reports, environmental monitoring records, equipment calibration and maintenance records, compliance audit reports, the enterprise's decision on external communication about its significant environmental aspects (see Subclause 4.4.3), P2 audit reports, equipment inspection reports, hazardous waste handling records, and the results of an emergency preparedness and response drill. *A record describes.*

A document prescribes. A document contains information, in varying amounts of detail, on what will or ought to be done. At one end of the document spectrum is the environmental policy statement of the enterprise — a broad, overarching statement of values, aspirations, and objectives. At the other end are highly detailed work instructions developed to govern the performance of workers with environmentally sensitive tasks. Because EMS documents provide guidance, they have to be updated regularly to remain effective. So, sooner or later, all EMS documents are superseded. In part because of that, documents are referred to often. Common EMS documents include the overall description of the EMS; the environmental policy statement; procedures related to environmental aspects; work instructions and standards (in ISO 14001 these are termed "operating criteria") related to environmental aspects; work plans for environmental management programs; legal and other obligations; and emergency response plans.

The EMS Committee has to stay alert to the danger of overdocumentation. It would be wonderful if everything connected with the EMS were immediately documented and stored for easy retrieval and use. But creating and maintaining documentation costs time and money; the more created, the more it costs and the more the documentation — rather than environmental management — tends to become the focus of the EMS.

Requirements of ISO 14001

Subclause 4.4.4 of the ISO 14001 EMS standard requires the enterprise to:

a) Establish and maintain, in hard copy or electronic form, information that describes the core elements of the EMS and the relationships among them
b) Establish and maintain, in hard copy or electronic form, information that gives directions for accessing related documentation

This is a very short subclause that represents a very large requirement. Although the subclause doesn't say what form the description of EMS core elements should be in, or even what the core elements are, one can easily imagine something like an overall EMS manual. How else could a certification auditor readily determine that an enterprise has met this requirement?

As for related documentation, this must include at least the documentation specified in the ISO 14001 standard:

- The environmental policy issued by top management (required by Clause 4.2)
- Procedures (requirements for documented procedures appear in several places in the standard)
- Operating criteria (detailed work instructions and standards required under Subclause 4.4.6)
- Records (requirements for records appear in several places in the standard)

That's not a small amount of documentation, and some would argue that documentation on legal and other environmental obligations of the enterprise should be added to the list. Again, one imagines a manual, this one either a collection of all the EMS documentation in one place, a directory to where the documentation is located, or a bit of both.

"Related documentation" can go way beyond the minimum requirements of the standard. In this chapter we have noted more than once that there really was no way but to create a certain document, even though it is not required by the ISO 14001 standard. Related EMS documentation could easily include training programs, work plans for environmental management programs, the results of P2 audits, and much more, including, as we said earlier, documentation on the legal and other environmental obligations of the enterprise. The language of this subclause means that the enterprise must provide written directions to all EMS documents, whether or not they are required by the standard.

A certification auditor will look for physical evidence that this requirement has been met; he or she may also assess the way instructions for accessing the related documents are provided: Are all related documents covered, and are the instructions user-friendly? A good certification auditor will also look for clear indication that the information is "maintained," that it's regularly updated. If an enterprise wants to make the auditor's job easier, it can have a written procedure for this; but then there would also have to be written directions for locating that procedure.

IMPLEMENTATION AND OPERATION: DOCUMENT CONTROL (SUBCLAUSE 4.4.5)

Controlling EMS Documents

This element of the EMS infrastructure ensures that EMS documentation is properly controlled. "Controlled" means that the enterprise has a deliberate process for tracking and managing the content and physical status of EMS documents. As a system based on continual improvement, the EMS frequently generates revised and updated versions of its documentation. Because documents provide guidance, it's absolutely imperative for useful operation of its EMS that the enterprise have in place a well thought-out mechanism for ensuring that the EMS documents people refer to are the ones currently in force.

The enterprise needs to have in place a well thought-out mechanism for ensuring that the EMS documents people refer to are the ones currently in force.

The document characteristics that can be used for controlling EMS document content and physical status can be many or few — as always, depending on the situation and the preferences of top management. ISO 14001 provides the basis for a list of such characteristics that offers a good starting point for any enterprise to develop its own. In brief, with respect to content, the control characteristics would be related to how the enterprise creates, approves, and revises its EMS documents;

with respect to physical status, the control characteristics would be related to how the enterprise stores, distributes, and ultimately disposes of its EMS documents.

Possible document *content control* characteristics include:

- Author's name
- Document title and subtitle
- Date of issue
- Date of revision
- Approver's name and approval date
- Interval or date of review
- Name of reviewer/reviser

Possible document *physical status control* characteristics include:

- Location
- Identifier/locator code
- Storage/retrieval procedures
- Locations in the enterprise where posted or stored
- Disposition when superseded

The danger, of course, is of going overboard with control. Again, the rule would be to have the minimum set of formal controls that allows for efficient and productive operation of the EMS.

Requirements of ISO 14001

Subclause 4.4.5 of the ISO 14001 EMS standard requires the enterprise to:

a) Establish and maintain procedures and responsibilities regarding how the different types of documents required by ISO 14001 are to be created and modified
b) Establish and maintain procedures for controlling required EMS documents, so that:
    - they are approved by designated approvers
    - they are reviewed, and revisions are considered periodically
    - they can be located easily
    - copies of the current versions of these documents are located and easily accessible at places in the enterprise where functions related to them are carried out
    - obsolete EMS documents are promptly removed from normal access throughout the enterprise, to keep them from being mistakenly used
    - any obsolete EMS documents retained for archiving are clearly identified as obsolete
c) Make sure that EMS documentation is:
    - legible
    - dated, and shows dates of revision
    - easily identifiable

- physically maintained in an orderly way
- retained for a specific period determined by the enterprise

This particular subclause of ISO 14001 is fairly detailed and self-explanatory. But please note that:

- The enterprise is required to regularly review and, if necessary, revise its procedures related to document control. A certification auditor will look for evidence of this.

- The requirements of this subclause apply only to EMS documents, not records (though one could argue that item c) refers to anything in writing related to the EMS because of the word "documentation" rather than "documents"). Records are not revised or updated or referred to regularly because they do not provide operational guidance. Procedures for taking care of records are discussed in Subclause 4.5.3, "Records."

- Many view the document control requirements of this subclause as applying only to EMS documents required by ISO 14001, though again, because of the language used, one could argue that item c) refers to anything in writing related to the EMS. In any case, as we have seen, any enterprise seeking ISO 14001 certification is bound to have more than the required documentation, both to make its EMS more effective and to facilitate the certification audit. It only makes sense for an enterprise that has established document control procedures as required by the standard to apply those procedures to all EMS documents, whether required by ISO 14001 or not.

- Subclause 4.4.5 says nothing about the physical form of the documents, so an auditor will be satisfied whether they are in print or electronic form. In fact, they do not even have to be typed, though the subclause does require explicitly that they be legible.

- In most cases, the enterprise will benefit the operation of its EMS and help the certification auditor by creating a "document control matrix" that lists the controlled documents as row headings, and control characteristics as column headings and corresponding information in the cells of the matrix.

IMPLEMENTATION AND OPERATION: OPERATIONAL CONTROL (SUBCLAUSE 4.4.6)

Controlling Enterprise Operations

This element of the EMS infrastructure has nothing to do with EMS and everything to do with EMS. It has nothing to do with EMS in the sense that the very essence of what *any* enterprise does is control operations — control the applications of materials, energy, equipment, and people in a process meant to produce something. The better an enterprise controls its operations, the better it can produce a product of consistently high quality at minimum cost. It has everything to do with EMS because what's true for sound operations in general is also true for managing the environmental aspects of enterprise operations, which at root is what an EMS is all about.

A significant portion of an enterprise's pollution and waste is the result of poor control of its operations; the biggest opportunities for cost savings in an enterprise tend to lie in better operational control.

Moreover, as explained in Part II of this book, a significant portion of an enterprise's polluting releases into the water, air, and land — or, you might say, a significant portion of its waste — is the result of poor control of its operations. The biggest opportunities for cost savings in an enterprise tend to lie in better control of its operations. Improving operational control can express itself in hundreds of ways, small and big, such as being sure a valve is turned off at the end of the day, installing a thermostat, or employing sophisticated automated process controls.

For its environmental management programs to achieve their targets and for its normal operations to run at peak efficiency and reliability, an enterprise has to have a system of controls in place, controls that enable the enterprise to obtain the desired performance from its operations with consistency and, therefore, predictability. Operational controls include specific equipment and specific procedures for using the equipment and for carrying out tasks in specific ways. The specification of how materials, energy, equipment, and people should be combined to prevent pollution or to reduce resource consumption by a particular work activity is called an EMS *operational control procedure*.

An EMS operational control procedure is a procedure related to an environmental aspect of the enterprise's operations. The idea is to have a procedure relating to every significant environmental aspect. Put differently, if a particular environmental aspect is important enough for the enterprise to call it "significant," then it is important enough for the enterprise to develop procedures to be sure its environmental impact is controlled as much as possible. An enterprise cannot afford to leave to chance the way its employees carry out an activity that could affect one of its significant environmental aspects. Because this is plain good management, it is possible that in a well-run enterprise many appropriate procedures will already be in place.

Although "procedure" has the ring of a bureaucratic burden, really it is nothing more than a specific way of doing things for a specific purpose. In the case of the EMS, the purpose is to help carry out the environmental management plan or otherwise minimize the environmental impact of the activity for which the procedure was written. With the possible exception of the very simplest ones, procedures have to be in writing, so there can be no mistake about the intent of management and the obligations of workers. A procedure describes the way a work activity is to be carried out: who does what; when and where they do it; what equipment they use and how they use it; what controls they apply; what training they must have; what sort of records they must make; and what purpose they're working to achieve. It's important for there to be a clear statement of the purpose of a procedure, not just because a worker who knows why things have to be done a certain way is more apt to follow the procedure, but also because that makes it possible for the worker to recommend ways of improving the procedure.

Although "procedure" has the ring of a bureaucratic burden, really it is nothing more than a specific way of doing things for a specific purpose.

Most enterprises prepare procedures on two levels: a broader level, covering an overall activity, and a more detailed level, covering procedures for specific tasks. For example, an enterprise might have a procedure for receiving-dock operations, and then detailed procedures for handling different types of materials (such as drums of caustic acid) as they arrive and are moved to inventory. Or, an enterprise might have a procedure for the final step in the production line for, say, small metal parts, and detailed procedures for their final cleaning, polishing, inspection, individual packaging, packing for shipment, and moving to inventory. Some enterprises use terms like "umbrella procedures" and "detailed procedures"; others distinguish between "procedures," and "work instructions or standards" for tasks that come under the procedure.

The term "work instructions *and* standards" best describes the set of step-by-step details of how a specific task covered by a procedure is to be carried out. These technical details state precisely what equipment or materials must be used, what exactly the worker should do, and what the standard for acceptable performance is. These are instructions for what to use and what to do to perform a specific task, and standards for performance and for results: "Use instrument A to measure compound B by following steps C through G; do this at X hour each day, taking Y amount of time or less, and record the results in Z record book." Though a procedure needs to be read and registered by everyone associated with an activity, work instructions and standards for specific tasks covered by the procedure only have to be known to the relevant individuals or teams.

There should be procedures covering all major elements of the enterprise's EMS as well as activities connected to its environmental aspects. Each procedure would be accompanied by whatever work instructions and standards for specific tasks are necessary to implement the environmental policy of the enterprise and its environmental management programs.

At this point many people throw up their hands and decide that they'll figure out a way to improve the environmental performance of their enterprise without all the paperwork. But it really is true that without well thought-out, written procedures and associated work instructions, it just won't work, at least not for very long. It is also true that an enterprise establishing an EMS does not have to do everything at once. The enterprise might begin with just its single most significant environmental aspect, which may mean just one objective and one target and not very many procedures to formulate. As top management, the EMS Committee, and everyone else in the enterprise get the hang of it, it will become easier and easier to expand the scope of the EMS to ever larger parts of the enterprise. In addition, some procedures will be for activities associated with more than one environmental aspect, and it will be possible to recycle parts of some procedures into others. And once it's done, it's done, except for regular updating and the new procedures needed for new activities.

Requirements of ISO 14001

Subclause 4.4.6 of the ISO 14001 EMS standard requires the enterprise to:

a) Identify the activities associated with its significant environmental aspects
b) Ensure that these activities (and their associated maintenance activities) are carried out in a fashion consistent with its EMS by:
    - establishing and maintaining written procedures that will cause the activities to be carried out in a manner that implements the environmental policy of the enterprise and helps achieve its environmental objectives and targets
    - specifying work instructions and standards ("operating criteria") in association with the procedures
    - establishing and maintaining such procedures also for identifiable significant environmental aspects of goods and services used by the enterprise
    - being sure that the procedures are known to those who provide the goods and services

Item a) requires that the enterprise identify discrete activities that contribute to the environmental impacts of its environmental aspects. These discrete activities are not to be confused with the much broader "activities, goods, and services" or "operations and products," or "activity or operation" that were mentioned earlier as the starting point for identifying environmental aspects. The actual language of ISO 14001 makes clear that this requirement refers only to activities associated with its significant environmental aspects, as identified by the enterprise in its EMS progression of environmental policy → significant environmental aspects → objectives → targets. Another way to look at it is that the standard requires the enterprise to identify activities associated with *all* its listed significant environmental aspects, and a certification auditor will look for this. Yet there is considerable flexibility here, when you take into account the latitude the enterprise has with its aspect significance criteria.

Item b) lists four requirements for ensuring that an enterprise carries out the activities identified in item a) in a manner consistent with its EMS. The first is to create, and regularly update, written procedures for the activities. The procedures should promote the environmental management programs of the enterprise and help implement its commitments to regulatory compliance, continual improvement, and pollution prevention. Quite often a key element of an environmental management program is, in fact, developing procedures to minimize environmental impact (see examples in Table 5).

The second requirement in item b) simply says that the enterprise has to be very explicit in its procedures, down to the level of specific work instructions and standards. It's important to know, and keep in mind, that the term used for work instructions and standards in ISO 14001 is "operating criteria," because that's the term certification auditors and colleagues from other enterprises may use.

The third and fourth requirements in item b) basically say that "they did it" is not an acceptable excuse. At the same time, ISO 14001 provides a sort of escape clause to ensure that an impossible burden is not placed on the enterprise: The enterprise has to prepare and communicate procedures only with respect to *identifiable* significant environmental aspects of goods and services it uses. While this requirement could possibly lead to disagreement between an enterprise and a certification auditor, the intent of the standard is only to be reasonable, and a worthy certification auditor will not have a problem with any interpretation of what is "identifiable" that is truly reasonable. That still leaves plenty of legitimate flexibility. And to make it even easier, these procedures do not have to be written.

IMPLEMENTATION AND OPERATION: EMERGENCY PREPAREDNESS AND RESPONSE (SUBCLAUSE 4.4.7)

Preventing and Responding to Environmental Emergencies

To be equipped to prevent and respond to environmental emergencies an enterprise has to identify potential emergencies, develop procedures to prevent them, and develop plans and procedures for what to do if they happen.

The final element in the EMS infrastructure is a set of procedures for dealing with environmental emergencies. Here we deal with the ultimate loss of control, such as happened in the cases of Three-Mile Island, Bhopol, Valdez, and Chernobyl. Incidents like these are never forgotten. Because of the scale and notoriety of such incidents, the organizations associated with them are forever tainted, have paid, and continue to pay dearly, for their loss of control; but these are the same *kinds* of loss of control that happen in enterprises every day. Improving controls to avoid environmental emergencies is extremely profitable, even thought the profits are "invisible," because they are losses that don't occur.

Most often, environmental emergencies result from loss of control during normal operations. This kind of loss of control can happen only if the relevant procedures were faulty (or there were no relevant procedures), or if the procedures were not followed. Sometimes environmental emergencies result from abnormal operating conditions — an extra shift responding to a sudden surge in orders, a rapid process change caused by sudden scarcity of a basic input, disruption of the production line as a result of moving operations to a new facility. And sometimes environmental emergencies are caused by events truly outside enterprise control, such as freak natural disasters or emergencies in the nearby facilities of other enterprises. No matter what the source of an emergency situation, how an enterprise prepares for and responds to environmental emergencies is ultimately a reflection of management competence. It is good strategic management to ensure that the enterprise is equipped to prevent and respond to environmental emergencies.

To be equipped to prevent and respond to environmental emergencies an enterprise has to identify potential emergencies, develop procedures to prevent them, and develop plans and procedures for what to do if they happen. Potential environmental emergencies include unintended air polluting releases, unintended polluting releases to the soil or water, and the environmental and human damage caused by such unintended releases. Accident prevention procedures and emergency preparedness and response plans might contain a wide range of information, including data on hazardous materials, responsible staff, and emergency service organizations; descriptions of avoidable emergencies and corresponding procedures for specific activities that will help avoid them; procedures to be followed in the event of different kinds of environmental emergencies; and classroom training requirements and emergency tests and drills to ensure employee competence in preventing and responding to environmental emergencies.

Requirements of ISO 14001

Subclause 4.4.7 of the ISO 14001 EMS standard requires the enterprise to:

a) Establish and maintain procedures to identify potential accidents and emergency situations
b) Establish and maintain procedures for responding to accidents and emergencies
c) Establish and maintain procedures for preventing and mitigating environmental impacts associated with potential accidents and emergencies
d) Review and revise its emergency preparedness and response procedures as indicated by experience when an accident or emergency has occurred
e) Periodically test its emergency preparedness and response procedures, if practicable

Items a), b), and c) together can be thought of as elements of an enterprise's overall emergency preparedness and response plan. Items d) and e) require that it be updated (especially after an emergency where some of its procedures should have come into play), and that the procedures be periodically tested if there is a meaningful and reasonable way of testing them. If there has been an environmental emergency at the enterprise, a certification auditor will check to see if the enterprise has reconsidered relevant procedures in its wake. Although this subclause of ISO 14001 does not require that emergency preparedness and response procedures be in writing, management that does not see to it that they are in writing does so at its peril.

At first, it might seem that *preventing* environmental emergencies is missing from the requirements of Subclause 4.4.7. But preventable accidents and emergencies are already addressed by EMS operational control procedures required under Subclause 4.4.6 of the ISO 14001 standard.

CHECKING AND CORRECTIVE ACTION: MONITORING AND MEASUREMENT (SUBCLAUSE 4.5.1)

We've formulated and publicized the enterprise's environmental policy; we've identified environmental aspects and their impacts, and determined which of them are significant and should be the focus of the EMS; we've set objectives and targets for reducing their environmental impacts, and devised environmental management programs to achieve those objectives and targets; and we've established an EMS infrastructure in the enterprise to support the environmental management programs and improved environmental performance of the enterprise generally, including a framework of roles and responsibilities, a training program, internal and external communication procedures, EMS documentation and document control procedures, operational control procedures and work instructions, and emergency preparedness and response procedures. We're ready to go. In fact, at this point much of the EMS is already going, at least on a preliminary basis. Now it's time to consider what the enterprise needs to track the progress of its EMS and the benefits that are accruing from it, and make changes that would make it even better.

> ***When an enterprise is first establishing its EMS it can expect to learn a lot from trial and error, so it needs to have a system from the outset for spotting the errors, and the successes, and for learning from them.***

Counting What Counts

Enterprises, governments, other organizations, in fact people, count what counts. If it's important to them, they count it. If they count it, that's a sure sign that it's important to them. Top management of an enterprise should not try to establish an EMS if it is not truly important to them; and if it is important to them they will want to "count it" — to measure, monitor, and evaluate the performance of the EMS and its results — and make improvements as required.

For an enterprise that is just establishing its EMS, this is extremely important for two reasons. First, if the EMS is being put in place properly, it will quickly produce very exciting results in terms of reduced environmental impacts and cost savings, especially through modest P2 initiatives. These successes need to be quickly documented and publicized in the enterprise (and perhaps beyond) to help everyone understand what the EMS is really all about — that it produces tangible and desirable results, and that top management is solidly behind it. Second, measuring and tracking from the very beginning of establishing the EMS allows mistakes, or even room for improvement, to be perceived quickly. The earlier the need for them is spotted, the easier it will be to make changes that improve the EMS and its performance. When an enterprise is first establishing its EMS, it can expect to learn a lot from trial and error, so it needs to have a system from the outset for spotting the errors, and the successes, and for learning from them.

Actually, those principles are just as important for a well-established EMS. As an integral part of its EMS, an enterprise needs procedures for regularly measuring and monitoring — measuring and monitoring if its EMS is operating as intended; if the EMS is producing the expected results; and, in general, if there is continual improvement in its environmental performance.

The term "measuring and monitoring" refers to a process that actually has five parts:

a) *Measuring* how often an event happens, the magnitude of an event, or the concentration or composition of a material
b) *Monitoring* the measurements and following the changes in them over time
c) *Recording* the findings of measuring and monitoring on paper or electronically
d) *Evaluating* the results of measuring and monitoring to determine if changes should be made in environmental management programs, EMS procedures, or other operational controls
e) *Revising* the EMS, environmental management programs, procedures, or other operational controls accordingly

The term "measuring and monitoring" refers to a process that actually has five parts: measuring, monitoring, recording, evaluating, and revising.

Some types of measuring and monitoring, particularly of certain process operations, are continuous. Others are carried out at regular intervals from numbers of minutes to numbers of months, the length of the interval depending on the nature of what is being monitored. Oven heat or boiler pressure, for example, would probably be measured and monitored continuously; effluent or emissions concentrations might be monitored continuously or at fairly brief intervals; adherence to maintenance procedures for a piece of equipment might be monitored on a weekly basis; conformance to the EMS training program that has been designed might be monitored on a monthly basis; a review of the enterprise's regulatory compliance status might be carried out every two months; an EMS internal audit might be conducted every six months. Whatever the exact schedules, overall EMS measuring and monitoring, like other EMS elements, is constantly at work. And once an EMS breakdown of some sort is identified through measuring and monitoring activities, it would be unconscionable to delay devising and implementing corrective and preventive action.

Here are examples of specific EMS areas for which an enterprise might want to develop measuring and monitoring procedures:

- Activities within its operations that can have a noteworthy negative impact on the environment (measuring and monitoring for these activities might be built into the procedures already covered in the earlier discussion on operational control)
- Environmental performance as related to each element of the EMS
- Environmental performance as related to each environmental management program, including its activity schedule and progress toward meeting targets and objectives
- The gap between the environmental objectives of the enterprise and actual performance
- Significant environmental aspects and their impacts
- Other environmental impacts of the operations of the enterprise
- Adherence of workers to EMS operational control procedures, work instructions and standards, and other operational controls
- Compliance with environmental laws and regulations
- Compliance with other environmental standards to which the enterprise has obligated itself
- Direct economic and other business benefits associated with specific environmental performance improvements

Some readers may contemplate abandoning the EMS venture when confronted with the idea of sets of procedures for a five-step process for measuring and monitoring all the things listed above, especially in addition to all the other procedures already discussed. At the risk of excessive repetition, we remind readers that our intent here is to provide material for fashioning an applied EMS model suited to the particular needs and circumstances of their respective enterprises. They can select part or all of what has been presented, or even none at all for the time being. As we've said before, it all depends on the size and nature of the enterprise, the scope of its EMS, and the enterprise's approach to establishing and operating its EMS. Being ready for ISO 14001 certification is another matter altogether.

Requirements of ISO 14001

Subclause 4.5.1 of the ISO 14001 EMS standard requires the enterprise to:

a) Establish and maintain written procedures for regularly measuring and monitoring the key characteristics of its activities that can have significant negative impacts on the environment
b) Include as an element of each of these procedures the recording of information for tracking
    - environmental performance
    - operational controls in relation to environmental performance
    - the relationship between actual performance and counterpart environmental objectives and targets
c) Calibrate and maintain the equipment it uses for measuring and monitoring
d) Keep records of equipment calibration and maintenance activities
e) Retain the calibration and maintenance records in accordance with its own procedures
f) Establish and maintain a written procedure for periodically assessing the enterprise's compliance with relevant environmental laws and regulations

Item a) does *not* require written procedures for periodically measuring and monitoring an activity that has environmental impacts — after all, how could you measure and monitor an activity meaningfully? — but instead requires periodic measuring and monitoring of *measurable characteristics of the activity* that have a bearing on its environmental impact (for example, the throughflow of water, air pollution releases, the quantity of chemical injected, the length of time in the oven, the number of units processed). In this requirement, the term "significant impact on the environment" means important, or major, impact on the environment, not significant in the sense of the significant environmental aspects for which the enterprise has environmental management programs under its EMS. Thus, the requirement applies to characteristics of *any* activity that has some form of serious impact on the environment, and not just of activities associated with significant environmental aspects in EMS programmatic terms.

Item b) requires that the written procedures from item a) cover, among other things, keeping records on the findings of measurement and monitoring activities, on potential causal relationships, and on relationships to target environmental measures, where relevant. ISO 14001 does not dictate a medium or format for the records. For the benefit of both the enterprise and a certification auditor, records should be in more or less permanent form and readily accessible where they are normally used.

Items c), d), and e) reflect the management principle that data are only as reliable as the methods and tools of measurement. The term "equipment" has to be taken as broadly as possible: it covers sampling and testing equipment, measuring and recording instruments, and computer hardware and software. There is no requirement that there be procedures for calibrating and maintaining the equipment, because these would be in the operator manuals. Item d) provides a certification auditor with a clear way of verifying if equipment calibration and maintenance is taken seriously by the enterprise, but keeping such records is obviously a smart idea in its own right. Item e) leaves it to the enterprise to decide how long to retain calibration and maintenance records, but implicitly requires that there be specific procedures governing those records (and maybe other records as well). This ensures that these records, like other EMS records, will not be dealt with casually.

Item f) reminds us of Subclause 4.3.2, "Legal and Other Requirements," which calls for the enterprise to track environmental regulations but says nothing about tracking the status of the enterprise's compliance with those regulations. Here in Subclause 4.5.1 the compliance side is filled in. However, unlike the earlier subclause, Subclause 4.5.1 requires that the procedure for tracking regulatory compliance be in writing (and thus be more easily verified by a certification auditor), requires only that the tracking procedure apply to "relevant" laws and regulations, and does not require that compliance or conformance be tracked with respect to "other environmental standards and practices" to which the enterprise has voluntarily obligated itself.

A good working interpretation of "relevant" can be inferred from the earlier subclause to mean legal requirements connected with the enterprise's environmental aspects. And required or not, why wouldn't an enterprise want to track compliance with *all* its environmental obligations, including those it has taken on voluntarily?

CHECKING/CORRECTIVE ACTION: NONCONFORMANCE AND CORRECTIVE AND PREVENTIVE ACTION (SUBCLAUSE 4.5.2)

Dealing with Breakdowns

This second element of checking and corrective action can be thought of as really belonging with the previous section, where "revising" was identified as the last of five components of the overall measuring and monitoring process. After all, what would be the point of measuring, monitoring, recording, and evaluating if the enterprise, its EMS Committee, and its top management were not going to promptly revise an EMS element or procedure to eliminate the cause of a breakdown? The matter of dealing with breakdowns is addressed separately here because ISO 14001 separately requires procedures on how they will be handled.

To keep its EMS running efficiently and productively top management of an enterprise has to ensure there are procedures for dealing with breakdowns uncovered through EMS measuring and monitoring activities, and be sure those procedures are incorporated into the EMS. The primary way of dealing with breakdowns is to revise the guidance that allowed the breakdown to happen. An enterprise might find that it needs to revise an element of its EMS, one of its environmental management programs, or an EMS procedure and associated work instructions, depending on the nature of the breakdown. A "revision" might involve refining, adding to, changing, or replacing an existing EMS element or procedure.

There are two essential types of breakdowns: a performance failure and a deviation from the EMS. A *performance failure* could be a breakdown in the progress of an environmental management program, the failure of a program to achieve the environmental management target, or the failure of an EMS procedure to produce the expected results. In this type of breakdown, everyone is trying to do what they were supposed to do — but it isn't working. The reason usually is because the design of the EMS element or program or procedure was unrealistic; perhaps it wasn't based on good information. Another common reason for this type of failure is that management has not devoted the necessary resources for effective implementation. Whatever the case, identifying and correcting the causes of EMS performance failures offer important learning opportunities, and the matter is best approached that way. The enterprise may need to revise the appropriate EMS element, program, or procedure; it may need to change the specific resource allocation, or the way it allocates resources to EMS operations.

A second type of breakdown is a *deviation from the requirements*, programs, and procedures of the enterprise's EMS — what in ISO 14001 language is called a "nonconformance." Examples of this type of breakdown range from top management not including a commitment to regulatory compliance in the environmental policy of the enterprise, to a worker not employing the measuring instrument prescribed in EMS work instructions. In this type of breakdown, the guidance is not being followed. Often the reason is because of neglect, but the reason for neglect almost always ultimately comes back, as does everything, to the effectiveness of management. However, this type of breakdown can also result from flawed EMS elements or procedures that cannot be followed properly or, again, lack of the necessary resources, such as time, needed to follow the guidance fully.

In some cases it can be hard to know if the breakdown is a performance failure or a deviation from the EMS, and the distinction is not always important. What is most important is a clear-headed assessment of the source of the problem and what needs to be done to correct it.

In some cases it can be hard to know if the breakdown is a performance failure or a deviation from the EMS, and the distinction is not always important. What is most important is a clear-headed assessment of exactly where the problem lies: Is there something wrong with the content of the EMS element, program, or procedure — with its scope, the assumptions or information on which it is based, its completeness? Is there something wrong with the guidance language used in it? Is it prominent and available where it is needed? Is there something wrong with the training for carrying it out? Have the right people been given responsibility for it? Is the right equipment available and properly maintained? Are the right types and quantities of other resources available to those responsible for carrying it out? Do they have competing or contradictory performance criteria? Are the workers that are responsible properly motivated?

Procedures for examining and correcting EMS breakdowns can be formulated by department or production process of the enterprise or by categories of activities, depending on how the enterprise and its EMS are organized. Each such procedure could include guidance concerning:

- Who is responsible for identifying performance failures and nonconformances
- Who is responsible for taking any immediate corrective action called for in response to a resulting environmental impact
- Who is responsible for
    - a careful examination to identify the cause of the breakdown
    - recommending actions to correct unacceptable environmental impacts that may have resulted from the breakdown
    - recommending actions to prevent a recurrence of the breakdown
- What is included in the process of examining the cause of the breakdown and in determining corrective and preventive actions needed (for example, review of procedures related to the breakdown, interviews with workers and supervisors, additional measurements, review of training records)
- The content of a report on the breakdown, and who the report is submitted to
- Who decides or approves the corrective and preventive actions; and, possibly, how (for example, does a certain committee have to meet?)
- Where the decisions are recorded and how they are publicized in the short run
- Who is responsible for drafting revisions in procedures or modifying other operational controls to incorporate the preventive actions decided upon
- Who is responsible for follow-up, and when, to ensure that the corrective and preventive actions have been carried out and to assess their effectiveness
- A timetable for these actions

Requirements of ISO 14001

Subclause 4.5.2 of the ISO 14001 EMS standard requires the enterprise to:

a) Establish and maintain procedures for determining responsibility and authority for:
    - responding to an actual or potential nonconformance
    - investigating to determine the causes of the nonconformance
    - taking appropriate immediate action to mitigate any environmental impacts caused by the nonconformance
    - initiating and carrying through the process of determining and implementing corrective and preventive action

b) Ensure that the corrective and preventive action taken is appropriate to the scale of problems linked to the nonconformance, and to the magnitude and nature of any associated environmental impact
c) Carry out any revisions in EMS procedures that are part of the corrective and preventive actions
d) Record any revisions in EMS procedures that are part of the corrective and preventive actions

The subclause requires that procedures for determining who deals with what aspects of nonconformances be established, and regularly reviewed and updated. It does not require, however, that these be separate procedures; so they could be incorporated with measuring and monitoring procedures, if the EMS Committee thought that was the right way to handle it. These procedures — again, like procedures for measuring and monitoring — could also be incorporated into operational control procedures. A certification auditor might instinctively expect to find separate procedures for corrective and preventive action, because they are required under a separate subclause of ISO 14001, but in the end an auditor will be satisfied with evidence that the requirements for these procedures are met in any fashion.

Subclause 4.5.2 mentions only nonconformances, not performance failures in the sense described earlier, and does not provide a definition for the term. An enterprise can interpret this clause in its narrowest sense as referring only to deviations from the requirements of ISO 14001, which would include unintended regulatory noncompliances. Or, it could take a more enlightened approach and interpret this subclause as referring also to the environmental management programs and procedures (and their associated work instructions) of the enterprise's particular EMS. Though a certification auditor cannot hold the enterprise to it, the enterprise could take an even broader approach and include also EMS and other environmental performance failures. The only place in the ISO 14001 standard where dealing with EMS performance failures clearly fits is in the management review covered under Clause 4.6. But nothing in the standard prevents the enterprise from first dealing with performance failures at lower levels and on a more immediate basis, and it certainly makes sense to do that. In fact, some interpreters of ISO 14001 insist that performance failures *are* implicitly covered in the present subclause because of the reference in it to mitigating environmental impacts. But, ultimately, the precise language of the standard rules.

A certification auditor might instinctively expect to find separate procedures for corrective and preventive action because they are required under a separate subclause of ISO 14001, but in the end an auditor will be satisfied with evidence that the requirements for these procedures are met either separately or as parts of other procedures.

Item a) actually does not require procedures for responding to and investigating nonconformances; it requires procedures for determining ("identifying" is the actual language of the standard) "responsibility and authority for" carrying out such procedures. Those who crafted the ISO 14001 EMS standard did not want to dictate what the contents of these procedures should be, or for what sorts of categories of activity they should be developed. In an effort again to provide as much latitude as possible to enterprises trying to conform, the standard in effect requires only that there be procedures for determining the appropriate procedures.

However, the procedures for determining procedures do not have to be in writing. That means that procedures for dealing with nonconformances themselves can serve as evidence that procedures for determining the procedures exist. This may at first sound silly, but it works: though ISO 14001 does not require them, the best way for an enterprise to prove to a certification auditor that the requirements of item a) have been met is to have written procedures for responding to nonconformances and for taking corrective and preventive action. In any case, if an enterprise uses the list we gave earlier that spells out what a procedure for examining and correcting EMS breakdowns should include, any certification auditor will be happy. That list covers items a), b), and c) under this subclause.

Item d) requires that the enterprise not only revise its EMS procedures as warranted to prevent a recurrence of the nonconformance, but that it create a separate record of the changes made. Written procedures are EMS "documents" that are handled according to the requirements of Subclause 4.4.5. "Document Control," discussed earlier. A record is handled according to the ISO 14001 requirements of Subclause 4.5.3, "Records," which we'll discuss next. So, the enterprise's revisions to EMS procedures that arise from nonconformances go on record in two different systems that are probably even physically in different places. In this way they are most likely to be preserved and are most easily accessible, either as guidance or as sources of research information for assessing the effects of different procedural changes.

CHECKING AND CORRECTIVE ACTION: RECORDS (SUBCLAUSE 4.5.3)

There is a discussion about EMS records and the difference between records and documents in the section on Subclause 4.4.4, "Environmental Management System Documentation" (page 60). The present subclause is not so much about records as such as it is about EMS information management, though the information happens to be in the form of records.

Requirements of ISO 14001

Subclause 4.5.3 of the ISO 14001 EMS standard requires the enterprise to:

a) Establish and maintain procedures for identifying, maintaining, and disposing of environmental records

b) Ensure that among the records are:
    - EMS training records
    - results of EMS audits
    - results of EMS reviews
c) Ensure that the records are:
    - readily identifiable
    - legible
    - easily associated with the activity, product, or service to which they are related
d) Store and maintain the records so that they are:
    - readily retrievable
    - protected against damage, deterioration, or loss
e) Establish retention times for the records and record them
f) Maintain the records to the extent possible in a manner that makes it easy to confirm that ISO 14001 requirements are being satisfied

What is an environmental record? Some environmental, or EMS, records are named in the standard, though not always with the word "record." Three of them are named in this subclause. If the report or documentation describes a decision, action, measurement, or finding, it is a record. Annex A of ISO 14001 recommends that enterprises concentrate on records needed for implementing and operating their EMSs and for recording progress toward meeting planned targets and objectives. In other words, the minimum set of records that a certification auditor will check for is those explicitly required by ISO 14001, but the authors of the standard think it would be a good idea for enterprises also to maintain records on actual environmental performance relative to target environmental performance. And that really means many kinds of records generated by environmental management programs. The standard allows for any type of record-keeping medium.

The minimum set of records that a certification auditor will check for is those explicitly required by ISO 14001; but the authors of the standard think it would be a good idea for enterprises also to maintain records on actual environmental performance relative to target environmental performance.

Environmental records could include:

- Training records
- Decisions regarding external communication about an enterprise's significant environmental aspects
- Information relevant to the EMS about contractors and suppliers
- Monitoring data

- Calibration and maintenance activities for measuring and monitoring, and pollution control equipment
- Regulatory compliance records (compliance audit findings)
- Revisions to EMS procedures in response to nonconformances
- Internal EMS audit results
- Environmental records retention times
- Results of management review
- Information on environmental aspects of the enterprise and on environmental performance generated by environmental management programs and EMS procedures
- Information on relevant environmental laws and regulations
- Information on other standards and practices to which the enterprise has obligated itself
- P2 audit results
- Environmental permits
- Emergency incident reports
- Information on emergency preparedness and response tests and drills
- Process and product information related to environmental aspects

Subclause 4.5.3, in short, says that the enterprise has to have a systematic and consistent way (a procedure) for managing its EMS-related records (and it names a few), so they are well cared for, easily retrievable, and easily identifiable in terms of what in the EMS they are about. Item f) is interesting in that it seems to be requiring that the enterprise organize its EMS records along the lines of the clauses and subclauses of ISO 14001 to the extent possible within its usual way of keeping records, so as to make life a bit easier for an ISO 14001 certification auditor. But this makes a lot of sense for purposes of the enterprise and its EMS as well.

CHECKING AND CORRECTIVE ACTION: EMS AUDIT (SUBCLAUSE 4.5.4)

Auditing Your EMS

Under elements of checking and corrective action that we discussed earlier, an enterprise would be constantly monitoring its environmental management programs, its EMS procedures and associated work instructions, and its overall EMS for performance failures and deviations from EMS requirements. With that, an enterprise that wants to be sure its EMS is always well-tuned to current needs, and functioning efficiently and productively, will from time to time want to audit its EMS specifically as an EMS. An audit comes at the EMS from a different perspective than regular measuring and monitoring, which answers the question, "Is environmental performance what it should be, and if not, where is the problem?" An EMS audit answers the question, "Does the structure and operation of the EMS match the plan and, if not, what can be done to correct the situation?"

Measuring and monitoring starts with performance and works back to the system; an EMS audit starts with the system and provides information for improving its overall performance.

Many people think of internal EMS audits as an annual event, but there's no real reason for an annual schedule unless it allows the EMS audit to tie in with other audits conducted by the enterprise, thereby reducing costs.

The purpose of an EMS internal audit is to develop information that top management can use as a basis for modifying the design of the EMS and how the EMS operates; this is in line with top management's commitment to continual improvement. Findings of an internal audit are not the only information top management considers when looking for ways to improve the EMS, but audit findings do represent an essential part of the necessary information, because audit findings provide a comprehensive overview of the EMS.

The basis for an EMS audit could be the detailed description of the EMS that appears in the EMS manual of the enterprise (in ISO 14001 this description is required under Subclause 4.4.4, "EMS Documentation"); or a comprehensive list of EMS requirements, such as the list that appears in this book for the requirements of ISO 14001; or one of the many internal audit checklists found in EMS literature; or a comprehensive checklist developed uniquely by the enterprise for its particular EMS (readers are offered an opportunity to try their hand at creating an ISO 14001 requirements checklist in Exercise 1 at the end of this chapter).

Whatever tool is used, the EMS audit is comprehensive, unlike information generated by routine measuring and monitoring activities. Only audit findings offer the EMS Committee and top management an opportunity to observe how all the pieces of the EMS are functioning as interrelated parts of a system, and how that system is integrating with the overall management system of the enterprise.

Many people think of internal EMS audits as an annual event, but there's no real reason for an annual schedule unless it allows the EMS audit to tie in with other audits conducted by the enterprise, thereby reducing costs. Enterprises could conduct EMS audits fairly frequently in the early stages of establishing the EMS, then gradually lengthen the period of time between audits if audit results are consistently good and the information the audits provide is not changing enough between audits to justify a high frequency. Of course, in a situation where a small failure in the operation of the EMS could lead to very bad environmental consequences — that is, in an enterprise with high environmental risk — frequent EMS audits would be the smart thing to do, whatever the case.

Some enterprises conduct their EMS audits all at once — say, over a two-week period once or twice a year. This approach creates a snapshot view of the EMS: All the elements can be assessed as they are, and as they interrelate, at a point in time. The disadvantage of this approach is that it usually has limited depth and certainly does not allow for an audit of each element in operation over a period of time.

At the other extreme, some enterprises spread the audits of core elements of their EMSs over 12 months — one or two core element are audited each month, some EMS auditing is always in progress, and the entire EMS is audited on an annual cycle. The disadvantage in this approach is that it can yield an imperfect picture of the way the interconnected EMS elements work together in real time.

There are many possibilities between these extremes — such as auditing each core EMS element over the course of a week, which would result in three auditing cycles annually. Each enterprise should use the approach that best suits its circumstances and its overall EMS approach. Whatever else it does, an enterprise pursuing ISO 14001 certification or recertification should conduct some form of pre-audit audit immediately before the third-party audit, as a sort of dress rehearsal.

Internal EMS audits can be carried out by enterprise personnel, by outside consultants, or by a combination of the two. If the enterprise can afford it, using a consultant, especially a trained EMS auditor, can result in a much more productive audit. Not only can a good consultant bring unequivocal objectivity and fresh eyes to the exercise, but he or she can share the experience of other enterprises, both in identifying nonconformances and in devising ways to improve the EMS. No matter how the enterprise chooses to go about it, at the very least the person in the enterprise with principal responsibility for the audit process should have formal training in the science and art of EMS auditing.

To be useful, EMS audits have to be thorough and conducted in a reasonably consistent manner. That means they have to be well planned, and the plan has to be in writing; in effect, the plan comprises one or more procedures for carrying out the audit. In a small enterprise, a single overall audit procedure may suffice. A larger enterprise will probably need to develop separate procedures for each EMS element or cluster of EMS elements, and these procedures together, perhaps under an umbrella procedure covering the entire audit, would constitute the audit plan.

The EMS audit plan would cover subjects such as:

- The scope of the audit — the range of enterprise operations, facilities, pollution media, and so on that the audit will cover, consistent with the current scope of the EMS as stated in the environmental policy of the enterprise
- The frequency of the audit, the overall auditing schedule (including reporting findings), and the schedule for each of its components
- Who will be responsible for managing the audit overall, and roles and responsibilities for conducting each component of the audit
- The training, experience, expertise, and rank requirements for auditors
- The authority of auditors
- How audits will be conducted — the process for each component, including techniques and tools (such as checklists) used, individuals interviewed, procedures reviewed, records reviewed, observations made, tests conducted, and other documents consulted
- Guidelines or standards for evaluating different elements of the EMS, and specifications for any scoring systems employed

- Specific equipment and other resources employed in the audit process, including software, instruments, other equipment, temporary auditing assistance, clerical support, and so on
- The form of the presentation of audit findings — that is, specifications for the EMS audit report, including
    - to whom in the enterprise audit findings are formally reported, as well as when and by whom
    - other directions for communicating audit findings

We'll digress for a moment to touch on the subject of major and minor nonconformances. Actually, the subject of major and minor nonconformances itself is quite relevant to this subclause of ISO 14001, but we've chosen to illustrate the subject by referring to the general practices of certification auditors. As a result, readers will also have the benefit of some useful information on the subject of major and minor nonconformances in certification and recertification audits.

Certification auditors will distinguish between major and minor nonconformances. A *major nonconformance* is a serious deviation from ISO 14001 requirements that could, or does, adversely affect the effectiveness of an enterprise's EMS, and possibly also its environmental performance. Examples of major nonconformances include:

- Incomplete environmental policy
- Absence of documentation regarding an internal audit
- Absence of evidence of a training program or of training records
- Absence of evidence of any procedures required by the ISO 14001 standard
- Inadequate document control
- Absence of records of calibration and maintenance of measurement and monitoring equipment
- A pattern of not creating separate records of changes made to EMS procedures in response to nonconformances

At the conclusion of a certification audit, the lead auditor will submit reports on the enterprise's major nonconformances to the certification company, or registrar. The registrar will notify the enterprise that its EMS does not conform to the ISO 14001 standard and will deny the application for ISO 14001 certification or recertification. Then it's up to the enterprise to make the necessary corrections to its EMS and arrange for a re-audit.

A *minor nonconformance* is an isolated deviation from the exact requirements of the ISO 14001 standard, but not one that causes major problems or cannot be corrected quickly and easily. Minor nonconformances are usually on the order of honest oversights to which all human beings are subject. Examples of minor nonconformances include:

- A piece of required document control information missing from a particular EMS document

- Lack of detail in a particular environmental management program
- Inconsistency or lack of precision between an environmental objective and a current target
- A procedure not posted in a particular place it is needed
- An obsolete procedure still on the bulletin board with the new one
- No record of a formal decision regarding external communication on significant environmental aspects
- A single instance of not creating a separate record of changes made to an EMS procedure in response to a nonconformance

Some very minor nonconformances can be corrected on the spot, and the audit continues without further reference to them. Some can be corrected during the course of the audit (which could last several days at a single facility), and the matter will be cleared up before an audit report is submitted. In some cases an enterprise might be given a little time to correct a nonconformance and an auditor will return to check that it has been corrected — again, before a final audit report is submitted. Generally, certification auditors will work with enterprises to overcome minor nonconformances when the enterprise is doing a good overall EMS job.

The same principles of major and minor nonconformances apply as well to internal EMS audits.

Requirements of ISO 14001

Subclause 4.5.4 of the ISO 14001 EMS standard requires the enterprise to:

a) Establish and maintain a program with explicit procedures (an audit plan) for carrying out EMS audits at regular intervals
b) Make sure that audit procedures are comprehensive and cover the audit's:
   - scope
   - frequency
   - methodologies
   - responsibilities
   - requirements
   - reporting
c) Ensure that the audits will enable the enterprise to determine whether or not the EMS:
   - conforms to what the enterprise intended
   - conforms to the requirements of ISO 14001
   - has been properly implemented
   - has been properly maintained
d) Ensure that the extent, depth, and frequency of the audit are appropriate to:
   - the nature and scale of the enterprise's interactions with the environment
   - how well the enterprise did on previous EMS audits
e) Ensure that management receives a report of audit findings

Subclause 4.5.4 of ISO 14001 requires the enterprise to develop, and regularly review and update, an audit procedure or plan (the actual language of the standard refers to one or more "programs and procedures") for carrying out a periodic internal EMS audit. The audit plan does not have to be in writing, but except for the very smallest enterprises, why wouldn't it be? Because this will almost certainly be a written EMS procedure, even though it may be called a "plan" and is not required by the ISO 14001 standard, it nevertheless must be handled according to the procedures for controlling EMS documents covered under Subclause 4.4.5, "Document Control."

The internal EMS audit addresses two fundamental questions: Is the EMS designed right? Is the EMS being carried out faithfully as designed?

Item b) refers to the audit procedures, or plan, being comprehensive in its coverage of audit activities, but these activities cannot be separated from the coverage of the audit itself. It's best to think of this requirement as referring both to the audit procedures and to the coverage of the EMS by audit. There should be a comprehensive audit plan for an audit that covers the EMS comprehensively.

The four points under item c) essentially say that the audit plan should be built around addressing two fundamental questions: With the insight of experience, is the EMS designed right? And, is the EMS being carried out faithfully as designed, including being regularly reviewed and updated? A certification auditor will try to discern not only if the procedural requirements of this subclause are being met, but also if the audit as designed clearly addresses these two overarching audit questions.

There's no mistaking the intent of ISO 14001 about what the ultimate point of the internal audit is. It is not about passing a certification audit; it's about supporting the process of continual improvement.

The point of item d) is appropriateness — for example, that a petrochemical refinery ought to conduct EMS audits more frequently than a software company because there is more potential environmental damage at stake in the event of an EMS nonconformance at a petrochemical refinery. Also, if an enterprise turned up a whole collection of nonconformances in each of its recent EMS audits, that's a clear sign that the management of the enterprise has a tendency to let things slip; to offset that, the enterprise ought to conduct EMS audits more frequently. The point is extremely well taken, though it's unclear how this could be dealt with from an auditing point of view: More frequent than what? How big is a big environmental impact? How bad is a bad audit? The best way for the EMS Committee to approach this is to conduct a meeting with top management to discuss this very question — At what intervals should an internal EMS audit be conducted, in light of the environmental implications of enterprise operations and recent EMS audit history? — and record the decision made and the rationale for it.

Item e) leads right into the next and final clause of ISO 14001. Here the standard says that information on EMS audit findings has to go to management of the enterprise, which includes top management. In Clause 4.6, it says that top management has to use audit findings when considering the need for changes in the EMS of the enterprise. There's no mistaking the intent of ISO 14001 about what the ultimate point of the internal audit is: It is not about passing a certification audit; it's about supporting the process of continual improvement.

MANAGEMENT REVIEW (CLAUSE 4.6)

Continual Improvement

Strategic environmental management, like strategic management in general, is based on the concepts of looking ahead and bringing the enterprise to where it will need to be even before it needs to be there, bringing long-term considerations into short-term decisionmaking, and fostering continual improvement. All of this comes together in the environmental policy of the enterprise, the strategic planning orientation of its EMS, and, most decidedly, in the EMS management review. Top management established the environmental policy that is the foundation and ultimate guidance for design and operation of the enterprise's EMS; the same top management must periodically review the design, operation, and performance of the EMS to determine what could be done to improve it and to help bring the enterprise to where it needs to be in terms of environmental performance, even before it is required by law to be there. A regular EMS management review is fundamental to strategic management and the EMS principle of continual improvement.

Although it's called a management review, it is not just management that participates in it, and certainly not just top management. Smart management will develop a procedure that includes EMS reviews at all levels of the enterprise. These reviewers will feed their conclusions into the review by top management, while input from outside parties — suppliers, contractors, citizen groups, local authorities, regulatory authorities — could also be invited.

Strategic environmental management, like strategic management in general, is based on the concepts of looking ahead and bringing the enterprise to where it will need to be even before it needs to be there, bringing long-term considerations into short-term decisionmaking, and fostering continual improvement.

The basic sources of performance information for the review are the enterprise's program of continual EMS measuring and monitoring (all five parts of the process), and the EMS internal audit. Additional information comes from the lower-level reviews and outside sources just mentioned, from past EMS audits and management reviews, and from such things as training reports, accident reports, notices of regulatory change, market studies, and studies of industry trends. The EMS management review has to be broader and more comprehensive than even the EMS audit, because it needs to consider not just the design and operation of the enterprise's EMS, but also specifically the environmental performance of the enterprise and trends in that performance, how the EMS is integrating with the overall management system of the enterprise, and the business benefits the EMS is delivering.

In the management review, top management has to carefully and thoughtfully examine every EMS element and environmental management program in terms of:

- The appropriateness of its design in light of experience and audit findings
- The appropriateness of its design in light of changing internal and external circumstances
- Its relationship to the current and anticipated regulatory framework
- The efficiency of its operation
- Its effectiveness
- Related environmental performance and performance trends
- Its contribution to reaching targets and objectives
- Implications of impending changes in the operations or products of the enterprise
- Advances in science and technology
- Its relationship to other elements of the EMS
- Its integration with the overall management system of the enterprise
- The short- and long-term business benefits associated with its operation

This rather thorough examination of each EMS element is ultimately aimed at answering the single question, "What changes should be made in the EMS so it operates more efficiently and effectively, and more forcefully promotes continual improvement in our enterprise's environmental and business performance?"

For the benefit of the enterprise, top management needs to review the design and operation of the EMS, and the environmental performance of enterprise operations, in relation to other management systems of the enterprise and its strategic business context.

The EMS review by top management typically takes place soon after completion of the EMS internal audit and, therefore, at the same frequency. Like the internal audit, the EMS management review does not have to consider everything at once: It can have an extended schedule, a compressed schedule, or a mix of both, attended by the same general advantages and disadvantages as the internal audit scheduling alternatives discussed in the previous section. A good and flexible middle ground in the case of the management review is to have one element of the EMS on the agenda of senior manager meetings at least once every two weeks, with a full-scale formal EMS management review one to three times per year. Of course, an enterprise in the early stages of establishing its EMS will want to conduct management reviews fairly frequently, to recycle lessons of experience back into the EMS implementation process quickly and to consider expansions in the scope of the EMS.

Just as an enterprise has to conduct its EMS audit in a relatively consistent manner to provide information that can be usefully compared over time, so it needs to ensure that the management review is conducted in a more or less comparable way each time, using similar information to make similar types of assessments. Therefore, the EMS Committee needs to think through a procedure for the periodic management review. The EMS management review procedure could cover:

- The scope of the management review
- The person responsible for overall coordination of the review process
- Roles and responsibilities regarding different parts of the review process
- Preliminary or preparatory meetings to be held
- Information to be prepared for review, including:
    - what types of information
    - how the information is generated (for example, through the EMS audit, or routine measuring and monitoring activities, or through a special effort)
    - who is responsible for developing and presenting it
    - the format in which it is presented
    - when it is presented
    - to whom it is presented
- Frequency of the EMS management review and a schedule for all activities in the overall review process
- Guidelines for conducting review deliberations (for example, the things to be explicitly considered with respect to each EMS element)
- The format, content, approval, and distribution of the management review report and other documentation resulting from the review process
- Responsibility for implementing decisions taken during the management review
- Reporting on implementation of management review decisions

The EMS management review procedure would be an EMS document, and its handling would therefore be governed by the EMS document control procedures discussed earlier (Subclause 4.4.5, "Document Control"). The management review report would be an EMS record, and its handling would be governed by the procedures for EMS records discussed earlier (Subclause 4.5.3, "Records").

Readers should not lose sight of the fact that, as always, the material presented here is intended as a menu from which an enterprise can select, and adapt in accordance with the scale and nature of its operations, the scope of its EMS, and its approach to establishing and operating its EMS. In a very small enterprise, the management review might be conducted by the owner, and his or her familiarity with the enterprise might be such that it can be a very short meeting. At the other end of the spectrum, the management review can be a very elaborate process, involving successions of meetings at different levels and facilities of the enterprise and ending with a formal presentation of the EMS management review report to the board of directors.

Requirements of ISO 14001

Clause 4.6 of the ISO 14001 EMS standard requires the enterprise to:

a) Conduct a top management review of the EMS, to ensure that the EMS continues to be suitable and adequate for the enterprise, and effective
b) Conduct the management review at regular intervals that it determines appropriate
c) Ensure that, as part of the management review process, all information that top management might need to conduct its review is assembled and made available
d) Ensure that the management review addresses the possible need for changes to environmental policy, environmental objectives, and other elements of the EMS
e) Ensure that, in making its decisions, top management takes into account EMS audit findings, changing internal and external circumstances, and its commitment to continual improvement
f) Make a record of the management review and decisions taken

Some say that this subclause of ISO 14001 requires only that top management periodically review the EMS, not the environmental performance of the enterprise, which is reviewed as part of the ongoing checking and corrective action element of the EMS. Some certification auditors accept this restricted view. But for the benefit of the enterprise, a broader interpretation is wiser. Not only should the environmental performance of enterprise operations also be included in the review, but the whole should be considered in relation to other management systems of the enterprise and its strategic business context. Moreover, the management review is explicitly for top management and represents an opportunity for them to review the EMS comprehensively as a system. And that means examining operation of the EMS and its individual elements *and* the individual environmental management programs and associated environmental performance results. The key is in the word "effective" in item a). What else could the EMS be effective for, if not for continually improving the environmental performance and the business performance of the enterprise? And how could that be evaluated if

Information needed for the EMS management review also is important for other business management purposes and should be collected as part of the enterprise's overall data collection and reporting program.

not by reviewing direct environmental performance data?

Item c) mentions a management review process. The authors of the ISO 14001 standard clearly had in mind the image of a process cycle with the management review meeting in the middle, information development preceding it, and reporting and implementing changes in the EMS following it. As for the information preceding it, they want to ensure that preparations for the management review meeting are not handled casually — that deliberate thought is given to the information top management will need to do the job they are required by the standard to do, and to the steps needed to be sure the information is delivered.

Item e) refers to audit findings and, implicitly, to information on changing circumstances, and that represents the auditable minimum of the information top management will need. But what are changing circumstances? Some examples mentioned earlier in this chapter include:

- The availability of new technologies
- New information from a P2 audit
- New findings of environmental health risk research
- Changes and new knowledge arising from the enterprise's experience implementing its environmental management programs
- Changes in the ISO 14001 standard
- Changes in government regulations
- New demands by local citizen groups or authorities
- Changing social values
- Changing environmental performance of the enterprise as a result of EMS operation
- An overall change in the business operations, facilities, equipment, processes, products, and markets of the enterprise.

If careful thought is not given to it, and a regular routine is not established, preparing the right information for an EMS management review can be a time consuming activity. But this information also is important for other types of business management purposes and should be collected as part of the enterprise's overall data collection and reporting program.

Lest there be any doubt, items d) and e) together make clear that the purpose of the management review is to examine every element of the EMS and consider changes that would improve it. In item f), ISO 14001 requires formal EMS documentation of the management review. The enterprise should create the documentation in a format that enables a certification auditor to determine that the other requirements of this subclause have been met. For example, there could be separate sections on scheduling, information, management review deliberations, and management review decisions, among others. If much of this is covered in a written procedure, then brief reference to it in the management review report will provide evidence that the procedure was followed, and auditors are always interested in that.

ONE MORE TIME

The basic elements of an EMS modeled after the ISO 14001 international standard are:

ENVIRONMENTAL POLICY

PLANNING

- Environmental Aspects
- Legal and Other Requirements
- Objectives and Targets
- Environmental Management Programs

IMPLEMENTATION AND OPERATION

- Structure and Responsibility
- Training, Awareness, and Competence
- Communication
- Environmental Management System Documentation
- Document Control
- Operational Control
- Emergency Preparedness and Response

CHECKING AND CORRECTIVE ACTION

- Monitoring and Measurement
- Nonconformance and Corrective and Preventive Action
- Records
- Environmental Management System Audit

MANAGEMENT REVIEW

The logic is as follows:

- First, top management establishes environmental policy, which provides the overall guidance for the EMS and environmental performance of the enterprise

- Then, the enterprise plans specific environmental management programs to improve its environmental performance and implement the policy

- Then, to support its environmental management programs and sound environmental practice overall, the enterprise establishes an EMS implementation and operation infrastructure

- Once the environmental management programs are underway, the enterprise monitors performance and takes corrective action as indicated to ensure that environmental targets are achieved

- Finally, top management reviews what it has wrought, and figures out ways to improve the EMS, starting with improvements in environmental policy

There is a sequential logic to the model, and initial EMS implementation activities may indeed follow a sequential course. In practice, however, the model is not of a sequence of activities that cycles from beginning to end, but of 17 elements, each of which repeatedly cycles through its own process, so that all are active and supporting each other at all times. For each of these 17 elements, the ISO 14001 EMS standard contains specific requirements that certification auditors will check.

An EMS is a tool for strategic environmental management. As such, it should integrate easily with the overall management system of the enterprise, and reinforce its general strategic planning and sound business practices and operations. One key to establishing an effective EMS at minimal cost is to build on the enterprise's existing systems, procedures, practices, and programs (if the enterprise has established ISO 9001/2, then to a great extent it can layer ISO 14001 onto that). Another key is to focus on establishing a productive EMS, rather than on certification requirements. Start small, with an initial EMS scope that is easily manageable, and gradually expand to more operations, more facilities, more pollution media, more environmental aspects of the enterprise. When an enterprise uses the ISO 14001 standard as a starting point for designing its own EMS, then in due course, as it extends its EMS, it will come to meet all the ISO 14001 requirements and be ready for a certification audit. At that point, it will have a great deal of experience with its EMS, will already have reaped substantial environmental and business benefits from it, and will really know what it is doing.

TWO EXERCISES

Exercise 1. The ISO 14001 EMS standard uses the word "shall" 51 times in Clauses 4.2 through 4.6, which cover the five basic elements of the planning and implementation process. The authors of ISO 14001 use this word to signal an absolute requirement of the standard; an enterprise that fails to meet *any* of the 51 "shalls" is considered not in conformance to ISO 14001. Actually, the number of discrete requirements is quite a bit higher than 51 because, as we have seen, many of the 51 have multiple parts. A list of the unconditional requirements of ISO 14001 serves as a basic checklist for determining what an enterprise has to do to implement ISO 14001, for creating detailed plans for implementing the EMS, and for EMS auditing — internal auditing, second-party auditing, or auditing for certification.

A list of ISO 14001 requirements written in your own words and organized by ISO 14001 clauses and subclauses will be an invaluable tool if you expect to be seriously involved in implementing the standard or in implementing a related type of EMS. In this exercise, you read through ISO 14001 Clauses 4.2 through 4.6

carefully and rephrase each of the explicit requirements of the standard in your own words. You can find out how to obtain a copy of the standard most easily at www.iso.ch. Use two-column paper, so you can write comments to the right of each requirement when you use your work as a checklist.

The following template will help you identify all the requirements of the standard. Each row with a dot at the left represents an ISO 14001 "shall." Parts or subrequirements or related points under a requirement are represented by additional rows under the row for the basic requirement.

Table 6. Template for Checklist of ISO 14001 Requirements.

ISO 14001 REQUIREMENT	COMMENTS
4.2 Environmental Policy	
•	
a)	
b)	
c)	
d)	
e)	
f)	
4.3 Planning	
4.3.1 Environmental aspects	
•	
•	
•	
4.3.2 Legal and other requirements	
•	

ISO 14001 REQUIREMENT	COMMENTS
4.3.3 Objectives and targets	
•	
•	
•	
4.3.4 Environmental management programs	
•	
•	
•	
4.4 Implementation and Operation	
4.4.1 Structure and responsibility	
•	
•	
•	
•	
a)	
b)	
4.4.2 Training, awareness, and competence	
•	
•	
•	
a)	
b)	
c)	
d)	

ISO 14001 REQUIREMENT	COMMENTS
•	
4.4.3 Communication	
•	
a)	
b)	
•	
4.4.4 Environmental management system documentation	
•	
a)	
b)	
4.4.5 Document control	
•	
a)	
b)	
c)	
d)	
e)	
•	
•	
4.4.6 Operational control	
•	
•	

ISO 14001 REQUIREMENT	COMMENTS
a)	
b)	
c)	
4.4.7 Emergency preparedness and response	
•	
•	
•	
4.5 Checking and Corrective Action	
4.5.1 Monitoring and measurement	
•	
•	
•	
•	
•	
4.5.2 Nonconformance and corrective and preventive action	
•	
•	
•	
4.5.3 Records	
•	
•	
•	
•	
•	
•	
4.5.4 Environmental management system audit	

ISO 14001 REQUIREMENT	COMMENTS
•	
a1)	
a2)	
b)	
•	
•	
4.6 Management Review	
•	
•	
•	

Exercise 2. Table 2 (page 33) contained examples of 10 enterprise activities and their aspects and impacts. In Tables 3 (page 41) and 5 (page 48), the first five examples were expanded to include corresponding objectives, targets, and environmental management programs. Table 7 below contains the last five examples, numbers 6–10. You are invited to try your hand at formulating possible objectives, first-year targets, and summaries of environmental management programs for these examples.

Table 7. Activities, Aspects, Impacts, Objectives, Targets, and Environmental Programs.

Activity or operation	Environmental aspects	Negative environmental impact	Objectives	Targets, 1st year (ending *[date]*)	Program summary
6. Maintaining vehicle fleet	Potential spills of hazardous chemicals	Potential soil and groundwater contamination			
	Exhaust emissions	Air pollution			
	Hazardous solid waste	Soil and groundwater contamination			

Activity or operation	Environmental aspects	Negative environmental impact	Objectives	Targets, 1st year (ending *[date]*)	Program summary
7. Front- and back-office operations	Use of electricity Use of paper Generation of solid waste	Upstream air pollution Forest resource consumption Contributes to need for landfills			
8. Maintaining grounds	Spraying and spreading of herbicides and pesticides	Soil and groundwater contamination			
9. Constructing new storage facility	Soil erosion	Degradation of the site and nearby areas			
10. Discarding used batteries supplied with our product (customer activity)	Placement of hazardous materials in household waste that may or may not be landfilled	Soil and groundwater contamination, human health risk			

QUESTIONS FOR THINKING AND DISCUSSING

1) What are the five biggest obstacles you would face in promoting adoption of an EMS in your enterprise, and how would you overcome each of them?

2) Which of the 17 basic elements (clauses and subclauses) of ISO 14001 described in this chapter do you think is/are most dispensable, at least in the initial stages of EMS implementation in your enterprise? Why?

3) Which of the 17 elements of ISO 14001 described in this chapter do you think would be the most difficult to establish and maintain in your enterprise? Why?

4) In developing a plan for implementing an EMS in your enterprise, where would assistance from a consultant fit in? What if the aim was an initial EMS without regard to certification? What if the aim was to be ready to have a successful certification audit within 12 months?

5) Based on what you know about your enterprise, what are the three environmental aspects of its operations that should be addressed first by environmental management programs?

6) Identify 10 discrete activities, operations, or services of your enterprise, or uses of its products, that have environmental aspects. Identify the environmental aspects and what you know about their environmental impacts.

7) Tell the story of the third example in Table 5: Use your imagination to describe what the core process is, why the particular objectives were decided upon, what some of the considerations of the EMS Committee were in outlining the environmental management programs, and what some of the give-and-take between devising the programs and setting the targets might have been.

8) Name five types of records and five types of documents that you use regularly in your work (see the section on Subclause 4.4.4, page 61, for a refresher on the distinction between records and documents).

9) How many procedures are required for conformance to the ISO 14001 standard? List them. How many of these must be written?

10) How many records are required for conformance to the ISO 14001 standard? List them. Remember, the standard may not explicitly say "make a record," but if it requires recording or reporting or documenting a decision, action, measurement, or finding, then an EMS record is required.

Chapter 3
EMS: TOOLS AND TECHNIQUES

USING THIS CHAPTER

This chapter provides tools and techniques for implementing ISO 14001, using the requirements of ISO 14001 discussed in Chapter 2 as the basis. Readers from enterprises for which ISO 14001 certification is not a priority in the near term can still benefit from what we present in this chapter — they can use the tools presented here as starting points for crafting similar tools consistent with their particular approaches to establishing and operating their EMSs. As the presentation in Chapter 2 was meant to illustrate, the fundamentals of all EMSs are similar, and so generally the implementation tools tend to be similar.

The material in this chapter will be most useful to readers who have reviewed the counterpart sections of Chapter 2, which provides a discussion of the details of an EMS, the clauses and subclauses (the EMS elements) of ISO 14001, the specific requirements for conformance to ISO 14001, and other information about implementing an EMS. It also provides a summary of the basic steps for implementing an EMS in general, and ISO 14001 in particular. None of that is repeated here, and obviously it is the material in Chapter 2 that gives meaning to the tools and techniques offered in this chapter.

In the sections of this chapter dealing with the first two ISO 14001 elements — Clause 4.2, "Environmental Policy, and Subclause 4.3.1, "Environmental Aspects" — many topics are touched upon that also are relevant for most other EMS elements. Examples include obtaining input from throughout the enterprise, tapping into other sources of information, using a standard EMS document header, using analytical matrices, and ISO 14001 requirements for the structure of a procedure. In the interest of text economy, after the first mention of a broadly applicable point, we try to avoid repeating it in this chapter. As a result, starting with the section on the third ISO 14001 element — Subclause 4.3.2, "Legal and Other Requirements" — the presentation becomes substantially leaner in terms of the amount of text that elaborates on the tools and techniques presented. For that reason, readers interested only in the tools and techniques for one or another specific ISO 14001 element will do themselves a service by nevertheless reading the sections on environmental policy and environmental aspects.

Throughout Chapter 3 we make frequent reference to the EMS Committee undertaking an implementation task. In small enterprises, the "EMS Committee" may be only the Management Representative working on the EMS part-time, while in very large enterprises there may be EMS committees at different levels that are all directly involved in EMS implementation and maintenance. Moreover, the EMS Committee carries out virtually all implementation tasks with the involvement of other employees. For example, regarding identifying significant environmental aspects (Subclause 4.3.1), many employees from different departments would probably be consulted by the EMS Committee in the course of defining discrete activities for analysis, articulating environmental aspects, assessing environmental impacts, and more. Throughout this chapter the phrase "EMS Committee" is meant to cover a wide range of possible EMS primary implementers in different enterprises, and to include involvement of other workers.

There are many other sources of information on tools and techniques for implementing ISO 14001. One of these is ISO 14004, a non-auditable companion standard to ISO 14001 that provides guidance from the ISO for implementing ISO 14001. The implementation guidance in ISO 14004 is thin, but it provides some insight into the thinking behind the individual clauses and subclauses of ISO 14001. The thinking behind the clauses and subclauses is not binding in the sense that a certification auditor can only hold an enterprise to the exact requirements of 14001, but it can be useful when an EMS Committee grapples with how to implement a particular element of the standard or the counterpart element of any other EMS. For information on purchasing copies of the ISO 14001 and ISO 14004 standards, visit www.iso.ch.

The "Additional ISO 14001 and P2 Resources" section (Appendix A) at the end of this book lists a number of ISO 14001 implementation handbooks. These books, as well as a number of the Web sites listed, offer:

- Illustrations of EMS documents, such as the environmental policy statement, that meet ISO 14001 requirements
- Templates for procedures, reports, records, and other EMS documentation
- Models for documentation that is not required but highly recommended, such as a document control matrix, or an EMS roles and responsibilities organization chart
- Checklists of all kinds for all elements of the ISO 14001 standard
- Step-by-step instructions for EMS implementation tasks, such as organizing the management review process
- Self-audit guides

The body of literature related to EMS and ISO 14001 is very dynamic because experience worldwide with this management tool is expanding rapidly.

An enterprise establishing an EMS needs to acquire and review some of these publications and surf the Web to develop knowledge of the different EMS implementation approaches, models, and tools available for each EMS element. Only then can it go about EMS implementation well informed. The aim in this chapter is to support that effort, not to complicate it or discourage readers from casting their nets widely. For that reason, material readily available from other sources is usually not included among the tools and techniques in this chapter. We've made an effort to restrict this chapter to versions of tools and techniques that are unique or that complement what can be obtained from other sources. Many sections of this chapter include lists of the sorts of tools and other implementation guidance materials available elsewhere.

For example, in the section of this chapter on environmental policy, we discuss matters related to formulating the EMS policy and creating the EMS policy document with the required content, but we don't include an example of an environmental policy statement. This is because an environmental policy statement that meets ISO 14001 requirements can be very simple — say, little more than a bulleted list of values, commitments, and overall objectives covering less than a page — or it can be relatively extensive, depending on the nature and size of the enterprise, the scope of its EMS, and whether or not elaborating annexes are attached to the policy statement. There are many examples of environmental policy statements, both real and imaginary, in other literature and resources. A committed and creative senior manager or ISO 14001 Management Representative would not only explore print literature but would also look for environmental policy statements on the Web sites of enterprises that are certified to the ISO 14001 standard. Names of certified enterprises can be found through Web surfing starting with Web sites listed in the "Additional ISO 14001 and P2 Resources" section of this book, and www.globalreporting.org will link researchers to even more environmental policy statements. Nothing would have been added by including one particular example of an environmental policy statement here.

In short, though the tools and techniques presented in this chapter stand on their own, they will serve better when they are put to work in combination with material from other ISO 14001 resources.

BEFORE POLICY

Meaningful EMS implementation always will come from within the enterprise, though consultants can help. As a practical matter, this means it is necessary to achieve awareness, interest, and commitment to some degree by both management and workers before any major EMS activities are undertaken. An enterprise that is serious about establishing an efficient and productive EMS may want to mount a program of environmental awareness either before beginning work on its EMS or during the early organizing stages. The program might aim to raise awareness among everyone in the enterprise of:

- The environmental aspects of enterprise operations and products
- The direct, indirect, and potential costs and cost savings associated with those environmental aspects
- The legal and regulatory requirements associated with different parts of enterprise operations, and the enterprise's record of compliance
- The growing pressure from customers, authorities, lenders, competitors, and other market forces to improve efficiency, productivity, and environmental performance
- The relationship between continually improving environmental performance and strategic business planning
- What, generally, P2 practices and an EMS can do for the enterprise and its future

As part of the process of working toward beginning EMS implementation activity, most enterprises will want to conduct an initial environmental review (IER) or a gap analysis. The IER is a general environmental survey of the enterprise; a gap analysis is a careful comparison of the requirement-by-requirement gap between the requirements of ISO 14001 and what the enterprise already has in place. A gap analysis might be carried out by an enterprise that already has an environmental program of sorts, maybe even its own EMS. Many enterprises that have already implemented ISO 9001/2 precede the ISO 14001 implementation process with a gap analysis. IER is discussed in Chapter 4 (see page 177); gap analysis is discussed further in this chapter — in the section on the internal EMS audit (page 157) — and is touched on again in Chapter 4, because both the gap analysis and the EMS audit use the same checklist of ISO 14001 requirements.

The Management Representative and members of the EMS Committee should be employees who are personally interested in improving the environmental performance of the enterprise. In other words, these should be people who want to create a successful EMS in the enterprise.

Conducting an IER or gap analysis can serve the awareness program because employees, through their participation in these exercises, may come to see the operations of their enterprise in a different light.

Some enterprises have undertaken P2 audits as a pre-implementation activity. P2 audits can serve both to raise awareness and generate important information for the EMS. A P2 audit can produce information on environmental aspects, their impacts, and possible P2 remedies for environmental problems, and it can demonstrate dramatically the linkage between good environmental management and good management of the enterprise's bottom line. P2 audits have been used by managers to convince top management of the merits of an EMS, which continually generates P2 innovations, and there are enterprises that have used the cost savings from initial P2 innovations to finance EMS implementation.

In addition to top management and worker awareness and commitment, an ISO 14001 Management Representative or overall EMS coordinator and, in many cases, an EMS Committee must be designated before any significant EMS implementation activity can take place. The EMS Committee should be made up of employees from various levels and functions throughout the enterprise. Both the Management Representative and members of the EMS Committee should be employees who are personally interested in improving the environmental performance of the enterprise. In other words, these should be people who want to create a successful EMS in the enterprise, not people who will participate only grudgingly or without real interest.

Once top management and worker awareness and commitment have been established, and a Management Representative and EMS Committee are in place — and, if needed, an IER or gap analysis have been carried out — EMS implementation planning can begin.

An initial decision will have to be made about the overall approach of the enterprise to EMS implementation and certification and, accordingly, the initial scope of the EMS. Much was already said about this in Chapter 2, and further help for establishing an EMS by degrees is provided in Chapter 4, so more will not be said more about it here. But perhaps we should make the point that small, consistently successful steps in EMS implementation will benefit the enterprise by raising the enthusiasm, energizing, involving, and maintaining the commitment of employees as these steps cumulate toward enterprise preparedness for an ISO 14001 certification audit.

In other words, there should be no rush to satisfy all the ISO 14001 requirements quickly. Moving too quickly foregoes not only the reinforcement given to employees by successes along the way, but also the opportunity to "own" and to learn how really to operate the EMS before certification auditors inspect it.

The plan for implementing the EMS will, of course, involve a schedule of activities and responsibilities for implementing each EMS element. If a consultant is used, each of the consultant's visits, and the terms of reference for the visit, would be included in the plan. It should be expected that the plan will be dynamic, that it will probably be revised and refined regularly by the EMS Committee and managers. When planning implementation, the EMS Committee will want to look for ways to take advantage of existing "platforms" in the enterprise. That means building ISO 14001 requirements onto existing employee training programs, document control procedures, performance monitoring programs, etc., to minimize the EMS implementation burden on the enterprise. A checklist of ISO 14001 requirements is a good point of departure for detailing tasks to be accounted for in the EMS implementation plan. A checklist of that sort is provided in the section of this chapter on the EMS internal audit (Subclause 4.5.4, Table 22 on page 160).

Be sure each increment of EMS implementation will be successful and a cause for self-congratulation.

Keep it as simple as possible. Plan carefully. Be sure each increment will be successful and a cause for self-congratulation.

ENVIRONMENTAL POLICY (CLAUSE 4.2)

Developing Environmental Policy

Although it can be, there's no reason that formulating environmental policy *must* be the first EMS implementation step. In fact, it's probably better to extend the policy development process so it overlaps with the environmental management program planning process: that way, the initial environmental policy statement and the first set of environmental management programs could be finalized at roughly the same time. This enables the programs to get up and running just after the policy is announced, providing immediate visible evidence that the enterprise's environmental policy will be reflected in action. No less important, by developing the policy at the same time as the environmental aspects, objectives, and management programs of the enterprise, each can inform the other, so that at the end of the day there is full consistency among them. Ideally, the current environmental objectives of the enterprise would be incorporated into its current environmental policy.

Everything in an enterprise, including an EMS, starts with top management. Or, if not, it doesn't go very far without top management getting behind it. Top management is the individual or group of senior managers that has overall responsibility, accountability, and authority for the resources of the enterprise, its operations, and its performance. These are the people who ought to care most about what an EMS can do for the enterprise, and who control the resources to implement a productive EMS. If you are not top management and you want to implement an EMS in your enterprise, your first job is to convince top management. Help with that can come from a P2 audit or actual P2 cost-saving experience in the enterprise; from stories of other enterprise EMS successes from the literature or the Web; and from the first section of Chapter 4, which describes the benefits of an EMS.

If you are not top management and you want to implement an EMS in your enterprise, your first job is to convince top management.

In developing environmental policy, top management needs input from all departments and levels of the enterprise, so the people closest to where environmental impacts take place, and those who supervise them, can bring their practical knowledge to the policy. In most cases, top management will obtain this input through comment on successive drafts of the environmental policy statement circulated by the Management Representative or by the EMS Committee. Most often, the Management Representative or EMS Committee will have overall responsibility for collating the input from across the enterprise, preparing successive drafts, and working closely with senior managers. The biggest challenge in the iterative process of formulating the enterprise's environmental policy may be keeping the policy drafts focused and succinct.

Probably the first issue managers will have to debate is the initial scope of the environmental policy and the EMS. This may be an issue whether or not a decision has been made to begin working toward ISO 14001 certification in the near term. But, particularly if the enterprise takes a more relaxed and learning-oriented approach, the real question is where and how to begin. Top management may want a first scope that will be the easiest to deal with, or they may want a first scope that addresses a known serious environmental problem, or there may be some other strategic determinant. Whatever the case, it's important to select an initial EMS scope that clearly can be handled successfully and without much disruption to normal operations. There is more on the subject of environmental policy and EMS scope in the counterpart section of Chapter 2 (page 22) and in the section of Chapter 4 on EMS Lite (page 188).

Other issues that senior managers and others might grapple with in the course of developing environmental policy include:

- The specific requirements of ISO 14001 Clause 4.2

- The environmental policy of the enterprise in relation to each of the elements of its EMS
- The importance of ISO 14001 certification and its relationship to the purposes of the EMS
- How far "upstream" and "downstream" the enterprise should be concerned with environmental impacts of its operations and products
- How environmental policy should relate to contractors and suppliers
- Policy concerning the relationship between improving environmental performance and improving the bottom line

The list could go on and on. Environmental policy development offers an opportunity to explore many environmental issues throughout the enterprise.

Annex A of ISO 14004 contains statements of guiding principles from the Rio Declaration on Environment and Development (June 1992) and from the International Chamber of Commerce's Business Charter for Sustainable Development, which could be useful in the process of formulating environmental policy for an enterprise. Many examples of environmental policy statements can be found in ISO 14001 literature and on various Web sites, as mentioned earlier. In addition, many enterprises will already have some mention of their environmental values in existing policy or mission statements, or annual reports or other publications; these can be researched and included in the collection of information consulted toward developing a consolidated and cohesive environmental policy for the enterprise.

The Environmental Policy Statement

The environmental policy statement is an EMS document required by the ISO 14001 standard, so, at least in file copies and in internally circulated copies, it would have the standard EMS document header that the enterprise uses. This header would contain all the document control information required by ISO 14001 Subclause 4.4.5, "Document Control," and is discussed under that subclause in this chapter (page 141).

Following the header there might or might not be an introductory overview paragraph, then the text of the policy. The policy statement could be a very simple bulleted list at one end of the spectrum, or a few pages of prose at the other, with many possibilities in between. Some enterprises have chosen to have a very basic environmental policy statement, accompanied by an annex providing an elaboration of each point. Some policy statement annexes also provide guidance on implementing the policy.

The environmental policy statement would be signed by the highest-ranking person in the enterprise.

The text of the policy statement should be organized logically and in a manner that makes it easy for a certification auditor to determine that it satisfies ISO 14001 environmental policy requirements. Table 1 shows one possible structure.

Table 1. Structure for an Environmental Policy Statement.

Structure for an Environmental Policy Statement
Mission What the enterprise's basic mission is, and how that mission relates to environmental concerns.
Values What the basic business and environmental values of the enterprise are, and the relationship between them.
Purpose What the purpose of this environmental policy statement is, and what its relationship is to the EMS of the enterprise.
Scope What part of the enterprise or its operations this policy and the EMS apply to and, if warranted, an explanation. If there is a limited initial scope, with the intent to systematically extend the scope as the enterprise gains experience, there should be a statement about this intent, about how the extension of scope will work, and about why the enterprise decided to start with this particular scope.
Commitments What general environmental principles the enterprise is committed to. Of course, this would include commitments to legal and regulatory compliance, and to honoring other environmental standards and practices that it has accepted; a commitment to continual environmental improvement; and a commitment to pollution prevention. But the enterprise might also want to declare other commitments, such as a commitment to obtaining and/or maintaining ISO 14001 certification, a commitment to maintain close working relationships with environmentally oriented citizen groups in the area, a commitment to support environmental education in the area, a commitment to encourage good environmental management among its peer enterprises, etc.
Objectives What the current environmental objectives of the enterprise are. These could be fairly broad or very specific; they could be the same as the objectives on which the enterprise's environmental management programs are based. This section needs to respond to the ISO 14001 requirement that the environmental policy of the enterprise provide the framework for establishing and reviewing the enterprise's environmental objectives and targets. That might suggest broader objectives for the longer term or more specific objectives for the shorter term; including both would not be a bad idea.

Related Practices
What the related practices adopted by the enterprise are. This section could list practices that top management does not want to raise to a level equal to the overarching environmental commitments or that apply broadly (not just to environmental concerns), and that should be restated here as they relate to environmental concerns. Examples might include how this environmental policy relates to the enterprise's contractors and suppliers; that environmental considerations will be taken into account in all enterprise activities; that the enterprise favors contractors, suppliers, and employees who demonstrate good environmental management practices; that the enterprise assumes environmental responsibility for its products (or the extent to which it assumes that responsibility); that the enterprise will maintain close working relationships with environmentally oriented citizen groups in the area (in cases where an enterprise feels this should be a related practice instead of a commitment).

Review
How often, by whom, and through what general process this environmental policy will be reviewed and reconsidered.

Communication
To whom, and how, this environmental policy statement will be communicated.

Most of what is contained in this structure, though not required by ISO 14001 to be incorporated into the actual policy, does reflect policy requirements of the standard that must be satisfied one way or another. Strength, clarity, and a stronger sense of commitment to the environmental policy is communicated to the reader when such things as additional concrete practices and statements about reviewing the policy and communicating it are built right into the policy document.

PLANNING: ENVIRONMENTAL ASPECTS (SUBCLAUSE 4.3.1)

The logical sequence of steps for identifying significant environmental aspects is:

1) Break down the operations and products (that are within the scope mentioned in the environmental policy) of the enterprise into activities for analysis
2) Identify environmental aspects of those activities
3) Establish the environmental impacts of those environmental aspects
4) Determine which of the environmental aspects are significant because they have significant environmental impacts

But, especially the first time around, as we will see, it may be easier and more useful to start near the end and work back.

Identifying Environmental Aspects

The most straightforward way of approaching the task of identifying the environmental aspects of enterprise operations and products is to begin by listing all the basic operations of the enterprise, and then breaking down those operations into individual activities in a way that makes sense as the basic units for analysis. The enterprise would handle the use and disposal of its products in the same way. At one end of the spectrum, the EMS Committee might begin simply by examining operations and product life cycles for activities that obviously have environmental connections, and focusing only on those activities. At the other end of the spectrum, the enterprise could employ scientific process-flow analyses. In between are acceptable methodologies that represent a wide range of variations in analytical detail and depth.

When considering activities to be analyzed for their environmental aspects, an EMS Committee and top management typically have to consider such questions as:

- How detailed should the activities be?
- How far upstream and downstream should we go in the analysis, keeping in mind that we must be able to exercise control or influence?
- What comes into the enterprise that we should take into account?
- Apart from our products and their use, what do we do or cause to happen outside the physical boundaries of the enterprise (for example, marketing, waste disposal, commuting)?
- What products does our enterprise circulate in the world, what does using them involve, and how are they disposed of?

How detailed should the activities be? How far upstream and downstream should we go in the analysis? What comes into the enterprise that we should be taking into account? What do we do or cause to happen outside the boundaries of the enterprise? What does using and disposing of our products involve?

Generally speaking, the range of activities examined will be those that fall within the scope of the EMS defined in the enterprise's environmental policy. If the process of developing the initial environmental policy for the enterprise was designed to overlap with the process of identifying environmental aspects and impacts, the nature of the environmental aspects identified may help define the initial range of activities to be covered by the environmental policy and EMS. In other words, each of the activities will inform the other, and in the end they are finalized simultaneously and in full coordination. As mentioned earlier, the process of simultaneous determination could include not only environmental policy and aspects, but also objectives, targets, and environmental management programs.

Continuing with the straightforward logical analysis model, once the EMS Committee articulates discrete activities of the enterprise, it would examine each of them to find how they interact with the environment during in the course of the activity. These interactions, or environmental aspects, could be direct: consumption of resources, releases to the environment, landscape alterations. Or, they could be indirect: consumption of goods made from natural resources, consumption of energy, use by others of the enterprise's products. And there you have the makings of an analytical matrix.

An analytical matrix is a tool to help think a matter through. The things you're interested in (in this case, activities) would be listed down the left as row headings. Something related to the quality you want to know about them (in this case, the environmental aspects they have, if any) would be listed along the top as column headings. In each cell of the matrix where a row intersects a column, an entry of some sort — a check mark, yes or no, or a piece of information — would be made if the thing of that row has the quality of that column. The analytical matrix is a basic information development tool for establishing and maintaining an EMS.

Table 2 lists the activities of the enterprise along the left, and categories of environmental interactions along the top. The EMS Committee, with the help of other employees, would work across each row, examining whether or not each type of environmental interaction takes place (or could potentially take place) in the course of each of the activities. If yes, then committee members would describe in a phrase the exact nature of the environmental aspect of the activity (for examples, see Table 2 in Chapter 2, page 33). When committee members have examined and commented on all the activities, the filled-in cells of the matrix would constitute the working list of environmental aspects of the enterprise.

Table 3 is another analytical matrix for identifying environmental aspects. In this case, the enterprise has used more specific column headings — that is, something closer to what an actual environmental aspect would be than was used in Table 2. The basic question is shown above the columns in Table 3. For each of the identified activities (rows), the specific questions are:

- Does that activity involve use of natural resources in the form of raw materials, indirect inputs, or energy?
- Does it produce emissions to the air?
- Does it produce releases to a nearby water body?
- Does it produce solid waste that is stored for a long period on the site?

And so on. The example in Table 3 was taken from an enterprise that did not examine its products to identify environmental aspects.

Tables 2 and 3, like all the tools presented in this chapter, are meant only to give readers a sense of different possibilities for ways the task (in this case, the task of identifying environmental aspects) can be approached.

Table 2. Starting the Analysis with Activities and Environmental Interactions.

Operation, Activity		Direct Environmental Interactions			Indirect Environmental Interactions		
		Consumption of resources	**Releases to the environment**	**Landscape alteration**	**Consumption of goods made from natural resources**	**Energy consumption**	**Use by others of the enterprise's products**
Manufacturing	**Generating process steam**	Coal, Water	NO_x, SO_x, and CO_2 emissions Fly-ash disposal				
	Activity B						
	Activity C						
Operation 2	**Activity D**						
	Activity E						
	Activity F						

ISO 14001 requires that a procedure to identify environmental aspects be put in place "in order to determine" which of them have significant environmental impacts. So a procedure is not actually required by the standard for identifying and quantifying environmental impacts or for determining which of them are significant (and therefore, which of their respective environmental aspects are significant). But for the process to hold up over time, the enterprise would have to do things fairly consistently so information generated over the months and years would be comparable. That means there has to be a procedure and, to ensure that the procedure is consistent, it really would have to be in writing. But, to repeat, the only procedure actually required here by ISO 14001 is a procedure for identifying environmental aspects; and that procedure does not *have* to be written.

If we continue further with a straightforward logical analysis model, the next step would be to set up a table like Table 2 in Chapter2, page 33, where the EMS Committee would record environmental impacts associated with each environmental aspect (which in turn would be associated in the table with a particular activity of the enterprise).

Table 3. Starting the Analysis with Activities and Specific Types of Aspects.

Operation, Activity		Does the activity involve or produce....?							
		Use of natural resources*	Emissions to air	Releases to water	Solid waste disposed of on site	Solid waste disposed of off site	Toxic or hazardous waste	Landscape alteration	Other**
Manufacturing	Generating process steam	Use of coal	NO_x, SO_x, and CO_2 emissions			Fly-ash disposal			
	Activity B								
	Activity C								
Operation 2	Activity D								
	Activity E								
	Activity F								

**Raw materials, indirect inputs, energy*
***Global, regional, local environmental issues*

Most enterprises need an EMS so that they can approach their environmental impacts systematically and strategically, not so that they can have a framework for identifying environmental impacts in the first place.

ISO 14001 does not require that environmental aspects or impacts be quantified. Eventually, of course, in most cases there needs to be some sort of quantification of environmental aspects or impacts, because it is the reduction of these quantities that represents improved environmental performance. The objectives and targets for four of the five examples in Table 3 of Chapter 2 (page 41) require baseline aspect or impact data to be meaningful and useful.

But it's important that the EMS Committee not get bogged down at *this* point with quantification. The emphasis here, especially when the EMS is first being established, should be on a free-wheeling effort, a brainstorming of sorts, to identify environmental aspects and associated impacts. If quantified information on the aspects identified or their impacts is readily available, so much the better. Otherwise, quantification can be part of the subsequent refinement of the list of aspects and impacts, or will occur during the process of setting objectives and targets.

The truth of the matter is, most enterprises are already well aware of the principal environmental impacts of their operations and, in many cases, are already monitoring them. In countries like the United States, where there has been an environmental regulatory framework established and enforced for some time, there are few enterprises that have never given any thought to their environmental performance. Many conduct regulatory compliance audits periodically and monitor at least some types of their environmental performance because of permit requirements. In many cases, it may only be the environmental officer who is fully aware of the environmental impacts of enterprise operations; but the environmental officer would surely be a prominent figure, probably the Management Representative, in an enterprise's EMS implementation activities. Most enterprises need an EMS so that they can approach their environmental impacts systematically and strategically, not so that they can have a framework for identifying environmental impacts in the first place.

For such enterprises the greater need is to put their environmental impacts into a framework of strategic management — so they can, for example, systematically reduce effluents in a sound business way rather than invest more and more money in end-of-pipe wastewater treatment as regulations get tighter and tighter. They need a framework that will help them see their environmental impacts as resulting from specific environmental aspects of specific activities; then they can focus on revising the activities that lead to pollution, rather than on the pollution itself, as the source of the problem. An enterprise that already knows its major environmental impacts might use an analytical matrix like the one shown in Table 4.

Because the enterprise knows its major environmental impacts, its examination can work by columns instead of rows — "We know we have air pollution; does it happen as a result of Activity A? Activity B? Activity C?" — and so on, for each of the enterprise's environmental impacts in turn. The first time around, the cells of the matrix could just be filled in with "Yes" where appropriate. Obviously, one environmental impact could be associated with several activities. For example, air pollution could be associated with generating process steam, with fugitive particulates, and with vehicle fleet maintenance. Then, in the second pass through the matrix each "Yes" would be translated into a specific environmental aspect of each related activity. As an illustration, Table 2 in Chapter2 (page 33), shows different environmental aspects for each of the three activity sources of air pollution just mentioned. When done with an analytical process, like that represented by Table 4, some enterprises may have something they have never had before: a set of causal relationships, from activities to their environmental aspects to their environmental impacts. Now, they can look at environmental performance in the context of operations, not just regulations.

Table 4. Starting the Analysis with Known Environmental Impacts.

	Operation, Activity	Known Major Environmental Impacts of the Enterprise					
		Air pollution	Water pollution	Soil and groundwater contamination	Landfilling solid waste	Forest resource consumption	Climate change effects
Manufacturing	Generating process steam	Yes NO_x, SO_x, and CO_2 emissions					Yes NO_x and CO_2 emissions
	Activity B						
	Activity C						
Operation 2	Activity D						
	Activity E						
	Activity F						

Another variant for a first probe into the question of activities and their environmental impacts is to begin with a matrix of known environmental impacts like Table 4, but with the leftmost column (the list of operations and activities) empty. The first question would be, what activities are associated with each of the known major environmental impacts of the enterprise? Through this line of inquiry, the EMS Committee would identify activities associated with known environmental impacts, and list only those activities (rather than all activities) in the leftmost column of the matrix. Then they would proceed, as before, to articulate the exact environmental aspect of each of the activities that results in the environmental impact.

Readers should be aware that a rich array of tools and alternative approaches for identifying environmental aspects, impacts, and environmental significance is found among the literature and Web sites listed in the "Additional ISO 14001 Resources" appendix at the end of this book.

Assessing Environmental Significance

In assessing significance of environmental aspects, an enterprise is free to include some strategic factors, such as cost or length of time needed to reduce the environmental impact. However, we will deal with all strategic factors together in the discussion on objectives and targets (Subclause 4.3.3), where they most logically fit.

The next step is to quantify the environmental impacts as much as possible. If that requires a lengthy process, start it or don't start it, but in either case move on fairly quickly to assessing environmental significance. Findings and decisions made on the fly can always be refined and revised later, if necessary.

Table 5. Matrix for Assessing Environmental Significance.

		Environmental Significance Factors							
Activity	**Environmental aspect**	**Environmental impact**	**Scale of the impact**	**Severity of the impact**	**Duration of the impact**	**Likely frequency of impact**	**Ecological health risk**	**Human health risk**	**Total score**
Generating Process Steam	NO_x, SO_x, CO_2 emissions	Air pollution	4	3	4	5	3	3	22
	NO_x, CO_2 emissions	Climate change effects	1	1	4	5	2	1	14
Activity B	Aspect B	Impact B							
Activity C	Aspect C	Impact C							
Activity D	Aspect D	Impact D							

Table 5 is an example of an analytical matrix that could be used to assess the significance of environmental impacts and, by association, the significance of the environmental aspects that cause them. In each cell of the matrix, the EMS Committee would enter an impact score — say, 1 to 5 — representing the range of environmental impact, 1 being slight, 5 being very heavy. These scores could be based on impressionistic assessments or, if available, on quantitative impact measures. Total scores for each environmental impact would show in the rightmost column.

An analytical process for assessing significance of environmental impacts, like the one represented by Table 5, has four built-in ingredients for determining significance. The first is the decision on which environmental significance factors to employ. After all, the EMS Committee implicitly excludes from its definition of significance the many factors that it could have included but did not. For example, one could take the position that the last three factors in Table 5 — frequency of the impact, and the ecological and human health risks — are sufficient for determining significance and that including any other factors unnecessarily complicates the scoring.

Second, the scores assigned obviously play a major role in determining the total significance. Third, significance factors can be assigned weights, to reflect different levels of importance in the view of the EMS Committee. For example, if the scale of the impact were assigned a weight of 1 compared to a weight of 2 for the severity of the impact, then the score of 4 for scale in the first row would remain at a value of 4x1=4 while the score of 3 for severity in the first row would be converted to a value of 3x2 = 6. Finally, the EMS Committee needs to set a numerical threshold for significance. That is, it has to decide on the total score below which an environmental impact, and therefore its associated environmental aspect, will not be considered significant by the enterprise for purposes of its EMS.

When the analysis is complete, the enterprise will have established the significant environmental impacts and aspects of its operations and products, and also will have prioritized those significant aspects by virtue of their relative scores. If the enterprise uses a similar approach for analyzing the factors that determine strategic importance of environmental aspects (discussed later in connection with Subclause 4.3.3), then it can easily incorporate relative environmental significance as a factor in determining relative strategic importance.

The science of environmental health risk assessment offers a fairly objective means for establishing significance of environmental impacts (to learn about this, search for "health, risk, assessment" on www.epa.gov). Most EMS Committees, however, will not be able to engage in scientific studies. Instead, how they determine which impacts are environmentally significant will be strongly subjective. That is perfectly fine from the ISO 14001 point of view, as long as the method used is consistent, systematic, and reasonable.

Most EMS Committees will determine which impacts are environmentally significant through a methodology that is strongly subjective. That is perfectly fine from the ISO 14001 point of view so long as the methodology used is consistent, systematic, and reasonable.

The Required Procedure

The required procedure must be employed consistently, so it probably would be written, though this is not required by ISO 14001. If it is written, the standard EMS document header (discussed under Subclause 4.4.5, "Document Control") would appear at the top of at least the first page.

A possible structure for an EMS procedure is presented in the discussion on Subclause 4.4.6, "Operational Control" (page 146). Essentially, a procedure for identifying significant environmental aspects would describe who does what, when they do it, and what tools they use to gather and process information and make decisions. There are no explicit ISO 14001 requirements for the structure and content of a procedure, only for what each procedure must be about and, sometimes, what it must accomplish. In this case, it is a procedure for an enterprise to identify environmental aspects of its operations — the rest is up to the individual enterprise.

One way to structure the procedure would be under headings that represent major steps in the process. For example, for a straightforward analytical approach to identifying environmental aspects, the core of the procedure might be structured with the following headings:

- ***Identifying Relevant Activities***
- ***Identifying Environmental Aspects of Relevant Activities***
- ***Establishing Environmental Impacts of Environmental Aspects***
- ***Determining Significance of Environmental Impacts***

The procedure would specify the analytical approach and steps, including details like the factors used to determine environmental significance, the scoring method used, and the like. Where appropriate, there would be an explanation for why the enterprise is taking a certain approach. The reason for the procedure is not to satisfy a certification auditor, but to ensure consistency of approach and establish a clear baseline for any changes that may be found necessary later.

With all that, every effort should be made to keep the procedure as simple and brief as possible.

PLANNING: LEGAL AND OTHER REQUIREMENTS (SUBCLAUSE 4.3.2)

To meet the requirements of this subclause, an enterprise needs a procedure for:

- Maintaining a list of all current laws, regulations, and other environmental standards and practices that it must honor that are related to its environmental aspects
- Keeping abreast of changes in these

- Ensuring that employees who need to be aware of legal and other obligations because of the nature of their work *are* aware
- Ensuring that these employees have ready access to the exact requirements and standards of the environmental obligations of the enterprise

This can all be encompassed in one procedure, or there can be separate procedures. Most enterprises maintain separate procedures for legal obligations on the one hand and "other" obligations on the other, because the means of tracking them are usually different. For example, many enterprises these days subscribe to commercial services that keep them abreast of relevant laws and regulations, and even provide distillations of essential requirements. A different approach would obviously be needed for the "other" category of obligations.

An enterprise needs to build into the procedures a process for communicating the requirements to employees who should know them. A certification auditor may interview workers to see if they are aware of relevant legal and other requirements, to see if they know where the requirements can be found, and to see if they can access the requirements easily.

This subclause of ISO 14001 does not say that the procedure has to ensure that legal and other obligations are maintained in writing. In principle, if the procedure specified a supervisor who would always be at hand and could provide the necessary information to workers, that would satisfy the ISO 14001 requirement — to access the appropriate legal and other requirements, a worker would have only to ask the supervisor.

At the heart of this procedure could be a table showing who is responsible for what functions related to the requirements of this subclause, how the functions would be carried out, and on what schedule the functions would be carried out. Along the left side of the table, as row headings, would be the names of statutes, regulations, and other obligations grouped by categories. The categories could be air, water, toxic waste, and so on, or they could be operations or facilities of the enterprise. Along the top as column headings would be functions performed, such as obtaining the latest version, reviewing and summarizing, storing, posting, and communicating. In the cells of the table would be the names or titles of those responsible for each function with respect to the item represented by each row, with an indication of when and how the function would be carried out.

PLANNING: OBJECTIVES AND TARGETS (SUBCLAUSE 4.3.3)

The Requirement and its Relationship to Environmental Significance

The first requirement of Subclause 4.3.3 is a written set of environmental objectives and targets. A table constructed like Table 3 in Chapter 2 (page 41) would satisfy this requirement, assuming that the objectives and targets documented in it reflect relevant levels and functions of the enterprise, are consistent with the enterprise's environmental policy, and reflect the enterprise's commitment to pollution prevention (as also required by the subclause). The discussions on environmental objectives and environmental targets and indicators in the counterpart section of Chapter 2 (page 37) provide practical information for establishing environmental objectives and targets in an enterprise. We encourage readers to review that information for purposes of EMS implementation planning.

A second set of requirements in this subclause is really a list of factors that must be taken into account by an enterprise when establishing and periodically reviewing its environmental objectives. One of those factors is the significant environmental aspects of the enterprise — those that were identified and prioritized under Subclause 4.3.1 ("Environmental Aspects").

An enterprise could have some environmental objectives that had nothing to do with environmental aspects: an employee training objective, or a community communication objective, or an objective to establish a foundation to support local environmental initiatives, for example.

An enterprise could have a set of environmental objectives with some objectives that have nothing to do with environmental aspects. For example, it might have an employee training objective or a community communication objective, or an objective to establish a foundation to support local environmental initiatives. Though none of these are directly related to the environmental aspects of enterprise operations, according the language of ISO 14001 such objectives are fine. All the standard requires is that the enterprise *consider* its significant environmental aspects, among other things, when deciding on its environmental objectives.

An enterprise can establish "management objectives" (as compared with "operational objectives") that are not necessarily in direct response to an environmental aspect but that make good sense in a context of strategic environmental management. However, most of the enterprise's environmental objectives would naturally address significant environmental aspects. For these, it's best to assess environmental significance and strategic importance in a two-step process. First, the enterprise should assess the significance of environmental impacts as described in the earlier discussion of Table 5 (page 116). Then, it should add factors of strategic business importance to the mix, as described below. The result: scores for significant environmental aspects that can account both for their relative significance and for their strategic importance, and that provide important information for establishing environmental objectives and targets.

Assessing Strategic Importance

ISO 14001 literature and Web sites feature large numbers of examples of approaches and analytical matrices for assessing the strategic importance of significant environmental aspects. Readers need to be aware, however, that in most cases these tools and techniques appear under a heading like "Assessing Environmental Significance" and are presented in connection with Subclause 4.3.1, "Environmental Aspects," rather than this subclause.

Why? Because even though the exact language of Subclause 4.3.1 speaks of identifying environmental aspects that "have or can have significant impacts on the environment," the (unauditable) practical implementation tip in ISO 14004 seems to forget the actual language of the standard and suggests a number of strategic business concerns that might be taken into account during the significance assessment process. To confuse matters even more, some of the strategic business concerns mentioned in ISO 14004 in connection with "environmental significance" are the same ones as listed here in Subclause 4.3.3 of ISO 14001 as things to be considered when establishing and reviewing environmental objectives. As pointed out in Chapter 2, from a certification audit point of view, the subclause under which these assessments are carried out doesn't really matter. And, once the enterprise has established its EMS, the whole question of sequence is greatly diminished because all elements of the EMS are operating more or less simultaneously.

An enterprise should first assess the significance of environmental impacts, and then add factors of strategic business importance to the mix.

The first step in assessing the strategic importance of significant environmental aspects is to select the factors to be used as the basis for assessment. A number of these strategic importance factors, or criteria, are specified in this subclause of ISO 14001, but many EMS Committees will want to include others, perhaps some of those listed in the counterpart section of Chapter 2 (see page 38).

The second step is to assess each significant environmental aspect in terms of these strategic importance factors. As an illustration, the analytical matrix shown in Table 6 lists ten possible strategic importance factors as column headings. The rows are the significant environmental aspects, as in Table 5. Information on the relationship between the strategic importance factors and each significant environmental aspect is entered in the cells of the matrix. In the emissions example in Table 6, the environmental aspect causes two environmental impacts as shown in Table 5, so their combined significance scores have been entered in the matrix.

The third step would be to rate the assessments of strategic importance on a scale. For simplicity, Table 7 is based on a low–medium–high scale of 1–3. The best way to establish the ratings is first to complete an analytical matrix like the one in Table 6; then observe the range of assessments under each factor (that is, the range of entries in each column); and then decide on the definitions of low, medium, and high for each strategic factor. In this way, the definitions of low, medium, and high are relative within the array of actual strategic assessment values for the enterprise. Another approach is to define the meanings of the rating levels in advance, to be used as a kind of further "independent" evaluation of the assessment entries. Either way, the result might look something like the table in Table 7.

Table 6. Assessing Strategic Importance of Significant Environmental Aspects.

	Activity	*Generating process steam*	*Activity B*	*Activity C*
	Environmental Aspect	*NOx, SOx, CO2 emissions*	*Aspect B*	*Aspect C*
Strategic Importance Factors	**Environmental Significance Score**	*22 + 14*		
	Regulatory Implications	*Violates regulatory standards*		
	Implications for Voluntary Standards & Practices	*None*		
	Future Environmental Liability Concerns	*None*		
	Technology Choices for Reducing the Env. Impact	*Low to Moderate Cost P2**		
	Cost of Reducing the Env. Impact by 25 %	*$ 100,000**		
	Employee Labor to Reduce Env. Impact	*30 days**		
	Economic Benefits of Reducing Env. Impact	*IRR** is 400 %*		
	Relationship of Impact Reduction to Business Needs	*None*		
	Community and Other Stakeholder Interests	*Organized complaints*		
	Factor K			
	Factor L			

** Assumes a program involving a number of P2 improvements completed in one year, some contractor installed, that will reduce total emissions by 80 percent (see "Programs to Achieve Environmental Targets," page 47, in the counterpart section of Chapter 2).*

*** Internal rate of return. Calculated as debt financing and five-year life; IRR is annual average. Economic calculations for P2 improvements are discussed at length in Part II of this book.*

Table 7. Establishing a Rating System for Strategic Importance Factors.

Strategic Importance Factors	Low 1	Med. 2	High 3
Environmental significance score	<10	10-25	*>25*
Regulatory implications	None	Very close to regltry. limit	*Violates regltry. standard*
Implications for voluntary standards and practices	*None*	Very close to limit	Violates standard
Future environmental liability concerns	*None*	Possible future liability	Likely future liability
Technology choices for reducing the env. impact	High cost non-P2	Others	*Low–mod. cost P2*
Cost of reducing the environmental impact by 25%	< $50,000	*$50,000-100,000*	> $100,000
Employee labor to reduce the environmental impact	< 30 days	*30 -60 days*	> 60 days
Economic benefits of reducing the environmental impact	IRR < 10%	IRR10–20%	*IRR > 20%*
Relationship of impact reduction to business needs	Contrary to needs	*None*	Reinforces business plans
Community & other stakeholder interests	None expressed	Isolated complaints	*Organized complaints*
Factor K			
Factor L			

The entries in italics in Table 7 reflect the assessments of strategic importance factors for the "Generating process steam" example in Table 6. On the basis of this rating system, the significant environmental aspect of generating process steam has a strategic importance rating of 23. But an enterprise could take a further step, and assign relative weights to the strategic importance factors, as discussed earlier in connection with assessment of environmental significance (see the discussion around Table 5, page 116).

The final ranking of significant environmental aspects according to their strategic importance would be a key to determining how aggressive in relative terms their environmental objectives and targets should be, at least insofar as a more aggressive objective translates into a commitment of more resources. But this is not a mechanical operation. The numerical strategic importance ranking is just a handy tool, not a determiner of decisions. The enterprise should take into account the content of the assessment that is behind the numerical ranking. Note that in three of the columns of Table 7, for example, "high" means everything above a certain value, so the actual values covered by "high" may need to be compared and taken into account. Or, there might be a decision that any environmental aspect that represents a regulatory violation goes to the top of the list, regardless of any other considerations.

PLANNING: ENVIRONMENTAL MANAGEMENT PROGRAMS (SUBCLAUSE 4.3.4)

Fortunately, environmental management programs is a subject much written about. To begin with, the discussion in the counterpart section of Chapter 2 (page 45) covers many practical matters related to this element of ISO 14001. Table 5 of Chapter 2 (page 48) offers summary models of different types of environmental management programs for different types of environmental objectives. In addition,

- examples of environmental management programs,
- templates for documenting environmental management programs,
- step-by-step checklists for the environmental management program planning process, and
- models for dealing with new developments as required by this subclause

abound in print literature and on the Internet. Annex A of ISO 14001 also contains a brief but useful elaboration on the subject.

Table 8 offers a handy summary of the principal elements of an environmental management program. This framework could be used both to guide the process of planning an environmental management program and to structure the documented work plan.

Table 8. Elements of an Environmental Management Program Work Plan.

Element	Comment
Required by ISO 14001	
Objective	By implication, the standard requires that the objective be incorporated into the environmental management program. It would make sense to also reference the associated activity, its environmental aspect, the environmental impact, and possibly even the element of environmental policy that the environmental management program is related to.
Target	The environmental target is also required by implication. Where appropriate, the enterprise should include the performance indicators and time frame associated with the target. The EMS Committee could also choose to include an explanation of why the particular target was selected, how it fits into a sequence of targets, and how it relates to other environmental targets of the enterprise.
Personnel	This refers to who is responsible for carrying out the overall program and its principal elements, and for achieving the target and objective. These responsible personnel should not be confused with employees who may carry out specific tasks under their supervision.
Tasks	A description of the basic tasks, in sequence — including, for each task, who does what, the time period in which it's done, reports and other documents (such as procedures and work instruction) to be generated, and the expected results.
Schedule	All the standard requires is a timeframe for achieving the target and objective. But the only way to figure that out is to estimate, task-by-task, the time required. Moreover, a task-by-task schedule will be needed to make revisions in the work plan later, so this information, summarized from the "Tasks" section, might as well be included. Schedule information could be represented as a bar chart or critical-path graphic.
Not Required by ISO 14001	
Communi-cations	Who is responsible for ensuring full communication, the e-mail group and who controls its membership, reports to be generated and how they are to be circulated, and specific reporting events, such as staff meetings.
Coordin-ation	Who is responsible for overall coordination of environmental management program implementation activities, the schedule for program team meetings, and the schedule for regular progress presentations to top management.
Resources	Time, personnel, equipment, and money committed to the program, and how they are to be utilized. Commitments of administrative and clerical personnel should be explicit.

Table 8. Continued.

Element	Comment
Not Required by ISO 14001	
Approach	A brief overall description of the program, with an explanation or justification of why the program is designed the way it is and perhaps how it represents implementation of the enterprise's environmental policy. Here, the EMS Committee can also include other comments about the program, its work plan, its aims, and so on, that do not fit elsewhere but are important to understanding what the program is really all about.
Baseline data	Additional baseline data that will help the enterprise evaluate the progress of the program and the progress of the enterprise toward the target and objective.

IMPLEMENTATION AND OPERATION: STRUCTURE AND RESPONSIBILITY (SUBCLAUSE 4.4.1)

As usual, we refer readers to the counterpart section of Chapter 2 (page 50) for useful information related to implementing this subclause of the ISO 14001 standard. ISO 14004 also contains some useful tips for creating a framework of people with specific ongoing EMS assignments and the authority, capacity, and resources to carry out their assignments. In addition, ISO 14001 publications and Web sites like those listed in the "Additional Resources" appendix at the end of this book offer or have links to

- examples of such frameworks,
- sample position descriptions,
- models of organograms,
- recommended tables of responsibility,
- model matrices for structuring resource allocations to EMS functions,

and more.

The centerpiece of the requirements of this subclause is a written description of EMS roles, responsibilities, and authorities for EMS positions in the enterprise, including the mandatory Management Representative position and its EMS functions. A written description structured to make it easy for a certification auditor to confirm that the requirements of this subclause have been met, but functional also for purposes of EMS operation, might appear like Table 9. Because it's a formal EMS document, the position description would have the standard EMS document header discussed in the section of this chapter on Subclause 4.4.5, "Document Control."

Table 9. Table of EMS Responsibilities and Resources.

Position Title; EMS Reporting	**EMS role**	**EMS Responsibility and Authority**	**Resources for EMS Work**
Vice President, Operations Reports to CEO	EMS supervisor and liaison for top management	Oversees and supports EMS implementation and maintenance on behalf of top management; has open door for the EMS management representative; approves reports from the Management Representative and EMS Committee; coordinates with other senior managers. Can require spot audits and reports on EMS operations; structures and leads top management EMS meetings.	Own time, 4 hours weekly Administrative Assistant, 8 hours weekly
Quality and Environment Officer Reports to VP, Operations	EMS Management Representative	Ensures that EMS requirements are implemented and maintained in conformance with ISO 14001 requirements; reports on the EMS to top management; *[detail specific functions, meetings, reporting, etc., and authorities as appropriate for the particular EMS of the enterprise].*	Own time, 100% for QMS/EMS Assistant, 100% for QMS/EMS
Production Manager Reports to EMS Management Representative	Member of EMS Committee (represents manufacturing operations)	*[Detail specific functions, meetings, reporting, etc., and authorities as related to participation in the general work of the EMS Committee; the level and function of the enterprise that this individual represents on the committee; and EMS elements for which this individual has lead responsibility in the enterprise.]*	*[Specify time, support personnel;, other personnel with specific functions or special skills; equipment; and funds allocated.]*
Property Manager Reports to EMS Management Representative	Member of EMS Committee (represents buildings and grounds)	"	"

Table 9. Continued.

Position Title; EMS Reporting	EMS role	EMS Responsibility and Authority	Resources for EMS Work
Office Services Manager Reports to EMS Management Representative	Member of EMS Committee (represents front-and back-office operations)	"	"
Position F			
Position G			
Position H			
Communication: - Who in the enterprise needs to have access to this table of responsibilities and resources, and why. (There may be different types of need for different purposes.) - Where the table will be kept, and how it can be accessed. - Where in the enterprise the table will be posted, and who is responsible for posting it. - How the information in the table will be actively communicated initially, and every time there is a change in it, and who is responsible for this.			

IMPLEMENTATION AND OPERATION: TRAINING, AWARENESS, AND COMPETENCE (SUBCLAUSE 4.4.2)

Again, readers will find useful ideas for implementing this subclause of the standard in ISO 14004 and the counterpart section of Chapter 2 (page 54). In print publications and Web sites they will find information about identifying environmental training needs, including

- training needs assessment guidelines,
- training needs analysis frameworks, and
- model training needs assessment questionnaires.

Regarding EMS training programs, they'll find

- information on training program models,
- recommended outlines for specific training or awareness topics,
- sample training evaluation questionnaires, and
- templates for training records.

A matrix is the basic tool that the EMS Committee should use to identify training needs, plan what training will be provided to what groups of employees, and track the training program. The committee might list on the left side of the matrix, as row headings, the different types of training being contemplated; along the top, as column headings, the committee could list the different groups of employees toward whom different types of training is targeted.

Table 10 is an example of a matrix like this. Employees have been divided first into new and veteran groups; the matrix for new workers would be exactly the same as the one for veteran workers (ideally, both groups would show on one matrix form). When the EMS is first being established, all employees can be considered new for EMS training purposes; after that, the distinction between new and veteran employees will be important.

In Table 10, senior and middle managers are considered employee groups for EMS training purposes, whatever their functional roles. Office and laboratory workers are categories that cross both functional and hierarchical boundaries. And shop-floor workers have been organized by function. Finally, contractors and suppliers are included in the training program. Of course, there are many ways that employees of an enterprise could be divided for EMS training purposes, and each enterprise will have to figure out its own approach. Many enterprises will have already grouped employees for other training purposes, and need only to consider carefully if the same grouping works for EMS training or needs to be adjusted for that purpose.

If the EMS Committee uses a standard matrix like the one in Table 10 to work through and record many different types of information about EMS training, that enables them over time to easily track and fine-tune both the needs assessment process and the training program. To say that a more-or-less standard analytical matrix would, over time, be used for the training program is not to say that it's unchangeable. Obviously, when the enterprise sees the need for change, it can combine or further divide different categories of employees, and reduce or expand the list of topics. But with a basic standard framework, these kinds of changes can be tracked and can inform further planning for EMS training. Table 11 illustrates how an analytical matrix like the one in Table 10 can be used to identify training needs, plan EMS training, and track EMS training.

Table 10. Analytical Matrix for EMS Training Needs, Planning, and Tracking.

	Veteran Employees								
	Managers		**Office and Lab Workers**		**Shop Floor Workers**				**Contractors and Suppliers**
Possible Types of EMS Training	**Senior**	**Middle**	**Front & Back Office**	**Researchers & Technicians**	**Machine Operators**	**Material Handlers**	**Function C**	**Function D**	
Overview of the enterprise's environmental policy, EMS, and relevant procedures (as relevant for each audience)									

Table 10. Continued.

Possible Types of EMS Training	Veteran Employees: Managers: Senior	Veteran Employees: Managers: Middle	Veteran Employees: Office and Lab Workers: Front & Back Office	Veteran Employees: Office and Lab Workers: Researchers and Technicians	Veteran Employees: Shop Floor Workers: Machine Operators	Veteran Employees: Shop Floor Workers: Material Handlers	Veteran Employees: Shop Floor Workers: Function C	Veteran Employees: Shop Floor Workers: Function D	Contractors and Suppliers
Environmental awareness, by function in the enterprise									
Environmental ethics									
Environmental policy and standards of the enterprise									
Operations of the enterprise as related to environmental performance									
Methods of environmental measuring and monitoring									
Applicable environmental regulations									
Environment-related work competence and skills, by function in the enterprise									
Emergency response procedures									
Principles of P2									
P2 auditing									
The enterprise's EMS									
EMS implementation in the enterprise									
Roles in the EMS									
EMS auditing									

Table 11. Analytical Matrix for EMS Training Needs, Planning, and Tracking.

Question for Each Training Topic (row) in the Analytical Matrix	Source of Information	Entry in the Matrix
Training Needs Questions		
Is training on this topic needed in principle for each group of employees?	Supervisor interviews	Yes; no; brief comments regarding content and depth
Is training on this topic needed in practice for each group of employees?	Supervisor interviews; employee interviews; employee questionnaires; training records	Yes; no; brief comments regarding content and depth
Planning Question		
What training on this topic will we provide, and what content and depth for each employee group, if any?	EMS Committee, in consultation with supervisors, senior managers, and possibly a training contractor	Dates of training sessions, and a phrase or two of description
Tracking Questions		
What training on this topic has been given? Record dates and a phrase or two of description.	Observation; training records	Dates of training sessions, and a phrase or two of description
How did trainees rate the training sessions on this topic?	Evaluation questionnaires; interviews with trainees	Evaluation questionnaire tabulations and additional comments

IMPLEMENTATION AND OPERATION: COMMUNICATION (SUBCLAUSE 4.4.3)

The requirements of this subclause of ISO 14001 are that the enterprise have procedures, not necessarily written, for conducting internal and external communication about its environmental aspects and its EMS; and that, with regard to its *significant* environmental aspects, the enterprise make and record a decision on how it will handle external communication.

On the subject of ISO 14001 communication requirements, in print literature and on the Internet there is an abundance of:

- Model procedures for internal and external communication
- Suggestions for internal and external communication methods
- Model logs for recording external communication received and the enterprise's response
- Examples of records of enterprise decisions on external communications regarding significant environmental aspects
- Model formats for recording enterprise decisions on external communications regarding significant environmental aspects

- Tips for establishing communication records that will readily demonstrate to auditors that the enterprise has met the requirements
- Tips on unexpected ways auditors might test the communication procedures

The record of the decision regarding external communication on significant environmental aspects of the enterprise could be a simple internal memorandum. Because the memorandum is a required EMS record, it would feature the standard EMS record header discussed under Subclause 4.5.3. "Records."

Whether or not the communication procedures are written, the key to following a consistent procedure and being able to demonstrate that to a certification auditor is to maintain communication logs. Among other things, the procedures would specify how the communication information is gathered and who is responsible for maintaining the different logs. There would be a log for each of the four directions of communication mentioned in the counterpart section of Chapter 2 (page 58). For both internal top-down communication and external enterprise-out communication, the logs would record essential information like date and EMS topic for each communication method and audience group involved in each communication event.

Tables 12 and 13 are illustrations of such logs. The column headings of Table 12, for internal top-down communication, are borrowed from those of Table 10. Using the same column headings, or list of audience groups, has practical value because training and communication, and the planning for them, are closely related. Table 13 reflects the same basic approach of Table 12, but for external enterprise-out communication.

Table 12. Internal Top-Down Communication Log.

Methods of EMS Internal Communication		**Managers**		**Office and Lab Workers**		**Shop Floor Workers**				**All Employees**
		Senior	**Middle**	**Front & Back Office**	**Researchers and Technicians**	**Machine Operators**	**Material Handlers**	**Function C**	**Function D**	
Briefing Sessions	Topic		EMS status report		EMS status report		Progress on objectives and targets			
	Date		May 13, 2002		May 14, 2002		April 9, 2002			
	Presenter		Jill Jamal		John Jones		Jane Jess			

Methods of EMS Internal Communication		Managers		Office and Lab Workers		Shop Floor Workers				
		Senior	Middle	Front & Back Office	Researchers and Technicians	Machine Operators	Material Handlers	Function C	Function D	All Employees
Internal Newsletter	Topic									EMS status report
	Date									July 2, 2002
	Author									Alex Ames
Enterprise Intranet	Topic									
	Date									
	Web-master									
EMS Award Ceremony	Topic									
	Date									
	Organizer									
Method E	Topic									
	Date									
	Leader									
Method F	Topic									
	Date									
	Leader									

Table 13. External Enterprise-Out Communication Log.

Methods of EMS External Communication		Local Governments	Regulatory Authorities	Other Fed. Agencies	Contractors/Suppliers	Labor Organizations	Consumer Groups	Stockholders	Trade Associations	Community Groups	Activist Groups	General Public	All
Annual Report	Date												
	Topic												
	Author												
Enterprise Web site	Date												
	Topic												
	Web-master												
Presentations: Public Meetings	Date												
	Topic												
	Presenter												
Method D	Date												
	Topic												
	Leader												

In addition to logs like those in Tables 12 and 13, an enterprise would have EMS communication logs for internal bottom-up communication and response, and for external outside-in communication and response. For both types of communication the logs would record:

- Who submitted the communication
- The medium of communication (letter, e-mail, Web site comment, telephone, fax, etc.)
- When the communication was received
- The general nature of the query, suggestion, or comment
- Who responded (per the procedure)
- The date of response

- The medium of response
- The general nature of the response

For bottom-up and outside-in communication, it may be best to give the log the form of a brief report for each communication exchange, rather than a table. The subheadings of the report would be the eight items of information listed above. Copies of the initial communication and the response could be attached to the report.

Ultimately, the best procedure any enterprise can use to deal with its EMS- or environment-related communication is the procedure it already uses for other communication. If the existing procedure is not adequate for EMS purposes, then the best approach is to upgrade that procedure and apply it to all communication. As with all ISO 14001 requirements, the emphasis should be on adequacy for the purpose, consistency, simplicity, and ease of implementation.

IMPLEMENTATION AND OPERATION: EMS DOCUMENTATION (SUBCLAUSE 4.4.4)

It's very important not to turn EMS implementation into an exercise in documentation. Focusing too much on EMS documentation diverts energy and creativity from the real purpose of the EMS — to improve environmental and economic performance of the enterprise — and adds enormously to the burden of EMS implementation and maintenance. The trick to minimizing the EMS documentation burden is threefold:

a) Create only documents that are required by ISO 14001 or are necessary for productive operation of the EMS. This will vary greatly from enterprise to enterprise.
b) Keep the documents and records that *are* created as simple as possible.
c) Build on existing documentation — that is, model EMS documents after existing documents; use elements of existing documents in new EMS documents; build EMS elements into existing documents; or incorporate existing documents and records fully into the EMS, as appropriate.

This subclause of ISO 14001 requires a written description of EMS core elements and written directions for locating and accessing related documentation. In available print and electronic literature, readers will find a wealth of information, such as:

- Annotated lists of "core elements," and what the descriptions of them should contain
- Annotated lists of "related documentation," and what the directions for locating and accessing them should contain
- Graphic representations and explanations of the "hierarchy of EMS documentation" (usually represented as a pyramid)
- Lists of documentation required by ISO 14001

- Lists of EMS documentation that enterprises are encouraged to create in addition to what the standard requires

Surprisingly, there are inconsistencies among the lists of documentation required by the ISO 14001 standard that different authors provide; and, of course, there is a wide range of voluntary documentation recommended by different authors. Because of this, we offer our own list of required and other possible EMS documentation in Table 14.

Table 14. Required and Other Possible ISO 14001 EMS Documentation.

ISO 14001 Element	Required Documentation		Other Possible EMS Documentation
	Documents	Records	
4.2 Environmental Policy	Environmental policy statement.		Procedure for periodic environmental policy review and revision
4.3.1 Environmental Aspects			Procedure for identifying environmental aspects. Procedure for determining which environmental aspects have a significant impact on the environment. Table of activities (and products and services), environmental aspects, and environmental impacts. Table of activities and *significant* environmental aspects and their impacts.
4.3.2 Legal and Other Requirements			Procedure for tracking relevant environmental laws and regulations. Procedure for tracking other environmental standards and practices to which the enterprise has obligated itself. List of applicable legal requirements. List of other applicable requirements.

ISO 14001 Element	Required Documentation		Other Possible EMS Documentation
	Documents	Records	
4.3.3 Objectives and Targets	Environmental objectives and targets.		Procedure for determining relative strategic importance of significant environmental impacts. Procedure for determining environmental objectives.
4.3.4 Environmental Management Programs			Work plans for environmental management programs. Checklist for new activities. Records of program changes for new activities.
4.4.1 Structure and Responsibility	EMS roles, responsibilities, and authorities.		Procedure for communicating roles, responsibilities, and authorities. Summary table of EMS responsibilities and resources.
4.4.2 Training, Awareness, and Competence		Training records — e.g., dates, sessions, topics, participants (required by Subclause 4.5.3)	Procedure for assessing EMS training needs. Training needs matrix. Training planning matrix. Training tracking matrix Training and awareness course materials.
4.4.3 Communication		Log of communication from outside parties and the responses. Record of decision concerning communication on significant environmental aspects to outside parties.	Internal and external communication procedures. Log of internal top-down communication. Log of internal bottom-up communication and responses. Log of external enterprise-out communication.

ISO 14001 Element	Required Documentation		Other Possible EMS Documentation
	Documents	Records	
4.4.4 EMS Documentation	Description of the core elements of the EMS and their relationships. Directions to other documentation related to the EMS.		EMS manual(s) containing the required documentation and perhaps more.
4.4.5 Document Control			Document control procedures. Document control matrix. Standard EMS header for controlled documents.
4.4.6 Operational Control	Procedures (with work instructions and standards as necessary) for all activities with a bearing on the environmental impacts of significant environmental aspects.		Procedure for identifying the activities for which procedures are required. List of activities for which procedures are required. Procedures for activities related to identifiable significant environmental aspects of goods and services used by the enterprise. Record of communicating relevant procedures to suppliers and contractors.
4.4.7 Emergency Preparedness and Response			Procedures for identifying potential accidents and emergency situations. Emergency preparedness and response plans, including procedures for responding to accidents and emergencies, and for preventing and mitigating environmental

ISO 14001 Element	Required Documentation		Other Possible EMS Documentation
	Documents	Records	
			impacts from the potential accidents and emergencies. Records of review and revision of emergency preparedness and response procedures after an accident or emergency has occurred. Record of periodic reviews, and tests of emergency preparedness and response procedures.
4.5.1 Monitoring and Measurement	Procedures for measuring and monitoring essential characteristics of activities that can have significant negative impacts on the environment. Procedure for monitoring legal and regulatory environmental compliance.	Records of calibration and maintenance of measuring and monitoring equipment.	Procedures for calibrating and maintaining equipment used for measuring and monitoring. Procedure for retaining calibration and maintenance records. Records of the findings of regulatory compliance inspections and audits.
4.5.2 Nonconformance and Corrective and Preventive Action		Records of changes in procedures that are part of corrective and preventive action in response to non-conformances.	Procedure for dealing with actual or potential EMS non-conformances. Records of nonconformances and of actions taken to correct them.

ISO 14001 Element	Required Documentation		Other Possible EMS Documentation
	Documents	Records	
4.5.3 Records		Retention times for EMS records (could be part of a records management matrix or record header, rather than a separate record).	Procedure for creating, identifying, maintaining, and disposing of environmental records. Record management matrix Standard EMS record header.
4.5.4 EMS Audit		Report of audit findings for management consideration. Results of EMS audits (required by Subclause 4.5.3; can be the same as the audit report or just a summary of the findings).	General audit procedure, including audit intervals. Audit plan. Audit checklist.
4.6 Management Review		Results of management review (required by Subclause 4.5.3).	Procedure for management review, including development and assembly of appropriate information.

An easy way to structure the EMS Manual that many enterprises assemble to satisfy the requirements of this subclause is to organize it by the elements (clauses and subclauses) of ISO 14001, and under each element provide:

- ***Description:*** the enterprise's approach, procedures, and generally how the enterprise handles the EMS element
- ***Relationships:*** how the element relates to other elements of the enterprise's EMS and with other management systems of the enterprise
- ***Personnel:*** who has what principal responsibilities for implementing and maintaining the EMS element
- ***Documentation:*** associated documentation, including both product documentation (policy, procedures, records) and process documentation (analytical matrices)
- ***Documentation directions*****:** how to locate and access related documentation

IMPLEMENTATION AND OPERATION: DOCUMENT CONTROL (SUBCLAUSE 4.4.5)

The requirements of this subclause of ISO 14001 are basically nothing more than what any enterprise should be doing to control the documentation that governs its operations. As we have said before, the key to reducing pollution and increasing efficiency in an enterprise — to achieving the greatest and quickest cost savings — is generally for management to improve control of enterprise operations. Improving control of operations always means revising procedures; that, in turn, always means revising the documentation that spells out and supports the procedures. On the face of it, the importance of controlling the documents that govern the procedures that control the operations is obvious. Similarly, with regard to its EMS, it is essential that an enterprise have in place a well thought-out mechanism for being sure that the EMS documents to which people refer are the ones currently in force, as noted in the discussion of this subclause in Chapter 2.

A well-run enterprise probably already has in place all the procedures necessary to satisfy this subclause of the ISO 14001 standard. Some tweaking may be necessary, but on the whole this element of ISO 14001 should integrate easily into the existing overall management system of the enterprise. And if it doesn't, meeting the requirements of the standard provides an opportunity to upgrade the document control element of its overall management system in a way that will enable an enterprise to operate much more dynamically and therefore extract more benefit from its EMS, which is based on continual improvement.

The principal evidence of proper document control procedures will be in the results: the consistency and adequacy of actual document control.

Table 15 is a worksheet for developing document control procedures based on the breakdown of document control elements discussed in the counterpart section of Chapter 2 (page 63). ISO 14001 does not require that document control procedures be written. Therefore, the principal evidence of proper document control procedures will be in the results: the consistency and adequacy of actual document control. If a certification auditor were to pull several EMS documents to examine whether or not the required document control procedures are in place in the enterprise, what are the questions he or she would ask about the sample documents? Those are the questions in the second column of Table 15.

The first column of Table 15 categorizes the requirements of the ISO 14001 standard represented by the questions first into content and physical status requirements. These are further subdivided into discrete document control variables. When a question in the second column is asked with regard to an EMS document, the answer, as described in the third column, should appear in the document control matrix and/or the standard document header when appropriate, as indicated in the two rightmost columns. The document control procedure is conceptually between the second and third columns. In other words, the procedure needs to ensure that the auditor will find what is shown in the third column, and will find it in the places indicated in the last two columns of the worksheet when appropriate.

Table 15. Worksheet for EMS Document Control Procedures.

Document control variable	Questions to be addressed by document control procedures	Result of the procedures (control characteristics)	Appears in control matrix	Appears in document header
Content				
Document Creation	Does the title of this document clearly identify what it is?	Title/subtitle is descriptive of what it contains and how it is used.	✔	✔
	Who is responsible for creating it (person, position, unit of the enterprise)?	Author's name, title, and unit appear.	✔	✔
	Who is responsible for editing, for typing, for formatting, and for physically producing it?	Document is legible and easily understandable.		✔
		Date of issue appears.	✔	✔
		Date of revision appears.	✔	✔
Approval	Who approves it?	Approver's name, title, unit, and date approved appear.	✔	✔
Revision	Who determines at what intervals or date it is reviewed and if necessary revised?	Interval/date of next review appears.	✔	✔
	Who is responsible for reviewing it and drafting proposed revisions or updates?	If revised, name of reviewer/reviser and title appear.	✔	✔
Physical Status				
Storage	Can the master copy of this document be easily found and retrieved?	Location or identifier/locator code (document number) appears.	✔	✔

Document control variable	Questions to be addressed by document control procedures	Result of the procedures (control characteristics)	Appears in control matrix	Appears in document header
	What is the storage/retrieval system for required EMS documents?	Written storage/retrieval procedures, including procedure for assigning identifier/locator codes.		
Distribution	At what locations in the enterprise are important EMS functions carried out for which the document contains guidance?	List of locations in the enterprise where it is stored or posted.	✔	
	Is a copy of the current version easily located and accessed at those locations?	Copy is easily located and accessed at the locations listed.		
Disposal	When the document is revised or superseded, who is responsible for assuring removal of obsolete versions from all locations?	Name of person or position responsible for assuring removal of obsolete versions appears.	✔	
	When it is revised or superseded, is the obsolete version destroyed or archived?	Statement of what to do with obsolete version appears.	✔	
	Who is responsible for destroying or archiving the obsolete version?	Name of person or position responsible for destroying or archiving appears.	✔	
	How long are archived documents retained?	Written enterprise archiving policy.		

Maintaining a document control matrix is fairly common practice among well-run enterprises. The document control matrix flows directly from document control procedures. It is basically an inventory of documents — in this case EMS documents — and a record of their status with respect to each of the prescribed document control characteristics. For new documents, it serves as a checklist of control characteristics that the enterprise must attend to. The document control matrix makes it possible for anyone (including a certification auditor) who needs it, to know the exact current title, location, and status of any controlled EMS document. It is the master record of EMS documentation. Table 16 shows what a document control matrix might look like if the document control characteristics are organized as they were in Table 15.

Along the left, as row headings, are all the controlled EMS documents. Because most controlled EMS documents will relate to operation and maintenance of the EMS itself or be directly related to the requirements of a specific EMS element, it's particularly convenient to organize the list of controlled documents by elements of the EMS. This will also make the job of the certification auditor easier (and therefore less expensive). For an EMS Committee setting up a document control matrix for the first time, the table of required and other possible EMS documents in Table 14 (page 136) may be helpful for creating an initial list of controlled EMS documents. Under ISO 14001 Subclause 4.4.6, "Operational Control," there might be listed a large number of procedures in the matrix, covering all environmentally sensitive activities related to significant environmental aspects. These EMS operational control procedures could be further categorized by type of operation, environmental aspect, or pollution and resource consumption categories (water, air, energy, etc.). Along the top of the Table 16 illustration, as column headings, are the now-familiar document control characteristics reflecting the requirements of this subclause.

Most often an employee will be reviewing an EMS document — say, an EMS procedure — without the document control matrix at hand. At the moment that he or she refers to the document, the employee needs evidence that this is the version of the document currently in force. And it may be important to have other control information about the document, like who approved it or when it was last revised. Certification auditors will look to see if this information appears on controlled documents — again, as evidence that adequate control procedures exist.

Taking the worksheet for EMS document control procedures in Table 15 as our guide for the necessary information, Table 17 is an example of a standard header that an EMS Committee might devise to appear at the top of every controlled EMS document that the enterprise issues. It contains selected items of information from the document control matrix. We should remind readers that neither the EMS document control matrix nor the standard EMS document header are required by ISO 14001. However, both represent meaningful evidence that the required document control procedures are in place. More important, both help enterprises make their EMSs as efficient and productive as possible.

Table 16. Example of an EMS Document Control Matrix.

Content							**Physical Status**					
Document Creation				**Ap-proval**	**Revision**		**Storage**		**Distrib.**	**Disposal**		
Title/Content	**Author**	**Date of issue**	**Date of revision**	**Approver**	**Review interval and date**	**Reviewer-reviser**	**Identifier/locator code**	**Location of master copy**	**Locations where posted**	**Resp. for removing obsolete versions**	**Disposition of obsolete versions**	**Resp. for dispos-ing of obsolete versions**
4.2 Environmental Policy												
Environmental Policy Statement												
Environmental Policy: Review and Revision Procedure												
4.3.1 Environmental Aspects												
Environmental Aspects: Procedure for Identifying												
Significant Environmental Impacts: Procedure for Determining												
Activities, Environmental Aspects, Environmental Impacts: Summary Table												
Significant Environmental Aspects and Impacts: Summary Table												
4.3.2 Legal and Other Requirements												
•												
•												
•												

Table 17. Example of an EMS Document Header.

ENVIRONMENTAL POLICY: REVIEW AND REVISION PROCEDURE	
Document Number: EMS/D 4.2-2	
Content: Procedure for periodically reviewing and revising the enterprise's environmental policy as overall guidance for our EMS	
Prepared by: Bob Brander, EMS Management Representative	**Date:** 01 January 2002
Approved by: Tony Ternall, CEO	**Date:** 15 January 2002
Effective Date: 15 January 2002	**Supersedes:** EMS/D 4.2-2 dated 15 January 2000
Review Date: 10 December 2002	**Review Resp:** Bob Brander, EMS management rep.

IMPLEMENTATION AND OPERATION: OPERATIONAL CONTROL (SUBCLAUSE 4.4.6)

Subclause 4.4.6 of ISO 14001 basically requires written procedures, with detailed work instructions and standards where called for, to control all environmentally sensitive activities related to significant environmental aspects of enterprise operations. In print and on the Internet, readers will find a wide array of:

- models and examples of procedures and associated work instructions and standards (operating criteria);
- analytical matrices and methods for identifying activities for which procedures should be written under the requirements of this subclause; and
- table formats for recording information about the status of development or revision of a procedure (column headings) with respect to the activity it addresses (row headings). ISO 14004 also contains some useful ideas in its "Practical Help" for this subclause.

An enterprise can design an EMS procedure any way it wants; many enterprises will have procedure formats that predate the EMS and are perfectly fine for EMS purposes. Table 18 offers a possible standard outline for an EMS procedure, and Table 19 offers the same for an associated set of work instructions and standards.

Table 18. Structure for an EMS Procedure.

EMS PROCEDURE
EMS Requirement: Quote the language of the related ISO 14001 element and provide a plain-English listing of the specific requirements.
Purpose: Describe the aim of the procedure, and why it's necessary.
Policy Reference: Reference the appropriate text from the enterprise's environmental policy.
Legal Reference: Reference the appropriate text from applicable laws and regulations.
Other References: Reference other environmental standards and practices that the enterprise has adopted.
Scope: Describe the activity that this procedure applies to, and explain if warranted.
Responsibility and Authority: Specify position titles assigned responsibility and authority for the tasks covered by this procedure and for ensuring the procedure is followed overall, specifying skill level, special training, or other professional requirements.
Tasks: Describe the component tasks of the procedure and, with respect to each, who does what, when and where, using what equipment in what way, applying what controls, with what training or qualifications. This is the heart of the procedure.
Related Documents and Information: Specify other documents and records that should be consulted in carrying out this procedure, including work instructions and standards.
Records: Specify data and records that should be generated from carrying out this procedure, and the control characteristics of the records so they can be easily identified and accessed.
Procedure History: Summarize recent revisions, dates and authors of revisions, and the reasons for the revisions.

Table 19. Structure for Work Instructions and Standards for an EMS Procedure.

SUBTASK #1: WORK INSTRUCTIONS AND STANDARDS (OPERATING CRITERIA)
Procedure Reference Reference the procedure that this work instruction is connected to.
Precautions Describe environmental impact, safety, and other risks from not executing this subtask properly, and special precautions and equipment to avoid or deal with these risks.
Equipment: Specify equipment to be used in carrying out this subtask.
Personnel Specify the position title or team members responsible for carrying out this subtask .
Qualifications Specify the skill level, special training, and other qualifications for carrying out this subtask.
Work Instructions and Standards Describe each step in the subtask in turn; give clear standards for proper performance of each step. If possible, represent the sequence of work steps and their performance standards graphically.

IMPLEMENTATION AND OPERATION: EMERGENCY PREPAREDNESS AND RESPONSE (SUBCLAUSE 4.4.7)

For implementing and maintaining this element of ISO 14001 readers will find a great deal of help in print and Internet sources, including:

- Analytical matrices for assessing potential accidents and emergencies connected with enterprise operations
- Sample emergency preparedness and response plans
- Recommended tables of contents or structures for such plans
- Model report forms for accidents or emergencies, including references to elements in the enterprise's emergency preparedness and response procedures that should be reviewed for possible revision in light of experience with an actual emergency event
- Formats for records of tests of emergency preparedness and response procedures
- Examples of emergency prevention and environmental impact mitigation methods that an enterprise can employ

For the benefit of readers for whom this subject area is a new one, Table 20 offers a list of some of the possible elements of an EMS emergency preparedness and response plan for an enterprise. This illustration focuses on emergency preparedness and response from an environmental point of view, but obviously EMS emergency preparedness and response procedures would be fully integrated with the enterprise's overall emergency preparedness and response plans. It would be a good idea to develop EMS procedures separately and document them (as in Table 20), so a certification auditor can access them easily, but then the enterprise should integrate them into the broader procedures.

Table 20. Elements of an EMS Emergency Preparedness & Response Plan.

Overview A review of enterprise operations broadly, identifying the specific potentials for environmental accidents and emergencies, what the environmental and business consequences could be, and how this plan relates to the environmental policy of the enterprise.
Legal References References to applicable laws and regulations.
Emergency Notification A listing of employees to whom different types of environmental emergencies should be reported, and a list of all emergency personnel and their functions.
Emergency Information A description of the information that should be collected and communicated to emergency personnel, and information to be entered in the record of the emergency.

Response Equipment Descriptions of the natures and locations of relevant emergency equipment throughout the enterprise, with the names and locations of the employees trained to use the equipment.
Identifying Potential Accidents and Emergencies Written procedures for identifying potential environmental accidents and emergencies.
Preparedness Measures Descriptions of preparedness measures for each identified potential environmental accident or emergency.
Prevention Procedures Written procedures for preventing the identified potential environmental accidents and emergencies, or reference to and excerpts from other procedures of which environmental emergency prevention activities are a part.
Response Procedures Written procedures for responding to environmental accidents and emergencies, or reference to and excerpts from other procedures of which environmental emergency prevention activities are a part.
Environmental Impact Mitigation Procedures Written procedures for mitigating environmental impacts from environmental accidents and emergencies, or reference to and excerpts from other procedures of which environmental emergency prevention activities are a part.
Site Map A schematic map of the facility and grounds, with descriptions of evacuation routes and locations of environmental emergency equipment and aid stations.
Training Requirements A listing of specific skill and training requirements for all personnel that have specific environmental emergency preparedness and response assignments, and environmental emergency preparedness and response training requirements for all employees.
Maintenance of this Plan A schedule and specification of responsibilities for review and revision of the environmental emergency preparedness and response plan.
Distribution A list of how and to whom this plan is distributed, and who is responsible for distribution.

CHECKING AND CORRECTIVE ACTION: MONITORING AND MEASUREMENT (SUBCLAUSE 4.5.1)

In response to the requirements of ISO 14001 Subclause 4.5.1, as part of its EMS an enterprise would have written procedures for measuring and monitoring key characteristics of its environmentally sensitive activities. By key characteristics, we mean characteristics of the activity that are principal determinants of its environmental impact: wastewater and air pollution releases, water use, waste generated, toxic chemical use, energy use, the number of units processed, and the like. As mentioned in the counterpart section of Chapter 2 (page 71), environmentally sensitive activities include those that are part of normal operations of the enterprise, as well as those that are specifically associated with its EMS.

One reason for developing the monitoring data is to assess the effectiveness of the enterprise's operational controls on the environmental performance of the activity. A second reason is to compare the environmental performance of the activity with environmental targets and objectives of the enterprise. A third reason is to provide data for tracking the environmental regulatory compliance status of the enterprise (for which there should also be a written procedure). Obviously, all this analysis would be a waste of time if the monitoring data were not accurate. To be sure that the data are accurate, the enterprise must calibrate and maintain the equipment used to take measurements, and keep records to verify and track the results of equipment calibration and maintenance. All of this is covered by the requirements of this subclause.

Fortunately, in the print and Internet sources like those listed in the "Additional Resources" section at the end of this book, there is a great deal of material to help enterprises implement and maintain this element of ISO 14001, as well as the next element (which deals with what is done with the monitoring information developed under this element). In keeping with our intent to avoid presenting material that is readily available elsewhere, we've listed below tools and other guidance that readers will find in other reference sources for implementing this subclause.

Reference materials about *measuring and monitoring the key characteristics of enterprise activities that can have significant impacts on the environment, and that relate the associated environmental performance to operational controls on the activity and to environmental targets and objectives, include*:

- Lists of recommended types of key characteristics to be covered by a measuring and monitoring program
- Examples of measuring and monitoring intervals for different types of key characteristics
- Recommended methods for tracking environmental performance associated with different key characteristics of activities
- Environmental performance tracking record forms
- Forms for recording and tracking indirect environmental performance measures, such as task completions or observations of occurrences of certain events, such as chemical spills
- Model procedures for measuring and monitoring
- Model procedures for evaluating operational controls in relation to environmental performance
- Worksheets for relating operational controls to environmental performance of different key characteristics

- Model operational control review report forms
- Model procedures for comparing the enterprise's objectives and targets to environmental performance of different key characteristics
- Worksheets for comparing the enterprise's objectives and targets to environmental performance of different key characteristics
- Model environmental objectives and targets review procedures
- Model environmental objectives and targets review report forms
- Model target or objective discrepancy report
- Guidelines for establishing a three-way relationship between operational controls on key characteristics, environmental performance of the key characteristics, and the enterprise's objectives and targets

Readers should also keep in mind that:

- Specific measuring and monitoring procedures for different key characteristics are described in environmental engineering literature, such as the *Standard Handbook of Environmental Engineering* (Robert A. Corbitt, McGraw-Hill Handbooks, 1999) and other literature cited in the "Additional Resources" appendix at the end of this book
- As a result of permitting requirements, many enterprises will already be generating some of the measurement data and monitoring records (such as records covering effluent composition) required by this subclause of ISO 14001
- Common statistical methods that enterprises would use for trend analysis, such as mean/median/mode, standard deviations, correlations, regression analysis, etc., are available from statistical handbooks and in computer programs

As a result of permitting requirements, many enterprises will already be generating some of the measurement data and monitoring records (such as records covering effluent composition) required by this subclause of ISO 14001.

Information regarding *calibrating and maintaining the equipment that the enterprise uses for measuring and monitoring, and about maintaining records of the calibration and maintenance, includes*:

- Model procedures for developing and maintaining a calibration and maintenance program (though this is not required by ISO 14001)
- Model calibration and maintenance programs
- Formats for calibration and maintenance reports
- Examples of calibration, and maintenance intervals for different types of equipment
- Examples of maintenance measures for different types of equipment
- Model EMS records for calibration and maintenance activities

In addition, calibration and maintenance procedures are provided in equipment operator manuals or can be obtained from equipment manufacturers, and records of equipment calibration and maintenance activities could be controlled and retained in accordance with the enterprise's normal procedures, as covered in the discussion on ISO 14001 Subclause 4.5.3, "Records" (page 156).

Information regarding the *establishment and maintenance of a written procedure for periodically assessing the status of the enterprise's environmental regulatory compliance* include:

- Model compliance monitoring procedures
- Examples of compliance auditing programs
- Matrices for tracking regulatory compliance
- Examples of noncompliance reports
- Recommended formats for noncompliance reports

In addition, in the United States and western Europe, enterprises can engage commercial environmental regulatory compliance auditing services to prepare a compliance auditing program for an enterprise and to conduct the compliance audits.

To tell the truth, most EMS Committees are probably capable of creating the tools and techniques they need to implement this EMS element without much outside help. The reason is that their enterprises already do much of what is required in this subclause, though perhaps not with a focus on pollution prevention, continual improvement, or an EMS. Most enterprises already have procedures for measuring and monitoring some characteristics of their operations; for comparing the findings with internally or externally established standards (benchmarks); for relating performance to operational controls; for recording the information and maintaining the records; and for calibrating and maintaining equipment used for measuring and monitoring. An enterprise needs to engage in this sort of activity to some extent if it maintains any semblance of control over its operations. Many EMS Committees will be able to build on procedures already in place to satisfy the requirements of this ISO 14001 subclause.

Most EMS Committees are probably capable of creating the tools and techniques they need to implement this EMS element without much outside help because their enterprises already do much of what is required in this subclause, though perhaps not with a focus on environmental concerns.

If that is the case, then the steps for gathering the information needed to plan implementation of this EMS element, and a part of the next element (corrective and preventive action) might well be to:

a) Identify the key characteristic of environmentally sensitive activities of the enterprise's operations and its EMS that should be monitored.

b) List these characteristics on the left as row headings in an analytical matrix, and along the top as column headings list the five parts of measuring and monitoring: measuring; monitoring; recording the findings; evaluating the results; and revising the EMS, environmental management programs, procedures, or other operational controls.
c) Indicate, in the cells of the matrix, whether each part of measuring and monitoring is routinely carried out with respect to each key characteristic; and, if so, include a phrase about the frequency, method, or other characterizing features of the way it's carried out.
d) Review other measuring and monitoring activities carried out by the enterprise as part of its overall operations management system, to determine if they provide a basis for the procedures required for its EMS.
e) Review current calibration and monitoring procedures of the enterprise, to determine if they provide a basis for the procedures required for its EMS.
f) Review current recording and record maintenance and retention procedures, to determine if they provide a basis for the procedures required for the EMS.
g) Review current compliance audit procedures, to determine if they provide a basis for the procedures required for the EMS.
h) Make a first rough plan of the steps needed to satisfy the requirements of this and the next subclause of ISO 14001.
i) Review what other enterprises have done with respect to this EMS element.
j) Refine the implementation plan for this EMS element.
k) Review print publications and Web sites for relevant tools and techniques.
l) Finalize the implementation plan for this EMS element.

CHECKING AND CORRECTIVE ACTION: NON-CONFORMANCE AND CORRECTIVE AND PREVENTIVE ACTION (SUBCLAUSE 4.5.2)

In accordance with the requirements of ISO 14001 Subclause 4.5.2, an enterprise, as part of its EMS, would have procedures for determining responsibility and authority for dealing with various aspects of an actual or potential nonconformance. Or, as explained in the counterpart section of Chapter 2 (page 75), as a practical matter (and evidence of conformance) the enterprise would have written procedures for dealing with nonconformances, and these procedures would very clearly include the responsibilities and authorities of designated employees. The enterprise would have to ensure that its response to a nonconformance was appropriate to the causes and environmental impacts of the nonconformance; and it would have to implement and record any changes to procedures that were part of the corrective and preventive actions taken in response.

As mentioned in the last section of this chapter, fortunately there is a great deal of material in print and Internet sources like those listed in the "Additional Resources" appendix at the end of this book to help the reader implement and maintain this element of ISO 14001. And again, in keeping with our intent to avoid presenting material that is readily available elsewhere, we've listed below tools and other guidance available in other reference sources for implementing this subclause.

Information regarding *procedures for dealing with nonconformances*, including the responsibilities and authorities for responding to a nonconformance, investigating to determine its causes, immediately taking action to mitigate its environmental impacts, and seeing that appropriate corrective and preventive action is taken, include:

- Examples of possible EMS nonconformances
- Lists of EMS nonconformances auditors find most often
- Lists of possible major and minor nonconformances
- Nonconformance report formats, and examples for both major and minor nonconformances
- Model procedures for assessing nonconformances and determining appropriate corrective and preventive action
- Worksheets for assessing causes of nonconformances to the EMS, environmental management programs, or environmentally related procedures and work instructions
- Worksheets for assessing causes of performance failures and regulatory noncompliances
- Rapid assessment techniques for assessing the causes of nonconformances and performance failures
- Lists of typical environmental impacts associated with different nonconformances and performance failures
- Lists of methods through which different nonconformances and performance failures are typically discovered
- Process diagrams for assessing nonconformances and determining appropriate corrective and preventive measures

Information regarding *how to determine appropriate corrective and preventive actions* include:

- Worksheets for assessing appropriate corrective and preventive actions for different types of nonconformances to the EMS, environmental management programs, or environmentally related procedures and work instructions
- Worksheets for assessing appropriate corrective and preventive actions for different types of performance failures and regulatory noncompliances
- Model decision trees for assessing nonconformances and determining appropriate corrective and preventive measures
- Examples of corrective and preventive actions for common types of nonconformances

- Corrective and preventive action report formats
- Examples of corrective and preventive action reports
- Model preventive action recommendation forms

Information regarding *how to record changes in EMS procedures* that are part of the corrective and preventive action include:

- Models and examples of procedure changes and records of procedure changes,
- See the discussion on records in the next section of this chapter.

In addition, in the section of Chapter 2 on the EMS internal audit (Subclause 4.5.4, page 81), we offer our own list of possible contents for a procedure for responding to performance failures and nonconformances, whether major or minor, potential or actual. We also offer our own illustrations of major and minor nonconformances. Moreover, the discussion in the previous section of this chapter regarding how an EMS Committee might use the tools and techniques available from ISO 14001 sources applies for implementing the corrective and preventive action element of the EMS as well.

If an enterprise has emergency preparedness and response procedures in place, these might provide a point of departure for developing nonconformance corrective and preventive action procedures. But it would make sense for an EMS Committee to want to have a feel for the nonconformances and performance failures the enterprise is most likely to encounter before setting out to prepare the procedures for dealing with them. A simple analytical matrix could help here. Down the left side, as row headings, would be listed the key characteristics of the enterprise's environmentally sensitive activities. This is the same list that was used in the analytical matrix for planning to create measuring and monitoring procedures, described in item b) at the end of the last section of this chapter. This time, along the top, as column headings, would be a list of possible types of breakdowns, both nonconformances (e.g., lack of required records, or not pursuing an environmental management program) and performance failures (e.g., emissions in excess of target, or regulatory noncompliance). In the cells of the matrix would be written phrase descriptions of the one or two specific nonconformances or performance failures that are most likely to occur at the enterprise (e.g., no training records, or emissions exceed target by 10 percent).

It would make sense for an EMS Committee to want to have a feel for the nonconformances and performance failures the enterprise is most likely to encounter before setting out to prepare the procedures for dealing with them.

When the EMS Committee has worked through this analytical matrix, it will have before it on paper a panorama of the most likely specific nonconformances and performance failures the enterprise could experience, and the key characteristics that each would be connected with. With this information, committee members can already begin thinking about the most probable causes of the nonconformances, and not only begin selecting the most useful tools for preparing procedures, but perhaps even begin planning some preventive measures.

CHECKING AND CORRECTIVE ACTION: RECORDS (SUBCLAUSE 4.5.3)

In brief, the requirement of this subclause of ISO 14001 is simply that an enterprise have one or more procedures in place that amount to good management of environmental records, so they can be readily identified for retrieval and use, are stored safely, and are retained and disposed of with forethought. The basic concerns about record identification, storage, and retention are similar to those discussed in connection with controlled EMS documents under Subclause 4.4.5, "Document Control." There is a difference, however. EMS documents represent authoritative guidance for how activities are to be carried out, so the document creation process is a major aspect of document control. But because records are essentially historical documentation of decisions, actions, measurements, or findings — in effect, the putting down of facts — the process of creation and approval is not so great a concern. In any case, responsibility for creating records would be included in the procedures governing specific activities that generate specific records.

Most enterprises, certainly well-run enterprises, already have some sort of records management system in place, and the EMS Committee can build on that system for its EMS records management procedures.

Most enterprises, certainly well-run enterprises, already have some sort of records management system in place, and the EMS Committee can build on that system for its EMS records management procedures. In addition, in print and Internet sources like those listed in the "Additional Resources" appendix at the end of this book, readers will find

- Lists of possible environmental records (see also the list in the counterpart section of Chapter 2, page 79)
- Guidelines for designing EMS records management procedures
- Models and examples of EMS records management procedures
- Formats for records management matrices

The principal issues an EMS Committee will confront when considering how to implement this element of the EMS are:

- What is included in the list of environmental records that must be managed in a controlled manner?
- Who has what responsibilities and authorities in the records management process?
- How is it assured that records both remain at the work site where they are

needed and are retained in a central records management location?
- Who decides on retention times?
- On the basis of what criteria are retention times determined?

The EMS Committee can develop a list of records that will be covered by EMS records management procedures by first listing records required by ISO 14001 (see Table 14, page 136, and other sources), and then adding environmentally related documents to this list (such as regulatory compliance audit results) that the enterprise already manages, and others that the EMS Committee thinks appropriate for the enterprise. It then can design a records management matrix along the lines of the document control matrix in Table 16 (page 145), but with fewer control characteristics and, therefore, fewer columns. The EMS Committee would design the records management matrix together with the procedures for managing all or different categories of environmental records, to ensure that the procedures and the matrix are coordinated. The committee would then detail the proposed procedures and design a standard EMS record header along the lines of the example in Table 21.

Table 21. Example of an EMS Record Header.

EMS AUDIT REPORT	
Record Number: EMS/R 4.5.4-1, 2002	
Content: Internal EMS audit process, findings, and recommendations	
Activity, date: Semi-annual EMS audit, February 2002	**EMS Element:** Checking & Corrective Action: EMS Audit
Prepared by: Betty Brown, EMS Management Rep.	**Date:** 15 March 2002
May be Archived After: 15 January 2005	**May be Destroyed After:** 15 January 2012

CHECKING AND CORRECTIVE ACTION: EMS AUDIT (SUBCLAUSE 4.5.4)

The best way for an enterprise to conduct its internal audit is through a two-track approach. On the first track it checks the presence, completeness, currency, and appropriateness of all required EMS documents, records, procedures, and work instructions and standards. Table 14 (page 136) provides the basis for a checklist of required EMS documentation for this purpose.

The list of procedures required by the EMS standard is as follows (* means the procedure must be written):

Subclause 4.3.1	Identifying environmental aspects
Subclause 4.3.2	Identifying and accessing legal and other requirements
Subclause 4.4.2	Employee EMS awareness training
Subclause 4.4.3	Internal EMS communications
Subclause 4.4.3	External EMS communications
Subclause 4.4.5	EMS document control
Subclause 4.4.5	Creating and modifying EMS documents
Subclause 4.4.6a*	Operational control of environmentally sensitive activities of the enterprise
Subclause 4.4.6c	Operational control of environmentally sensitive activities related to goods and services provided by suppliers and contractors
Subclause 4.4.7	Identifying and responding to potential emergencies
Subclause 4.5.1*	Monitoring and measuring key characteristics of environmentally sensitive activities
Subclause 4.5.1*	Evaluating compliance with environmental laws and regulations
Subclause 4.5.2	Establishing responsibility and authority for dealing with EMS nonconformances, and corrective and preventive action
Subclause 4.5.3	Identifying, maintaining, and disposing of EMS records
Subclause 4.5.4	Carrying out periodic internal EMS audits

Checking all the EMS documentation and procedures together allows the verification process to be very efficient — but it foregoes a focus on much of the content, on how each EMS element operates as an integral element and in relation to other elements. Hence the second track, which involves checking for conformance with the requirements of ISO 14001 in terms of operational content, element by element.

The best way for an enterprise to conduct its internal audit is through a two-track approach. On one track, it checks all required EMS documents, records, and procedures; on the other, it checks conformance to the requirements of ISO 14001, element by element.

The counterpart section of Chapter 2 (page 81) includes discussions about the frequency, timing, content, and overall approach for an enterprise's internal EMS audit. These discussions provide information directly useful for planning audit procedures. In addition, again, there is an abundance of tools and techniques in the print literature and on the Internet that EMS Committees can build on, including:

- Checklists of ISO 14001 requirements
- Recommended EMS audit techniques
- Model EMS audit programs
- Model EMS audit procedures
- Nonconformance report formats and examples for both major and minor nonconformances
- Worksheets for planning an audit schedule
- Audit tests for completeness of a procedure, a record, environmental policy, an environmental management program, and other specific ISO 14001 EMS components

Readers who explore other ISO 14001 resources will find in particular a large number and wide variety of checklists of ISO 14001 requirements. This is fortunate, because the standard itself is written as normal text, sometimes in ambiguous language, and sometimes with several requirements nested within a single paragraph. The ISO does not provide a "translation" in checklist format, yet a checklist of discrete requirements is obviously an indispensable tool for an enterprise conducting an EMS internal audit (not to mention that it is the basic tool in second-party audits and certification and recertification EMS audits as well).

In fact, a checklist of ISO 14001 requirements is the centerpiece of a gap analysis (see the section on IER and gap analysis in Chapter 4, page 175), the most basic direct tool for EMS implementation planning, and it is an indispensable reference for actually implementing and maintaining an EMS. Because of its unique and important roles in pre-planning, planning, implementing, and self-auditing the EMS, in Table 22 we provide our own checklist of ISO 14001 requirements, despite the ready availability of these checklists elsewhere.

The aim of Table 22 is, as the title suggests, to provide a list of discrete ISO 14001 requirements in plain English. The requirements have been structured as questions; a "yes" answer means conformance to the requirement of the ISO 14001 standard. The requirements are listed in relatively logical sequence within each EMS element, in turn. Opposite the list of requirements are columns for indicating whether or not the enterprise conforms to the standard, the evidence of conformance, and additional comments the internal auditor may want to note, such as what it would take to achieve conformance. This makes the checklist useful to an EMS Committee as a tool for gap analysis *and* EMS implementation planning *and* EMS implementation *and* EMS maintenance *and* internal EMS auditing.

Someone conducting a gap analysis, for example, could fill in the column for evidence as he or she progresses in the analysis, noting where and what sort of evidence was found that indicated conformance. Or, the EMS Committee could fill in the column in advance of an internal audit, indicating what the evidence is intended to be for each requirement on the basis of the EMS implementation plan; then the internal auditor would make note in the comments column of whether or not the evidence existed in the place and form intended. Establishing evidence of conformance can involve use of specific tools and techniques, such as checklists of standards or targets, interviews with workers or managers, reviewing procedures, reviewing records, making observations, conducting tests, consulting other documents, and more.

The list of ISO 14001 requirements in Table 22 is based on information in the "Requirements of ISO 14001" subsections of Chapter 2. It is interesting to note that while reference is commonly made to the 51 or 52 "shall" requirements of the ISO 14001 EMS standard, Table 22 reveals that these in fact incorporate approximately 130 discrete requirements.

Table 22. Plain English Checklist of ISO 14001 Requirements.

ISO 14001 Requirement ("yes" for conformance)	**Yes**	**No**	**Evidence**	**Comments**
Clause 4.2, Environmental Policy				
1. Does the enterprise have an environmental policy?				
2. Is the environmental policy issued by top management?				
3. Is the policy appropriate to the nature, scale, and environmental impacts of the enterprise's operations and products?				
4. Does the policy include a commitment to comply with applicable environmental laws and regulations?				
5. Does the policy include a commitment to conform with other environmental standards and practices to which the enterprise has obligated itself?				
6. Does the policy include a commitment to continual improvement?				
7. Does the policy include a commitment to pollution prevention?				
8. Does the policy provide the framework for establishing and reviewing the enterprise's environmental objectives and targets?				
10. Is the policy in writing?				
11. Has the policy been communicated to all employees?				
12. Is the policy being implemented?				
13. Is the policy reviewed regularly and kept up to date?				
14. Is the policy available to the public?				
Subclause 4.3.1, Planning: Environmental Aspects				
1. Has the enterprise established and does it maintain one or more procedures to identify environmental aspects of its operations or products that it can influence or control?				

ISO 14001 Requirement ("yes" for conformance)	Yes	No	Evidence	Comments
2. Has the enterprise determined which of the aspects have or could have significant impacts on the environment?				
3. Does the enterprise consider its significant environmental aspects when setting environmental objectives?				
4. Does the enterprise keep the information on aspects and impacts up to date?				
Subclause 4.3.2, Planning: Legal and Other Requirements				
1. Has the enterprise established and does it maintain a procedure to track environmental laws and regulations that relate to the environmental aspects of its operations and products?				
2. Has the enterprise established and does it maintain a procedure to track other environmental standards and practices to which it has obligated itself?				
3. Has the enterprise established and does it maintain a procedure to ensure that employees are aware of and have access to the requirements of its legal and other environmental obligations?				
Subclause 4.3.3, Planning: Objectives and Targets				
1. Has the enterprise established and does it maintain environmental objectives and targets?				
2. Are they in written form?				
3. Have they been established for each relevant operational function and level of the enterprise?				
4. Are they consistent with the enterprise's environmental policy?				
5. Do they reflect a commitment to pollution prevention?				
6. Does the enterprise consider environmental laws and regulations with which it must comply when establishing and reviewing its environmental objectives?				

ISO 14001 Requirement ("yes" for conformance)	Yes	No	Evidence	Comments
7. When establishing and reviewing its environmental objectives, does it also consider other environmental standards and practices to which it has obligated itself?				
8. Does it also consider its significant environmental aspects?				
9. Does it also consider available technology options for achieving the objectives?				
10. Does it also consider its financial requirements?				
11. Does it also consider its operational requirements?				
12. Does it also consider other business requirements?				
13. Does it also consider the views of trade associations, citizen groups, and other outside interested parties?				
Subclause 4.3.4, Planning: Environmental Management Programs				
1. Has the enterprise established and does it maintain one or more environmental management programs to achieve its objectives and targets?				
2. Do the plans for these programs specify the people responsible for carrying them out and achieving the objectives and targets at each relevant function and level of the enterprise?				
3. Do the plans for these programs specify the means and timeframe by which objectives and targets are to be achieved?				
4. Does the enterprise amend these programs as needed to ensure that it is applying principles of environmental management to new developments and new or modified operations or products?				
Subclause 4.4.1, Implementation and Operation: Structure and Responsibility				
1. Has the enterprise determined EMS roles, responsibilities, and authorities?				
2. Are they described in writing?				
3. Are they communicated?				

ISO 14001 Requirement ("yes" for conformance)	Yes	No	Evidence	Comments
4. Does the enterprise provide essential resources in the form of people, special skills, equipment, and money to implement and control the EMS?				
5. Has the enterprise appointed one or more Management Representatives who, regardless of their other responsibilities, have defined roles, responsibilities, and authority for ensuring that EMS requirements are implemented and maintained in accordance with ISO 14001?				
6. Are Management Representatives also responsible for reporting on the EMS to top management, so top management can review EMS performance and operations and make informed decisions about ways to improve it?				
Subclause 4.4.2, Implementation and Operation: Training, Awareness, and Competence				
1. Does the enterprise make a deliberate effort to identify its EMS training needs?				
2. Does the enterprise require all employees whose work could have a significant impact on the environment to receive appropriate training?				
3. Has the enterprise established, and does it maintain, procedures (such as a training and competence program) to make its employees aware of the importance of performing in a manner that supports its environmental policy and procedures, and that supports the requirements of its EMS?				
4. Do its programs also make its employees aware of the significant environmental impacts and potential environmental impacts of their work activities, and the environmental benefits of improving their performance?				

ISO 14001 Requirement ("yes" for conformance)	Yes	No	Evidence	Comments
5. Do its programs also make employees aware of their roles and responsibilities in achieving conformance with the policy, procedures, and requirements of the enterprise's EMS, including emergency preparedness and response requirements?				
6. Do its programs also make employees aware of the potential consequences of not following specified operating procedures?				
7. Does the enterprise take steps to ensure that, as a result of appropriate education, training, and/or experience, workers performing tasks that can cause significant environmental impacts are competent at their jobs?				
Subclause 4.4.3, Implementation and Operation: Communication				
1. Has the enterprise established, and does it maintain, procedures for conducting internal communication on its environmental aspects and EMS among all levels and functions of the enterprise?				
2. Has the enterprise established, and does it maintain, procedures for conducting external communication, including receiving, documenting, and responding to relevant communication from outside parties?				
3. Has the enterprise made and recorded a decision regarding how the enterprise will handle external communication on its significant environmental aspects?				
Subclause 4.4.4, Implementation and Operation: EMS Documentation				
1. Has the enterprise established, and does it maintain in hard copy or electronic form, information that describes the core elements of the EMS and the relationships among them?				
2. Has the enterprise established, and does it maintain in hard copy or electronic form, information that gives directions for accessing documentation related to the EMS and its operation?				

ISO 14001 Requirement ("yes" for conformance)	**Yes**	**No**	**Evidence**	**Comments**
Subclause 4.4.5, Implementation and Operation: Document Control				
1. Has the enterprise established, and does it maintain, procedures and responsibilities regarding how the different types of documents required by ISO 14001 are to be created and modified?				
2. Has the enterprise established, and does it maintain, procedures for controlling required EMS documents so they are approved by designated approvers?				
3. Do the document control procedures include periodic reviews and revisions (if necessary) of EMS documents?				
4. Do the procedures allow EMS documents to be located easily?				
5. Do the procedures ensure that copies of the current versions of these documents are located and easily accessible at places in the enterprise where functions related to them are carried out?				
6. Do the procedures ensure that obsolete EMS documents are promptly removed from normal access throughout the enterprise, so they will not mistakenly be used?				
7. Do the procedures ensure that obsolete EMS documents retained for archiving are clearly identified as obsolete?				
8. Is the enterprise's EMS documentation legible?				
9. Does the date appear on the documentation, with dates of revision?				
10. Are the documents easily identifiable?				
11. Are the documents physically maintained in an orderly way?				
12. Are the documents retained for a specific period determined by the enterprise?				

Subclause 4.4.6, Implementation and Operation: Operational Control				
1. Has the enterprise identified the activities associated with its significant environmental aspects?				
2. Has the enterprise established, and does it maintain, written procedures that will cause these activities (and associated maintenance activities) to be carried out in a manner that implements the environmental policy of the enterprise and helps achieve its environmental objectives and targets?				
3. Are work instructions and standards (operating criteria) specified in association with the procedures?				
4. Has the enterprise also established, and does it maintain, such procedures for identifiable significant environmental aspects of goods and services used by the enterprise?				
5. Has the enterprise communicated these procedures to those who provide the goods and services?				
Subclause 4.4.7, Implementation and Operation: Emergency Preparedness and Response				
1. Has the enterprise established, and does it maintain, procedures to identify potential accidents and emergency situations?				
2. Has the enterprise established, and does it maintain, procedures for responding to accidents and emergencies?				
3. Has the enterprise established, and does it maintain, procedures for preventing and mitigating environmental impacts associated with potential accidents and emergencies?				
4. Does the enterprise review and revise its emergency preparedness and response procedures, as indicated by experience when an accident or emergency has occurred?				
5. Does the enterprise periodically test its emergency preparedness and response procedures if practicable?				

Subclause 4.5.1, Checking and Corrective Action: Monitoring and Measurement				
1. Has the enterprise established, and does it maintain, written procedures for regularly measuring and monitoring the key characteristics of its activities that can have significant negative impacts on the environment?				
2. In its measuring and monitoring procedures, has the enterprise made provision for recording information for tracking environmental performance?				
3. In the procedures, has it made provision for recording information for tracking operational controls in relation to environmental performance?				
4. In the procedures, has it made provision for recording information for tracking the relationship between actual performance and counterpart environmental objectives and targets?				
5. Does the enterprise calibrate and maintain the equipment it uses for measuring and monitoring?				
6. Does the enterprise keep records of equipment calibration and maintenance activities?				
7. Does the enterprise retain those records in accordance with its own procedures?				
8. Has the enterprise established, and does it maintain, a written procedure for periodically assessing its compliance with relevant environmental laws and regulations?				
Subclause 4.5.2, Checking and Corrective Action: Nonconformance and Corrective and Preventive Action				
1. Has the enterprise established, and does it maintain, procedures that determine responsibility and authority for dealing with actual and potential nonconformances?				
2. Do these procedures establish who responds to an actual or potential nonconformance?				
3. Do these procedures establish who investigates to determine the causes?				

4. Do they establish who takes appropriate, immediate action to mitigate any environmental impacts caused?				
5. Do they establish who initiates and carries through the process of determining and implementing corrective and preventive action?				
6. Has corrective or preventive action taken by the enterprise been appropriate to the scale of problems linked to the nonconformance, and to the magnitude and nature of any associated environmental impact?				
7. Does the enterprise expeditiously implement revisions in EMS procedures that are part of the corrective and preventive actions?				
8. Does the enterprise make a record of any revisions in EMS procedures that are part of the corrective and preventive actions?				
Subclause 4.5.3, Checking and Corrective Action: Records				
1. Has the enterprise established, and does it maintain, procedures covering identifying, maintaining, and disposing of environmental records?				
2. Does the enterprise include among those records EMS training records?				
3. Does it include results of EMS audits among the records?				
4. Does it include results of EMS reviews among the records?				
5. Are the enterprise's environmental records readily identifiable?				
6. Are the records legible?				
7. Are the records easily associated with the activity, product, or service to which they are related?				
8. Does the enterprise store and maintain its environmental records so that they are readily retrievable?				
9. Are the records protected against damage, deterioration, or loss?				

10. Has the enterprise established retention times for the records and does it keep a record of retention times?				
11. Does the enterprise maintain environmental records to the extent possible in a manner that makes it easy to confirm that ISO 14001 requirements are being satisfied?				
Subclause 4.5.4, Checking and Corrective Action: EMS Audit				
1. Has the enterprise established, and does it maintain, a program with explicit procedures (an audit plan) for carrying out EMS audits at regular intervals?				
2. Are the audit procedures of the enterprise comprehensive?				
3. Do the audit procedures cover the audit scope?				
4. Do the procedures cover audit frequency?				
5. Do the procedures cover audit methodologies?				
6. Do the procedures cover audit responsibilities?				
7. Do the procedures cover audit requirements?				
8. Do the procedures cover audit reporting?				
9. Do its EMS audit procedures enable the enterprise to determine from its EMS audit if its EMS conforms to what the enterprise intended?				
10. Do the audit procedures enable the enterprise to determine from the audit if its EMS conforms to the requirements of ISO 14001?				
11. Do the procedures enable the enterprise to determine from the audit if its EMS has been properly implemented?				
12. Do the procedures enable the enterprise to determine from the audit if its EMS has been properly maintained?				
13. Are the extent, depth, and frequency of its EMS audits appropriate to the nature and scale of the enterprise's interactions with the environment?				

14. Are they appropriate in light of how well the enterprise did on previous EMS audits?				
15. Does top management of the enterprise receive a report of EMS audit findings?				
Clause 4.6, Management Review				
1. Has top management of the enterprise conducted a review of the EMS to ensure that the EMS continues to be suitable and adequate for the enterprise and effective?				
2. Is a management review conducted at regular intervals that management has determined is appropriate?				
3. Is there a process for assembling and providing top management with the information it needs to conduct its EMS review?				
4. Does the management review address the possible need for changes to environmental policy, environmental objectives, and other elements of the EMS?				
5. In making its decisions on changes in the EMS, does top management take into account EMS audit findings, changing internal and external circumstances, and its commitment to continual improvement?				
6. Are the management review and the decisions taken as it recorded?				

Before planning the procedures for internal audits, the EMS Committee may want to review its work on measuring and monitoring (Subclause 4.5.1), and on nonconformance and corrective and preventive action (Subclause 4.5.2). Much of what was developed for these two EMS elements will help the committee develop an internal audit plan.

As part of its internal audit planning, the EMS Committee may find it useful to create report forms for reporting major and minor nonconformances discovered during the audit. The completed forms would be included in the audit report, giving top management the full story on each nonconformance. As noted earlier, model formats for these forms can be found in many ISO 14001-related print and Internet resources, but basically the nonconformance report forms would have blocks of space set aside for:

- The name of the auditor
- When the audit of this element was conducted
- What the auditing procedure was

- A description of the nonconformance
- How the nonconformance was determined or the evidence of conformance that was lacking
- A statement of the cause of the nonconformance, and how the cause was determined
- Questions and issues related to the nonconformance or its cause
- Recommendations for correcting and preventing the nonconformance (for examples, changes in procedures or environmental management programs), and other associated measures to improve the enterprise's EMS (for example, changes in EMS roles and responsibilities, or frequency of internal audits)

MANAGEMENT REVIEW (CLAUSE 4.6)

The counterpart section of Chapter 2 (page 87) includes discussions about:

- Who participates in the management review
- The frequency of the management review
- What might be covered in the review
- The contents of the (optional) management review procedure
- Changing circumstances that would be incorporated into the review
- The structure of a management review report

All of this is directly useful for planning to satisfy the requirements of this clause of ISO 14001.

In addition, other print and Internet sources offer a variety of tools and techniques for implementing and maintaining this EMS element, including:

- Model agendas for the management review
- Lists of the types of information needed for addressing different topics in the management review
- Lists of the types of changes in the EMS that might come out of a management review, especially as related to continual improvement of the EMS
- Formats and examples of records of the management review
- Examples of management review procedures

Beyond all that, there is very little mystery to the purpose and what is required for the management review. It must ultimately be conducted by top management (and coordinated by the EMS Management Representative), and produce top management decisions; these decisions must be aimed at continually improving the efficiency and performance of the EMS; the review must be based on a full set of relevant information, like the results of the internal EMS audit, measuring and monitoring, compliance audits, and ISO 14001 certification audits; it must take account of changing internal and external conditions; it must be conducted at regular intervals; and a record must be made of the review and the decisions taken.

ONE MORE TIME

Here are the keys to establishing and maintaining an efficient and productive EMS with minimal burden on the enterprise:

a) Implement the EMS in a manner and at a pace that suits the enterprise and its approach to its EMS. Ensuring that each EMS implementation step taken is successful and rewarding is more important than trying immediately to satisfy all ISO 14001 requirements throughout the enterprise.

b) Build on existing general management features and systems in the enterprise, such as environmental policy elements, training programs, ISO 9001/2 systems, document control procedures, measuring and monitoring procedures, record management procedures, and the like.

c) Examine a variety of ISO 14001 resources in print and on the Internet, such as those listed in the "Additional Resources" appendix at the end of this book, for implementation ideas, tools, and techniques suited to your enterprise. It's also a good idea to explore the Web sites of enterprises already certified to ISO 14001.

d) Before planning your EMS implementation, conduct an initial environmental review (IER) or gap analysis (see the section on this subject in Chapter 4, page 177). This is not just to identify what needs to be done to establish an EMS; it's also to explore what is already in place in the enterprise that can serve as a basis for EMS requirements.

e) Plan the EMS with an eye to overall strategic management of the enterprise, and ways that the EMS can contribute to improving business performance.

f) Involve everyone in the enterprise in EMS implementation and maintenance, and visibly reward good EMS performance.

g) Keep the EMS, its environmental management programs, its procedures, its records — everything — as simple as possible.

EXERCISES

1. From EMS literature or the Web, identify four examples of environmental policy statements that conform to ISO 14001 requirements. Which of them comes closest to what would be appropriate for your enterprise?

2. Adapt the environmental policy statement structure in Table 1 (page 108) to your needs, and try your hand at writing a draft environmental policy for your enterprise.

3. Work through the analytical matrix in Table 4 (page 115) to identify activities and environmental aspects related to known environmental impacts of your enterprise.

4. Based on your knowledge (not research), complete a row of Table 6 (page 122) for a known significant environmental aspect of the operations of your enterprise.

5. In Table 22 (page 160), a checklist of ISO 14001 requirements, fill in the "Evidence" column as it might be for your enterprise once its EMS is in place. A good place to start is in the "Requirements of ISO 14001" subsections of Chapter 2, which often note things certification auditors might look for. The present chapter, Chapter 3, also frequently refers to what a certification auditor might look for. Good ideas for objective evidence of conformance can be found throughout the print and Internet literature on ISO 14001. Beyond that, your knowledge of your enterprise and your imagination will serve well.

QUESTIONS FOR THINKING AND DISCUSSING

1. Where in its policy statements, annual reports, or other public documents are there statements of your enterprise's policy on environmental matters? Does your enterprise effectively have an environmental policy, written or unwritten? If it does, can you describe it?

2. What analytical approach would you use for determining significant environmental aspects of your enterprise's operations and products? How would you break enterprise operations down into activities for analysis? What criteria would you use for significant environmental impacts? Do you know now what the significant environmental impacts of your enterprise are?

3. What method does your enterprise employ to keep abreast of environmental regulatory requirements? How does it communicate relevant information about regulatory requirements to employees who, because of the nature of their work, need to know?

4. What would be the single most important environmental objective you would recommend for your enterprise, based on what you already know about enterprise operations and products? What would be the outlines of an environmental management program to achieve it?

5. Who in your enterprise would likely be assigned the role of ISO 14001 Management Representative, and why? If you think someone would have to be hired from the outside for this function, please explain.

6. Consider Table 10 (page 129). Into what categories do you think it would be best to group the workers in your enterprise for EMS training purposes? What kinds of EMS training for which categories of employees do you think are most immediately important in your enterprise?

7. Does your enterprise currently have a document control system? If so, what are the principal document control characteristics under that system? Would document control procedures that conform to ISO 14001 requirements integrate easily with the existing document control system?

8. Does your enterprise use written procedures to maintain control over operations? If so, how does management communicate those procedures to employees whose work should be guided by them?

9. Does your enterprise periodically conduct any form of general environmental audit or environmental compliance audit? If so, how does it compare with what would be required for an internal EMS audit? If not, is there any sort of audit conducted regularly by your enterprise (for example, quality assurance, or occupational health and safety) that would provide some base of experience upon which internal EMS audit procedures could be built?

Chapter 4
EMS: FIRST STEPS

THE BENEFITS OF AN EMS

If you've reached this point in this book you probably don't need convincing about the benefits of an EMS. But you may need more arguments to convince your colleagues throughout the enterprise that implementing an EMS would be a good investment. As noted early in Chapter 3, if you want to implement an EMS in your enterprise, your first job is to convince top management. Lists of the benefits of implementing an EMS, and especially of implementing the ISO 14001 EMS standard, abound in ISO 14001-related print literature and Web sites. In addition, some enterprises certified to ISO 14001 offer testimonials on their Web sites about how they have benefited from implementing the standard. Finally, in the "Additional Resources" appendix at the end of this book, readers will find one book devoted entirely to the subject of how to build a business case for implementing ISO 14001. As a convenience to readers, but also because making the case for an EMS is the first of the first steps toward establishing an EMS, we offer our own list of the benefits of implementing an EMS.

The U.S. Environmental Protection Agency (EPA) is sponsoring a continuing study of the benefits that accrue to enterprises from establishing EMSs, many of them based on the ISO 14001 standard. For the latest findings of this multi-year study, see www.eli.org/isopilots.htm.

The question CEOs sometimes ask when someone proposes establishing an EMS in the enterprise is, "What will it cost?" But that's the wrong question. The right question is, "What is the net benefit?" The net benefit is the sum of the likely benefits, minus the likely cost of EMS implementation and maintenance. Sure, there can be endless arguments over the monetary value of non-market benefits and non-market costs, but rough estimates of cost and revenue streams associated with EMS implementation and operation *can* actually be made for a specific period of time — say, the first five years. Once these streams are estimated they can be analyzed and presented to top management in readily understandable business terms, like a range of possible benefit/cost ratios, net present values, or internal rates of return under a range of different assumptions.

> ***Sure, there can be endless arguments over the monetary value of non-market benefits and non-market costs, but rough estimates of cost and revenue streams associated with EMS implementation and operation*** **can** ***actually be made, and presented to top management in business terms.***

However it's handled, the first step of all is to develop a firm grasp of the possible benefits of a well-run EMS. The 10 most important broad benefits are listed below in three necessarily overlapping categories: benefits associated with the EMS as a management system; benefits associated with improved environmental performance as a result of the EMS; and benefits associated with responding to the growing demand everywhere for increased environmental responsibility on the part of enterprises.

Because of the EMS emphasis on principles of good business management, strategic environmental management, and continual improvement, enterprises that implement and maintain an EMS enjoy the benefits of:

1) *Environmental management cost savings,* by integrating environmental management with quality management systems and occupational health and safety management systems, and the overall management system of the enterprise.

2) *Improved environmental performance at lower cost,* as a result of building capacity within the enterprise for strategic environmental management, including anticipating future environmental performance demands and continually identifying pollution prevention opportunities, rather than simply reacting to legal requirements.

> ***EMSs yield high rates of return on pollution prevention and energy efficiency investments, returns that often far exceed rates of return that the enterprise encounters in the normal capital marketplace.***

3) *Better overall enterprise management and better overall business performance,* as a result of EMS requirements to clearly allocate responsibility throughout the organization, establish monitoring systems, ensure adequate operational controls, minimize waste and energy consumption, and establish and maintain other elements of sound enterprise management.

4) *Readiness for the coming trends in environmental regulation,* such as the gradual movement from media-based to systems-based compliance enforcement approaches that require strategic environmental planning by top management (for recent evidence of this, see EPA's new National Environmental Performance Track program at www.epa.gov/perftrac/program/program.html).

Because an EMS leads to continually improving environmental performance through pollution prevention and other measures, enterprises that implement and maintain an EMS enjoy the benefits of:

5) *Increased likelihood of long-term sustainability,* as a result of minimizing the demands of enterprise operations for natural resources, minimizing the risk of environmental liability, and complying with environmental legal requirements through proactive management.

6) *Reduced operating costs,* as a result of minimized waste and efficiencies in energy consumption. These reductions in costs — from an energy-saving improvement, for example — continue to accrue year after year.

7) *High rates of return* on pollution prevention and energy efficiency investments (especially when reduced risk of environmental liability is factored in), returns that often far exceed rates of return the enterprise encounters in the normal capital marketplace.

Because an EMS, and most visibly an EMS that is certified to the ISO 14001 standard, responds to growing demands from consumers, business customers, other business partners, communities, governments, and regulators for demonstrated environmental responsibility, enterprises that implement and maintain an EMS enjoy the benefits of:

8) *Improved access to capital and lower capital and other costs.* Because investors and lenders tend to favor enterprises with an EMS, lenders will sometimes offer them lower interest rates, insurance companies will sometimes offer them lower premiums, and some state and local jurisdictions are preparing to waive certain environmental fees and requirements for enterprises with an EMS. And the list goes on.

9) *Improved market access,* because governments, major business customers, and consumers are increasingly demanding, as a condition of doing business, that their suppliers demonstrate a commitment to environmental responsibility through conformance to the ISO 14001 or other EMS standard.

10) *Better public, community, and government relations,* through an environmental management approach that encourages dialogue and responsiveness to the concerns of outside parties.

INITIAL ENVIRONMENTAL REVIEW (IER) AND GAP ANALYSIS

There is a discussion of initial steps toward implementing an EMS, the steps that need to be taken before implementation can get underway, in the second section of Chapter 3, "Before Policy" (page 103). That discussion provides the context for this presentation of IER and gap analysis.

Initial Environmental Review

An enterprise that until now has paid little attention to environmental matters, or has not paid attention in any organized way, will want to carry out an initial environmental review of itself before it does much else. One purpose of an IER is to bring together, for the first time, available information on the enterprise's environmental interactions, its environmental performance, its regulatory compliance status, its environmental programs, and the overall way that it manages the environmental aspects of its operations. An IER would also touch on such related subjects as quality management and health and safety management, because another purpose of the IER is to identify existing systems or programs in the enterprise that the EMS could build on. Ultimately, the purpose of the IER is to provide information for planning the process of implementing the enterprise's EMS, so the IER might also include a first pass at identifying the significant environmental aspects of the enterprise's operations and products.

Like the EMS itself, for it to be really useful the IER has to be carried out by employees of the enterprise. Consultants can help a great deal in both planning and conducting the IER, but the actual work of it and developing the findings has to be done by people who will use the information to plan and implement the EMS, and who will live with the results of their work. Conducting the IER and carrying out an initial environmental awareness program could be the first two initiatives of the EMS Committee. Planning and carrying out an IER offers an opportunity for the EMS Committee to develop its working systems and style, which will make planning and implementing the EMS later much more efficient.

One purpose of an IER is to bring together, for the first time, available information on the enterprise's environmental interactions, its environmental performance, its regulatory compliance status, its environmental programs, and the overall way that it manages the environmental aspects of its operations.

Table 1 is an example of an IER information collection form. It has been pieced together from several such forms developed and used by different enterprises. The first section of the form contains basic information about the enterprise. The reason for including this information is to establish, in writing, a common understanding of the enterprise, its relevant operations, and its business purposes. The second section is a questionnaire for gathering the basic IER information described earlier, while the third section is a guide and record of observations for a walk-through inspection. An EMS Committee can use this information collection form as a starting point for designing a form that is uniquely suited to its enterprise and IER approach.

Table 1. IER Information Collection Form.

Dates of IER: ______________________

IER team leader and title: ____

IER team members and titles:

____

____

____

Top management liaison for the IER and EMS:

Name and title:

____

Telephone: Email:

<u>I. Basic Enterprise Information</u>

1. Name of the enterprise:

2. Location:

3. Type of business:

4. Relevant enterprise ownership history:

5. General makeup of the enterprise (type, size, and number of facilities):

6. Setting of the enterprise (rural, urban, industrial park, near river, near housing, etc.):

7. Enterprise mission statement or equivalent:

8. Principal products of the enterprise:

9. Principal markets:

10. Principal customers/clients:

11. Principal suppliers and contractors, and the goods/services that they provide:

12. Types of production processes:

13. Enterprise organization chart:

14. Scope of the IER (i.e., parts of the enterprise to be included in the IER):

II. IER Questions

A. General, Quality, and Occupational Health and Safety

1. Does the enterprise have written policy related to:
☐ Quality? (attach)
☐ Occupational health and safety? (attach)
☐ Other?

2. Does the enterprise have a:
☐ Written detailed structure of responsibilities and authorities? (attach)
☐ Training program? (attach description or program)
☐ Internal communication procedure? (attach or describe)
☐ External communication procedure? (attach or describe)
☐ Documentation system? (attach procedure or describe)
☐ Document control system? (attach procedure or describe)
☐ A system of written operational control procedures? (describe)
☐ A system of written work instructions and standards for controlled tasks? (describe)
☐ Emergency preparedness and response plan? (attach or describe)
☐ General measuring and monitoring program? (attach procedures or describe)
☐ Procedures for dealing with out-of-spec performance? (attach or describe)
☐ A records management system? (attach procedure or describe)
☐ Schedule of regular senior management meetings? (attach schedule or describe)

3. Does the enterprise have a:
☐ Quality manager?
☐ Occupational Health and Safety Department?
☐ Occupational health and safety officer?
☐ Safety engineer?

4. To what standards is the enterprise certified?
☐ None
☐ ISO 9001/2
☐ Other (explain)

5. What methods does the enterprise use to ensure the quality of its products?

6. Does the enterprise measure and monitor the quality of its products?
☐ No
☐ Yes (explain)

7. Does the enterprise keep records of its quality performance?
☐ No
☐ Yes (explain)

8. Does the enterprise conduct quality audits to determine the effectiveness of its quality controls and programs?
☐ No
☐ Yes (describe; how frequently?)

9. Has the enterprise identified the occupational health and safety hazards to its employees?
☐ No
☐ Yes (explain)

10. Has the enterprise identified legislation and regulations relevant to the health and safety of its employees?
☐ No
☐ Yes (where is the information recorded, who tracks it, how is it tracked?)

11. Does the enterprise have an employee health monitoring program?
☐ No
☐ Yes (explain)

B. Environment

12. Does the enterprise have environmental policy?
☐ No
☐ Yes
    ☐ Environmental policy statement (attach)
    ☐ Environmental policy components of other policy statements (attach or describe)
    ☐ Other (e.g., environmental statement in annual report) (attach or explain)

13. Has the enterprise assigned overall environmental responsibility?
☐ No
☐ Yes
    ☐ Environmental Department
    ☐ Environmental officer/engineer
    ☐ Other (explain)

14. Has the enterprise analyzed if its operations or products have any impact on the environment?
☐ No
☐ Yes (explain; describe environmental aspects identified)

15. Has the enterprise identified and quantified the natural resources it consumes (water, energy, fuels, raw materials)?
□ No
□ Yes (explain; provide data)

16. Does the enterprise use any specific methods, programs, or practices to minimize the environmental impacts and resource use of its operations?
□ No
□ Yes (explain; provide specifics)

17. Does the enterprise track legislation and regulations relevant to its operations?
□ No
□ Yes (How? Where is the information recorded? How is it communicated?)

18. Has a general environmental legal and regulatory compliance audit been conducted at the enterprise?
□ No
□ Yes (When? Who performed it? What were the findings?)

19. Has the enterprise been cited for any environmental noncompliance?
□ No
□ Yes (list dates, noncompliances, penalties, and corrective actions)

20. Does the enterprise have any environmental performance objectives or targets?
□ No
□ Yes (explain; attach objective and target documentation)

21. Has the enterprise adopted any other environmental practices or standards?
□ No
□ Yes (explain; describe the practices and standards)

22. Does the enterprise monitor and make records of its environmental performance in relation to external and internal environmental requirements, practices, and standards?
□ No
□ Yes (explain; provide the comparison data)

23. Does the enterprise maintain any other records of its environmental performance?
□ No
□ Yes (explain; attach records or provide data)

24. Does the enterprise use any specific method to track the quantities of hazardous chemicals it processes?
□ No
□ Yes (explain; provide data)

25. Are any major hazardous wastes generated at the facility?
☐ No
☐ Yes (explain; provide information on wastes and their disposal)

26. Does the enterprise have a written emergency response plan?
☐ No
☐ Yes (attach; explain where it is kept, how often it is reviewed, how it is made known to employees)

27. Are emergency response procedures practiced by employees on a regular basis?
☐ No
☐ Yes (explain)

28. Does the enterprise monitor the effectiveness of its methods, programs, and techniques for managing its environmental performance?
☐ No
☐ Yes (explain; provide information, attach records)

29. Has the enterprise carried out an internal environmental audit?
☐ No
☐ Yes (explain who conducted it, when it was conducted, what the findings and resulting actions were; attach audit report)

30. Has there been a P2 audit at the enterprise?
☐ No
☐ Yes (explain; provide information, attach audit report)

31. Has the enterprise encountered any community pressure or public relations problems connected with its environmental performance?
☐ No
☐ Yes (explain)

32. Does the enterprise have a procedure for responding to environmental complaints from the community?
☐ No
☐ Yes (explain)

33. Has any environmental awareness training been conducted for the employees of the enterprise?
☐ No
☐ Yes (explain; attach documents)

34. Does the enterprise have a program to motivate good environmental performance among employees?
☐ No
☐ Yes (explain; attach documents)

35. Are there any motivators, requirements, or other circumstances in the enterprise that work against good environmental practices?
☐ No
☐ Yes (explain)

III. Walk-Through Inspection *(photograph recorded observations when possible)*

Observation	**Comments**
☐ General housekeeping problems	
☐ Evidence of chemical releases	
☐ Inadequate or incorrect hazardous materials labeling	
☐ Leaking valves, lines, containers	
☐ Improperly stored chemicals, wastes	
☐ Inadequate, incorrect, or unmaintained emergency equipment	
☐ Potential environmental accidents or emergencies	
☐ Energy waste or inefficiencies	
☐ Water waste or inefficiencies	
☐ Improper handling of solid waste	
☐ Improper handling of toxic waste	
☐ No-cost/low-cost pollution prevention opportunities	1. 2. 3. 4.
☐ Moderate-cost pollution prevention opportunities	1. 2. 3.

Observation	Comments
☐ Major pollution prevention investment opportunities	1. 2.
☐ Apparently significant environmental aspects	1. Aspect: Activity: Operation: 2. Aspect: Activity: Operation: 3. Aspect: Activity: Operation:
☐ Apparently less significant environmental aspects	1. Aspect: Activity: Operation: 2. Aspect: Activity: Operation: 3. Aspect: Activity: Operation:
☐ Other observations	

After collecting all this information, the EMS Committee would review and analyze it carefully, and distill it into a report for top management. Following is a structure of possible section headings for the IER report:

- ***Significant Environmental Aspects of Our Operations and Products***
- ***Other Environmental Aspects of Our Operations and Products***
- ***Legislative and Regulatory Requirements***
- ***Our Regulatory Compliance Status***
- ***Findings of Compliance and Environmental Audits***

- ***Environmental Performance and Performance Comparisons***
- ***Interactions with External Parties on Environmental Issues***
- ***Pollution Prevention Opportunities***
- ***Our Environmental Management Practices, Programs, Procedures, and Internal Standards***
- ***Other Environmental Programs that We Conduct or Participate In***
- ***How We Manage Our Environmental Aspects***
- ***Circumstances in Our Enterprise that Work Against Good Environmental Performance***
- ***Existing Systems or Programs We Could Leverage for EMS Purposes***
- ***Recommended Initial Priorities, Approach, and Scope for Our EMS***
- ***Proposed Next Steps***

Gap Analysis

If an enterprise already has an environmental management system or program of some sort in place, or if it has a quality management system certified to ISO 9001/2, then it would conduct a gap analysis before embarking on EMS implementation planning. An EMS gap analysis compares the system or program in place at the enterprise with the demands of ISO 14001, requirement by requirement. To develop the full range of information to support good EMS implementation planning, the EMS Committee would also want to conduct an IER or incorporate elements of an IER into the gap analysis, focusing especially on elements that provide information on existing programs or procedures that can be built upon, like question number 2 in Section II of the IER information collection form above.

> ***An EMS gap analysis compares the system or program in place at the enterprise with the demands of ISO 14001, requirement by requirement.***

As with an IER, a gap analysis would be conducted by the organization's own personnel, but could be usefully assisted by an external consultant. And, as with the IER, the first step in a gap analysis would be to define its scope.

The discussion in Chapter 3 on the internal EMS audit (Subclause 4.5.4, page 157) contains important practical information for conducting a gap analysis, including a checklist of ISO 14001 requirements (page 160). When this checklist is used for gap analysis, the comments column would be for notations on what needs to be done regarding each requirement to create an EMS in the enterprise that conforms to the ISO 14001 standard. Or, readers may do several other things: they can choose to use the checklist in their own words that they created in the first exercise at the end of Chapter 2; they can assemble a checklist from the "Requirements of ISO 14001" sections of Chapter 2; or they can review the print literature and surf the Web, where dozens of ISO 14001 requirements checklists and gap analysis guides are available.

Once the EMS Committee has gap analysis information in hand, it needs to prepare it for presentation to top management. A good structure for a gap analysis report would be first to borrow some appropriate sections from the IER structure presented earlier to convey information on the environmental context of the enterprise; then to provide a simple summary table of gap analysis findings and EMS implementation needs, like the one in Table 2; then, to offer recommendations regarding the EMS initial scope and implementation planning process. Appendices could be used for detailed environmental information and gap analysis findings. A summary table like the one in Table 2 serves both as a format for presenting gap analysis findings and as a tool for the first broad step in the EMS implementation planning process.

Table 2. ISO 14001 Gap Analysis Findings and EMS Implementation Needs.

ISO 14000 Section/Clause	**Conforms?**		**Implementation Needs**
	Yes	**No**	
4.2 Environmental Policy			
4.3 Planning			
4.3.1 Environmental Aspects			
4.3.2 Legal and Other Requirements			
4.3.3 Objectives and Targets			
4.3.4 Environmental Management Programs			
4.4 Implementation and Operation			
4.4.1 Structure and Responsibility			
4.4.2 Training, Awareness, and Competence			
4.4.3 Communication			

ISO 14000 Section/Clause	Conforms?		Implementation Needs
	Yes	No	
4.4.4 EMS Documentation			
4.4.5 Document Control			
4.4.6 Operational Control			
4.4.7 Emergency Preparedness and Response			
4.5 Checking and Corrective Action			
4.5.1 Monitoring and Measurement			
4.5.2 Nonconformance and Corrective and Preventive Action			
4.5.3 Records			
4.5.4 EMS Audit			
4.6 Management Review			

EMS LITE

It has been nearly a mantra in this book that an enterprise can establish its EMS in convenient increments. There is no need to try to meet all the requirements of ISO 14001 to certification standards from the very beginning — though, if an enterprise is certain it is up to the job, there is nothing wrong with it. But an enterprise looking to reduce the scope of its initial EMS can limit:

- The number of environmental aspects it deals with
- The portion of the enterprise, or even of a single facility, it will cover
- The functional areas of the enterprise it will cover

- The environmental "sectors" it will cover (energy, water, wastewater, solid waste, air pollution, and so on)
- The extensiveness of the EMS as a system

These limits can be combined — for example, an enterprise might start with one or two environmental aspects in one portion of the enterprise, with very limited EMS documentation. This is fine, as long as the enterprise approaches this initial EMS as a step in a learning process and part of a plan to gradually expand outward to other parts of the enterprise, and as long as the initial EMS is designed as a good foundation for building, by degrees, an EMS that eventually *does* conform to the ISO 14001 standard. Table 3 offers an example of a limited-scope initial EMS that would provide that foundation.

Table 3. A Limited-Scope Initial EMS

1. Develop an environmental policy statement

The policy statement must:

- Make a commitment to regulatory compliance
- Make a commitment to continual improvement
- Make a commitment to pollution prevention
- Be issued by top management of the enterprise

2. Describe the enterprise's principal operations and products

This description should include:

- The types of products or services the enterprise produces
- Its principal markets and customers
- The general nature of its operations (type, size, and number of facilities, use of suppliers and contractors, types of production processes)
- Other general descriptive information

The purpose of this element is to establish a common understanding — among management, employees, and others — of the basic characterization of the enterprise. Much of this information may be available in annual reports, marketing materials, or on the Web site of the enterprise.

3. Conduct a "P2 Lite"

Conduct a walk-through audit as in Section III of the IER Information Collection Form (page 184) as a guide. Look closely and with a fresh eye at small-scale ways energy could be saved, materials use can be reduced, waste can be reduced or recycled. Develop and carry out a program to make small-scale P2 improvements in the enterprise based on findings of the walk-through.

4. *Identify the enterprise's significant environmental aspects*
   This identification should include:
   - Releases to air, water, and soil
   - Solid and toxic waste
   - Use of energy
   - Use of water
   - Use of other environmental resources

Each environmental aspect would be associated with the activity of which it is a part and, in turn, with the larger operation of which the activity is a part. This information can be developed with the help of analytical matrices like the ones in Chapter 3 - Tables 2 (page 112), 3 (page 113), or 4 (page 115). It can then be represented in a simple table, like the one in Table 2 of Chapter 2 (page 33).The information could be developed through a two-step process: first, environmental aspects would be identified and recorded nonquantitatively, with the emphasis on the identification process; then, all or only the most significant aspects would be quantified.

5. *Identify applicable environmental regulations and the enterprise's compliance status*
   Again, a simple analytical matrix would be useful for assembling and presenting the information. Such a matrix could be organized like this:
   - Along the left of the matrix, as row headings, the enterprise would list operations and activities, and their respective environmental aspects
   - Along the top, as column headings, it would list the titles of applicable regulations
   - Cells would contain the applicable emission standard or allowance; point or means of measurement; and the date and finding of the most recent compliance check

6. *Prepare an environmental action plan*
   This plan should address one or more of the enterprise's environmental aspects and regulatory compliance needs, with priority going to compliance needs. The action plan would include four elements:
   (a) Setting near-term environmental performance targets related to selected environmental aspects and regulations. Targets would be expressed:
   - as quantities and percentages of reduction in specific emissions or resource use
   - within explicit periods of time
   - with respect to each selected aspect or regulation

These targets would represent the best that the enterprise can realistically expect to achieve, however modest.

(b) Devising programs for achieving the targets. Programs for achieving the targets would include:

- details of tasks in sequence
- schedule for carrying out the tasks
- employees assigned
- resources allocated
- the means and schedule for measuring results

(c) Planning an internal EMS audit. The audit plan would specify:

- when the audit will be carried out
- who is responsible for planning, coordinating, and conducting the audit
- what exactly will be audited
- what evidence will be examined (observations made, documentation reviewed, interviews conducted)
- how and to whom the findings will be reported

(d) Planning an EMS review by top management. The EMS review plan would include:

- when the review would be carried out
- who would participate
- what information in addition to the results of the internal audit would be reviewed
- on what subjects decisions will be made (changes or next steps with respect to the five basic elements in this limited-scope EMS, and next steps toward conformance with ISO 14001)

7. *Implement the action plan*

ONE MORE TIME

Here are the 10 main benefits of implementing and maintaining a productive EMS, especially one that conforms to the ISO 14001 standard.

Because the EMS is a good management system, it fosters:

- Environmental management cost savings
- Improved environmental performance at lower cost
- Better overall enterprise management and better overall business performance
- Readiness for the coming trends in environmental regulation

Because the EMS leads to improved and constantly improving environmental performance:

- It increases likelihood of long term sustainability
- It reduces operating costs
- It gives the enterprise high rates of return on pollution prevention and energy efficiency investments

Because the EMS responds to the growing demand for increased environmental responsibility:

- It improves access to capital, and lowers capital and other costs
- It improved market access,
- It improves public, community, and government relations

In conjunction with the "Before Policy" activities mentioned in Chapter 3 (page 103), an enterprise that had been paying little attention to environmental matters would conduct an IER to assemble environmental information to support planning to implement an EMS. If an enterprise has already established an environmental management system of some sort, or if it has a quality management system certified to ISO 9001/2, then it would conduct a gap analysis, though it would serve itself well to also carry out selected elements of an IER.

In any event, the EMS Committee would use the IER or gap analysis report to present the information it had gathered to inform top management of its findings, and to use as a foundation for detailed EMS implementation planning. If top management approves the committee's recommendations, the next step is for the committee to develop an EMS implementation plan that details specific tasks, responsibilities, an implementation schedule, resources allocated to do the job, and a timetable for specific EMS elements to get up and running.

Enterprises should not lose sight of the fact that certification to the ISO 14001 standard is not an end state. At the core of any EMS is the principle of continual improvement, so there* is *no end state. Continual improvement is a process that can start well before certification to ISO 14001, just as it will continue long after certification to ISO 14001.

An enterprise can establish its EMS in increments, to ensure the success of each step and to apply any lessons learned to the next step. Most enterprises will want to aim at having their EMSs certified to the ISO 14001 standard eventually, but they can set their own schedules for accomplishing that. They can reap many benefits from their EMSs along the road to certification. Enterprises should not lose sight of the fact that certification to the ISO 14001 standard is not an end state. At the core of an EMS is continual improvement, so there *is* no end state. Continual improvement is a process that can start well before certification to ISO 14001, just as it will continue long after certification to ISO 14001.

EXERCISES

1) Modify the information collection form in Table 1 (page 179) so it works better for your enterprise.

2) Adapt the IER report structure presented on page 185-186 to a structure that you think would be more suitable for your enterprise.

3) Develop a full gap analysis report structure for your enterprise, using the brief discussion on page 187 as a starting point.

4) Select one element of ISO 14001 and break it down into detailed implementation tasks for implementation planning purposes.

5) Devise a table for planning out the implementation process. It will require space for descriptions of discrete implementation tasks for each EMS element, for who is responsible for carrying out each task, and for the scheduling, resources allocated, and outcome of each task.

6) Develop a recommendation for an initial EMS scope, to present to your enterprise's top management.

7) Develop a recommendation on whether or not to pursue an EMS that conforms to the ISO 14001 standard from the outset, to present to your enterprise's top management. Justify your recommendation.

8) Simplify or expand the limited-scope initial EMS in Table 3 (page 189) to a model that would be appropriate for your enterprise.

Road Map to Part II: POLLUTION PREVENTION

The focus of the second half of this book is on pollution prevention, or P2. The subject is covered in four chapters. Collectively these chapters have the following objectives:

1) If your enterprise is not doing P2 on a regular basis, we want to try and convince you that it should have a dedicated P2 program
2) To convince you that there are many types of cost-savings benefits associated with P2, but also that pollution prevention should only be implemented when it makes financial sense
3) To show you how to implement a P2 program

The last objective needs some qualification. We maintain that the elements of a successful pollution prevention program (P3) are:

1) Identification of P2 opportunities
2) Proper justification though the use of standard financial indicators
3) A thorough P2 action plan
4) Roll-out throughout every phase of an enterprise's business operations

We cover the first three elements only. The four chapters show you how to identify the opportunities, how to develop proper justification for P2 recommendations, and then how to develop the action plan. Implementation and roll-out are only discussed in passing, simply because once the first three elements are in place, implementation of P3s are no different than any other management- and/or technology-driven project in an enterprise.

The reader will find that the majority of tools needed for the first three elements are contained in only two chapters, Chapters 7 and 8. But there are a number of concepts, background information, and general methodology considerations in the Chapters 5 and 6 that the reader should review before learning to work with the tools that we discuss.

Chapter 5 covers general principles, providing the reader with our point of view and the terminology used in describing pollution prevention. We define P2, discuss it in terms of waste minimization and energy efficiency programs, which are both different from and similar to P2, and we discuss in very general terms why an enterprise like yours should be doing P2. At the end of this chapter, there are questions for thinking and discussing, which will help you orient your thoughts for the more in-depth discussions that follow.

Chapter 6 covers industry-specific P2 practices. The chapter starts by reviewing some of the general principles and ideas behind P2 discussed in the first chapter, but then quickly gets into industry-specific case studies. The authors are firm believers that practical examples are among the best tools for training, but that actual exercises are even better. As such, in this chapter you'll find a synthesis of industry-specific P2 practices in concise tables, as well as exercises at the end of the chapter that will get you thinking about how P2 solutions from industries similar to your own might apply to your enterprise. There are some concepts introduced in this chapter that you are likely not to run across in other P2 publications. These include simple but effective performance-tracking tools, such as the *P2 case-study matrix* and the *P2 benefits matrix*.

Chapter 7, which covers the P2 audit, is in many ways the primary route in your tour through P2. The audit enables you to identify the need for, develop much of the justification for, and begin creating, a P2 action plan. We present a three-phase approach to implementing the audit. This systematic, step-wise approach is presented in great detail, along with sample audit spreadsheets and engineering calculation methods for performing necessary material balance analyses. There are questions for thinking and discussion at the end.

Chapter 8 is organized into seven sections, with each section addressing different but related project-financing tools. This is an important chapter but it must be read only after the steps and methodology of the P2 audit are understood. This last chapter contains tools, calculation methods, and a step-by-step procedure for developing the project-financing justification needed to decide about the merits of a P2 recommendation. The chapter discusses life-cycle analysis, but only in terms of the essential elements needed to evaluate the cost-attractiveness of P2 projects.

An underlying premise throughout all four chapters, and particularly the case when reading Chapter 8, is that we should approach P2 from the standpoint that there are only two categories, namely P2 projects that require little to no investments (those that result from following best practices) and those that require investment. Among the latter category, for which the cost-accounting tools in Chapter 8 apply, we consider P2 recommendations based largely on off-the-shelf technologies. In other words, although some industry situations certainly demand development of entirely new technologies, others merely require *proof of principle*, and this volume focuses on making use of existing methods and equipment. In short, the reader will find that the cost-accounting tools and, indeed, the entire approach to P2 we present assumes that pollution reductions are achievable with standard technologies. It is the application of off-the-shelf equipment and best practices to unique industry settings that constitutes unique P2 solutions.

Chapter 5
POLLUTION PREVENTION: PRINCIPLES & CONCEPTS

THE BASICS OF P2: WHAT IS IT, AND WHY DO IT

Defining Pollution

Although the word *pollution* is self explanatory, different stakeholders view it according to their standpoints. To environmental regulators, pollution is viewed from the standpoint of enforcement, where there are strict fines, penalties, forced shut-downs and interruptions of business operations, and even imprisonment of the CEOs and responsible parties that willfully violate federal, state and local laws. *Environmental legislation* was devised first and foremost to protect the health and well-being of the public, and then to protect our environment; therefore, in industrialized countries, environmental legislation (and the regulations that enforce it) is a matter of national policy, and a focus of governmental agreements between countries.

The impacts of pollution vary depending upon different stakeholder's points of views — regulators, the public, and industry.

To the public, the term is viewed from the same standpoint, as well as from a standpoint of consumer confidence. Why should we, the public, allow an enterprise or a particular industry to get rich by sacrificing the quality of the air we breathe, the water we drink, and the environment in which we live?

Today's captains of industry view pollution from the same vantage points of government and the public, but they also are concerned with the costs of pollution. There are, in fact, a number of costs, including:

- The cost of complying with large amounts of complex and, in some cases, ambiguous legislation
- The cost of non-compliance, because even innocent or unintentional violations of environmental laws can carry with them strict and costly penalties
- Lost revenues and capital from the loss of consumer and investor confidence when environmental damages or catastrophes occur

- High insurance premiums or even loss of insurance policies for the assets and operations associated with enterprises that poorly manage their environmental affairs
- Higher operating and maintenance (O&M) costs, due to the end-of-pipe hardware needed for waste control, treatment, and disposal
- Higher direct operating costs, because pollution-control hardware is integral to plant operations, and requires utilities, sophisticated controls, and a trained and dedicated workforce
- Higher health costs for employees who may potentially face both acute and long-term chronic exposures to pollutants
- Legal costs, as a result of third-party or off-site incidents involving pollution from a plant operation

So, what do all these costs add up to? There are no reliable estimates. According to the "Economy and Environment" portion of the U.S. Environmental Protection Agency's Web site (www.epa.gov/economics), American industry spends about $210 billion annually on compliance with all environmental statutes. On a macroeconomic scale, this represents only about 2.1 percent of U.S. gross domestic product, which the EPA argues is minuscule. At even at twice this amount, EPA argues, complying with environmental regulations represents only small operational changes to the way an enterprise handles its overall operations.

Industry representatives disagree, countering that the costs for meeting today's environmental challenges in this country alone are staggering. They point out that EPA's estimates of the cost of compliance only account for direct costs, not for indirect costs, which can vastly increase the cost of compliance.

> ***The reason for doing pollution prevention is to eliminate operating costs and liabilities. Pollution has direct, indirect and hidden costs that can affect profits and even sustainability of enterprises.***

A number of the cost factors we cite in this book are indirect and, in some cases, may result in social/economic impacts that go far beyond individual enterprise profits. For example, what about plant operations being shut down because of environmental problems, throwing thousands of workers onto the unemployment ranks? Each of the costs of environmental compliance can collectively or separately have a dramatic negative impact on an enterprise's bottom line and, indeed, on the sustainability of its business operation.

Because every enterprise's objective is to make a profit, and to maximize that profit, it is a matter of common sense to eliminate or minimize costs. Practicing pollution prevention (P2) and applying an environmental management system (EMS) to its operations enables a business to cost effectively manage its environmental affairs and, ultimately, to add to its bottom line.

Defining Pollution Prevention

Webster's Dictionary defines *prevention* as "the act of keeping something from happening." P2 practices are activities aimed at keeping an operation from polluting. Examples of P2 practices include recycling wash water in the bottle-cleaning stages of a beverage bottling plant, and converting a coal-fired electricity generating utility plant to one that burns natural gas. In the first case, a polluting stream (wastewater) is reused in the process operation, thereby reducing the overall discharge from the plant. In the other case, both a technology change and a feedstock substitution reduce the pollution.

In the case of the coal-fired electric utility plant, air pollution emissions include unburned hydrocarbons (UHC), particulate matter (PM), nitrogen oxides (NO_x), sulfur oxides (SO_x), carbon dioxide (CO_2), and mercury. In contrast, the combustion of natural gas only creates NO_x and lower amounts of CO_2 (there are other pollutants, depending on the specific gas turbine technologies used).

By converting the plant to natural gas, we displace the use of coal and eliminate the hazardous air emissions associated with its combustion — but we need to invest in a new technology. The new process is generically referred to as "clean production," simply because it is inherently cleaner or less polluting, and the technology is referred to as "green technology."

The potential direct savings and benefits from P2 include:

- ***Raw materials***
- ***Energy***
- ***Product recovery***
- ***Valuable by-products***
- ***Reduced pollution fees***
- ***Improved product quality***
- ***Higher throughputs and productivity***

In both P2 examples, there are cost savings carried along with the P2 activities. In the case of the bottling plant, there will be savings associated with lower treatment costs of the wastewater in an off-site municipal wastewater treatment facility, savings from the use of less raw feedwater, and savings derived from the use of less make-up wash chemicals.

In the case of the electric utility operation, the savings categories include lower direct and O&M costs, because the P2 measure eliminates the need for such air-pollution control devices as electrostatic precipitators, cyclone separators, baghouses, and NO_x and SO_x scrubbers; savings from not having to treat and dispose of wastewaters and solids wastes from gas scrubbing and coal combustion operations;

and lower energy costs, because, when considered over its full cycle of use, natural gas is more economical and has a higher burn efficiency than coal.

As indicated by these two examples, the cost savings can go beyond the list of compliance-related financial losses associated with the act of pollution itself. There can be raw materials savings, energy efficiency savings, materials substitution savings, and — as we will show by way of examples later on — there can be savings associated with improved product quality and higher levels of productivity and throughput. In some cases, pollution streams can have valuable by-products that can be recovered, recycled, or reconstituted and sold into secondary markets, thereby generating new revenue streams.

A good P2 program is built around a set of well-planned activities that involves re-engineering the way an operation works, so that pollution does not occur or is at least kept to a minimum. By doing this, the enterprise saves money. The re-engineering can take the form of process changes, technology changes, operational changes, and, in very simple cases, nothing more than good housekeeping practices.

WHAT POLLUTION PREVENTION IS ALL ABOUT

P2 is quite simply the act of eliminating or minimizing the production of pollution at its source. Because this production costs an enterprise money, there should be financial rewards. This is not to say that all P2 activities carry with them savings that pay for the re-engineering. The old saying is, if you see a dollar bill on the ground, you'll probably bend over to pick it up — but not if you have to break your back to do it. The same rule of thumb applies to P2. P2 makes sense when it carries along with it cost benefits that are attractive enough for an enterprise to change the way it currently operates. To illustrate this further, let's return to the example of the coal-fired utility plant conversion.

Sorting through Tradeoffs and the Objectives of P2

In the electric utility example, the incentives for investing in natural gas-fired plants are both environmental and economic. From the environmental standpoint, both smog precursors and climate-change species production (or "greenhouse gases") can be directly attributed to coal use, and hence investments into clean technologies are a matter of global sustainability.

From an economic standpoint, electricity production from a new natural gas plant versus a newly built coal-fired plant heavily favors natural gas, because natural gas-fired plants are cheaper to build. However, older coal-fired plants built 20 to 30 years ago are often more profitable than newly built gas plants, because — among other reasons — coal is cheaper than natural gas, and the older plants have long since paid for their capital investments.

Emissions-control requirements can also be less stringent when it comes to older coal-fired plants. In the United States the great majority of NO_x emissions from the power sector come from existing coal-fired plants, many of which were built between 1950 and 1980. Due to a process known as "grandfathering," most existing plants are subject only to NO_x standards imposed by Title IV of 1990's Clean Air Act Amendments, which only requires the installation of low-NO_x boiler technology or the equivalent. The standards were set at 0.45-0.50 pounds per million British thermal units, or MMBtu (equivalent to 1.5–1.7 lb/MW-hr) for wall-fired and tangentially fired coal boilers during Phase I of the Title IV program (1996–99), and from 0.40 to 0.86 lb/MMBtu for a wider group of boilers beginning in 2000. However, because companies are legally allowed to average the emissions of all their units, and because some units are exempt in Phase II (i.e., cyclone units less than 155 MW), some units continue to operate without controls while emitting NO_x at levels as high as 2.0 lb/MMBtu (roughly 630 parts per million, or ppm).

The Title IV standards for NO_x produced minor innovation, because they were set at levels that could be achieved through low-cost boiler modifications. The intent of the U.S. Congress on this issue was quite clear, as the law allows units that could not meet the numerical limit though use of low-NO_x burner technology to meet an alternative emission limit set at the lowest level they could reach using this technology. Because the standard simply required the use of an existing and known technology, the only innovations were to reduce the cost of low-NOx burner technology for some boiler types. In contrast, the standards for existing oil- and gas-fired plants are significantly more stringent than those for existing coal plants, as most were built after 1971 and were subject to new-source standards that establish a minimum NO_x emission limit of 0.20-0.30 lb/MMBtu (40 CFR 60.44, 60.44a). New-source standards for new gas-fired facilities are even more stringent — in some states as low as 0.02 lb/MM Btu, or an order of magnitude lower than the levels mentioned above.

There are several problems with these standards for older coal-fired plants that limit a sensible NO_x-reduction policy. First, they are relatively lenient. Because they require only that low-NO_x burners be used at coal-fired plants, relatively inexpensive NO_x-reduction technologies are not being used, because the plants already meet the minimum standards. Second, standards are set on a technology-by-technology basis, and have resulted in more lenient requirements for dirtier technologies. This has created little or no incentive to switch to cleaner processes in the past decade. Thirdly, these standards are input-based, which means they provide no incentive for efficiency within any technology category.

This example leads us to a very important observation. While pollution prevention may seem like the obvious thing to do, it is not always cost effective, and investments into re-engineering processes must be carefully scrutinized. In fact, in this example, we see environmental legislation that actually discourages investments into green technologies.

In the early days of implementing P2 in the United States, all too often activities were implemented without carefully assessing the economic attractiveness (or unattractiveness) of a P2 project. Very often, enterprises implemented P2 programs because it seemed like the right thing to do, yet in practice these implementations would cost an enterprise more than the status quo, or cause the enterprise to ignore other opportunities with a greater potential for cost savings (such as simple end-of-pipe technologies).

We could argue that protecting the environment must take priority over financial savings. After all, investments into clean technologies, such as a gas-fired plant, have global environmental implications, so why shouldn't a utility owner/operator take a little less profit?

The problem with this is two-fold. First, even if the owner/operator of an older coal-fired plant decides to make the conversion, the investments required are so high that the conversion must be done in stages. As a result, the owner/operator will inevitably dispatch electricity more frequently from the dirtier coal-fired portion of the operation, to recover the capital-investment costs of the conversion. This means that the environmental benefits that could be achieved by clean production will not take place. A second problem is that NO_x-emission standards from natural gas plants are being ratcheted down to as low as 2 ppm in many states. This ultra-low emission target is in fact at the extreme limits of today's combined-cycle natural gas technology, and to achieve these limits there are not only tradeoffs in efficiency (making gas turbines less financially attractive than coal-burning technology), but a worsening of emissions because machines produce other pollutants beyond NO_x. The lesson here is that force-fitting P2 into an enterprise or industry sector is simply not the right thing to do.

P2 programs can only be financially justified when they meet the criterion of win-win investments (WWI). A WWI means that the re-engineering not only eliminates the pollution, but the economics are equal to or better than break-even.

This brings us to what the real objectives behind P2 are about — or at least what they should be about. Pollution prevention is a carefully planned investment aimed at reducing an enterprise's operating costs through the elimination of harmful pollution. A successful P2 activity is a win-win type of investment — that is, the enterprise not only eliminates pollution at the source, but does so on the condition that, at the very least, the activity pay for itself and, more favorably, provides attractive financial returns. The re-engineering considered for the pollution reduction and/or elimination must meet a set of well-defined financial goals within the enterprise; otherwise it is not a worthwhile P2 practice.

ESSENTIAL ELEMENTS OF POLLUTION PREVENTION

The coupling of emission-target reductions with attractive financial credits or savings constitutes a pollution prevention program. To implement a P2 program, four elements are brought together. These are *engineering* or *process know-how*; *auditing practices*; *project cost accounting*; and *monitoring/roll-out*. All four are essential elements of a successful P2 program.

Process Know-How

Engineering, or process know-how, requires intimate knowledge and understanding of enterprise operations. The old saying that "there is no substitute for experience" is never truer than here. This part of the P2 formula can't be taught in a course or explained in a textbook. The expertise must be derived from the technologists, the process engineers, and the operational personnel from within the enterprise and, in some cases, with the help of outside consultants and equipment and technology providers. The use of industry-specific case studies can also help. The technical solutions achieved within industry sectors often provide a roadmap and general guidance on the levels of emissions-target reductions and on the cost savings achievable for common problems with certain P2 technologies and practices. This book provides some examples and case studies, but more importantly, it directs

The Elements of a P2 Program:

1) Process Know-How

2) The Audit

- ***2.1) The Protocol***
- ***2.2) Material and Energy Balances***
- ***2.3) Financial Screening***

3) Project Cost Accounting

- ***3.1) Total Cost Accounting***
- ***3.2) Life Cycle Analysis***

4) Continual Monitoring and Roll-Out

the reader to resources that can assist in formulating industry-specific technical solutions for achieving pollution reductions with financial benefits.

The Audit

The P2 audit is made up of three sub-elements. These are the protocol; engineering principles that make use of material- and energy-balance methods; and financial screening. The objectives of the audit are:

1) To identify pollution reduction (and waste reduction/energy savings) opportunities
2) To quantify the emissions reductions that different re-engineering options are likely to achieve
3) To screen technologies and practices on the basis of financial attractiveness, to recommend a cost-effective P2 activity

A P2 audit goes far beyond problem definition. Its objective is analysis — namely, it provides an evaluation of the pollution problem, provides options for preventing pollution, and defines for management what the options will cost.

The audit procedure is rigorous, and in this book we explain a step-by-step procedure that all P2 audits can follow. There is sufficient flexibility in the procedure for enterprises to adapt it to specific practices, but we feel that the steps outlined in this book have been tried and tested in P2 programs throughout the world, and therefore represent a good protocol for most organizations to follow.

Project Cost Accounting

Project cost accounting uses total cost accounting to develop the financial justification that management needs to champion an activity.

P2 projects compete for budgets in the same arena as other capital investments in an enterprise's future operations. As such, a proposed P2 activity must not only make financial sense, but it must be competitive with other investment priorities.

We recommend that all P2 investments be organized into three investment categories: low-cost/no-cost, moderate, and high. The division between these categories is subjective, and really depends on an individual enterprise's overall business views. To facilitate discussions in this book, we assume two threshold levels. For small business operations (let's say, enterprises with gross yearly revenues of less than $10 million), the low-cost/no-cost investments are those less than $ 25,000; moderate-cost investments fall between $ 25,000 and $ 250,000; and high-cost investments are those that cost more than $250,000. For large businesses (enterprises with annual gross revenues of more than $10 million), the thresholds we discuss in the book's case studies and examples assume that low-cost/no-cost investments cost less than $50,000; moderate-cost investments are those up to $500,000; and high-cost investments fall above the $500,000 mark.

To properly determine the cost of a project, we first need to establish a baseline for comparative purposes. If nothing else, a baseline defines for management the option of maintaining the status quo. Changes in material consumption, utility demands, staff levels, and other options being considered can be measured against the baseline.

In this book, we follow the methodology of R.T. McHugh, author of *The Economics of Waste Minimization* (McGraw-Hill Book Publishers, 1990). McHugh defines four tiers of potential costs that are integral to pollution prevention:

- *Tier 0:* Usual or normal costs, such as direct labor, raw materials, energy, and equipment
- *Tier 1:* Hidden costs — examples include monitoring, reporting and record keeping, permitting requirements, environmental impact statements, and legal expenses
- *Tier 2:* Future liability costs — examples include remedial actions, personal injury under OSHA regulations, and property damage
- *Tier 3:* Less tangible costs — examples include consumer response/confidence, employee relations, and corporate image

Tier 0 and Tier 1 costs are direct and indirect costs. They include the costs of engineering, materials, labor, construction, any contingencies, etc., as well as waste collection and transportation services, raw material consumption (increase or decrease), and production costs. Tier 2 and Tier 3 represent intangible costs. Much more difficult to define, they include potential corrective actions under the Resource Conservation and Recovery Act (RCRA); possible site remediation at third-party sites under Superfund; liabilities that could arise from third-party lawsuits for personal injury or property damages; and the benefits of improved safety and work environments. Though these intangible costs often cannot be accurately predicted, they can be very important and should not be ignored when assessing the merits of a proposed P2 project.

Monitoring and Roll Out

Rarely can one expect significant savings from a single activity. A P2 program won't really be effective unless its principles can be rolled out to all parts of an enterprise's operation. To do this, the enterprise needs to highlight the benefits from successfully implemented activities. Monitoring the benefits achieved by P2 activities provides the momentum needed to identify opportunities in other operations and to develop P2 strategies.

Monitoring the benefits and implementing corrective actions to maximize the benefits derived from a P2 activity enable an enterprise to develop a record of lessons learned from the activity. This can then become a model by which the enterprise can integrate P2 into all aspects of its business practices.

WHAT ABOUT MATERIAL WASTE AND ENERGY?

There are still many enterprises that view pollution control as simply a normal operational cost — that is, pollution control and compliance are simply among the normal costs of doing business. In these enterprises, P2 practices are sometimes performed on a very small scale and are implemented inconsistently. P2 projects are generally given the same low priority as any other environmental project.

This leads us to modifying our definition of P2 for those enterprises that operate in countries where the direct costs of environmental compliance do not provide sufficient financial incentives, beyond the occasional P2 opportunity.

Most people equate pollution with the terms harmful and toxic; they see pollution as harmful to the environment and poisonous. For the most part, that is true. But from the industrial stakeholder's point of view, pollution is a valueless by-product — or, more simply, it is waste. This valueless by-product can be a waste of materials, a waste of energy, or a waste of staff time. Yet, as we know, waste is not valueless. All forms of waste represent lost revenues.

What is Waste Minimization?

When we view pollution as a waste of potential resources, then the reasons for investing in P2 practices become clearer, even to those enterprises that don't believe that environmental compliance is a high priority for their organization. A simple example of this is a story from one of our fathers, who was a famous pollution control engineer.

In the early 1970s, he worked for a precious-metal refining operation where a part of the process generated PM air emissions, which were viewed as more of a nuisance to the surrounding community than a harmful and regulated air emission. The discharge was collected for 10 years in a baghouse operation and, once every month, the company collected the PM and sent two to three truckloads to a local municipal landfill.

It cost the company about $70,000 per year to operate the baghouses, transport the waste, and deposit it in the landfill. Then, an analytical test showed that the PM contained a high percentage of platinum, worth about $500,000 in material losses per year in today's dollars. Over 10 years, the enterprise lost close to $6 million (the lost platinum, plus the O&M and direct costs for pollution controls and disposal). These losses would not break this large enterprise, but imagine 10 or more such losses in various stages of the plant operations. Now the financial losses are not $6 million but $60 million! So, regardless of whether or not there is a strong regulatory climate, industry already has incentives to apply P2.

Some enterprises may embrace the term "waste minimization" more easily than P2. The term "minimization" encompasses avoidance of waste generation when practical, as well as the productive utilization (recycling) of any wastes that are generated. Waste minimization is simply one of a number of related terms and concepts that, despite having similar overall goals and often being used interchangeably, may differ significantly in basic principles and in emphasis. It is therefore important to have clear definitions that establish why enterprises need to make such investments.

As discussed above, waste minimization comprises avoidance and utilization. Avoidance refers to actions by the producer to avoid generating waste. Utilization includes the range of actions that make the waste a useful input to other processes, eliminating the need for its disposal. Processes that reduce the toxicity or potentially harmful impacts of a waste can, in some cases, be regarded as minimization, although in other circumstances such changes represent treatment before ultimate disposal.

Although the terminology used may vary, we can distinguish a number of important activities. *Reuse* refers to the repeated use of a "waste" material in a process (often after some treatment or makeup). *Recycling* refers to the use by one producer of a waste generated by another, or reuse of a waste as a raw-material component within an existing manufacturing process. *Recovery* is the extraction from a waste of components that have value in other uses.

With few exceptions, most enterprises are concerned to one degree or another with sustainable development. Waste avoidance and utilization can be viewed as part of a broader hierarchy of approaches to achieving sustainable development. At the highest level of the hierarchy are approaches that seek to satisfy human needs and requirements in ways that do not waste resources or generate harmful by-products or residuals. These approaches include changing consumer behavior, and reexamining the range and character of the products and services produced. At a slightly lower level are efforts to redesign products and services, and to raise consumer awareness about the impacts of their decisions.

Application of techniques such as life-cycle analysis (LCA) is part of the difficult process of determining the overall impacts of products and services on the environment. For the most part, industries in more-advanced countries adopt such approaches, while industries in developing countries focus on improving production processes. These approaches include cleaner production, pollution prevention, and waste minimization (all of which are more or less related to better management), as well as improvements in production processes, substitution of hazardous inputs, reuse and recycling of wastes, and so on. Enterprises in highly developed countries also focus on improving production processes, but perhaps to a lesser extent. This is especially true in some industry sectors — for example, many heavily polluting U.S. industries made the investments to green technologies more than a decade ago.

Looking at these concepts in terms of a hierarchy, the next step — which should be minimized but certainly not neglected — is treatment and proper disposal of wastes. The lowest level in the hierarchy, and the one that all the other levels strive to eliminate, is remediation of the impacts of wastes discharged to the environment. Cleanup is simply costlier than prevention.

There are differences in philosophies between enterprises and even countries about what really constitutes waste minimization and pollution prevention. Consumers in some of the wealthier countries are moving toward a greater awareness about the need for waste reduction, as shown by participation in recycling schemes and the demand for environmentally friendly products. However, progress is often slow, and there is a need for ongoing education and awareness, as well as for careful analysis of options and incentives. In developing countries, the demand for resources often leads to significant recycling of materials such as glass, metals, and plastics. These recycling systems have important social and economic consequences at the local level, and their improvement must be approached with care.

On a microeconomic scale — the scale that really matters for each enterprise — there are not great differences between P2 and waste minimization, because they both share one common objective: to reduce financial losses.

What is Energy Efficiency?

Efficient use of energy is one of the main strategic measures not only for the conservation of fossil energy resources but also for abatement of air pollution and the slowing down of anthropogenic climate change. Accordingly, economic and technical measures to reduce specific energy demand should be priorities across all sectors of an economy. Many opportunities exist for improving efficiency, but progress has been disappointingly slow in many cases. The phrase "efficient use of energy" includes all the technical and economical measures aimed at reducing the specific energy demand of a production system or economic sector. Although implementation of energy-saving techniques may require initial investments, short-term financial returns can often be achieved through lower fuel costs due to the reduced energy demand.

Industrial production processes often show a high specific energy demand. In the United States, industry is estimated to account for between 25 and 35 percent of total final energy consumption. Although great progress has been made in the rational use of energy in the industrial sector during the last two decades, improvements in cost-effective energy utilization have not nearly been exhausted. This holds true for both new and existing plants. Improvement in energy end-use efficiency offers the largest opportunity of all alternatives for meeting the energy requirements of a growing world economy.

Many of the technical options for energy saving require only small investments and are easy to implement. In a number of cases, even simple organizational changes bring about considerable energy savings, yielding not only environmental benefits but also financial returns. Energy-saving measures often show very short payback times, especially in industrial applications. However, as in the case of cleaner production, it is often difficult to generate management interest in, and support for, the identification and implementation of energy-saving measures. Without such support, success is almost always limited.

The first step in identifying the energy-saving potential within an enterprise — say, an industrial facility — is to carry out an *energy audit*, taking into account the specific conditions at the plant and the local conditions at the production site. The energy audit, which we consider to be a subset of the P2 audit, is required to determine the scope of the energy efficiency project, to achieve a broad view of all the equipment installed at the production site, and to establish a consistent methodology of evaluation. Preparation of an improved energy utilization scheme starts with an inventory of the equipment, its energy demand, and the flow of energy through the plant. Electrical energy and heat should be recorded separately, and energy demand should be charted over time, to determine patterns of usage.

A few key areas can be identified on which to focus conservation efforts:

1) Electricity production typically requires three times as much primary energy as direct heat use. Therefore, electricity should only be used if it cannot be replaced by other, more direct energy sources.
2) The chemical energy contained in fuels should be utilized as efficiently as possible. When a process or an industrial plant uses combustion processes to meet energy demand, it should achieve high combustion efficiencies by utilizing as much as possible of the thermal energy contained in the flue gases; by minimizing heat losses (through, for example, use of insulation); and by recovering the thermal energy contained in combustion by-products, such as ashes and slag.
3) The enterprise should pay special attention to separation processes for recovering and purifying products, which account for up to 40 percent of the total energy demand of chemical processes. For example, industrial

> facilities can achieve energy savings of 10 to 40 percent through heat integration of the reboiler and the condenser of distillation columns, by using heat pumps or water-compression systems. In several applications, it may also be possible to replace the common but very energy-intensive distillation process with advanced separation processes, such as membrane processes, that significantly reduce energy demand.

The first step in breaking the energy-environment link is to capture the opportunities for reaping environmental benefits through economically attractive solutions at no additional cost (i.e., the no-cost/low-cost measures). These opportunities include, at the very least, improvements in energy efficiency on the supply and demand sides, and a switch to less-polluting energy sources. These can be referred to as "win-win" measures and can go a long way toward reducing local environmental degradation.

HOW ARE EMS AND P2 RELATED?

The reader might very well ask why the subjects of environmental management systems and pollution prevention are treated in a single volume. Indeed, much of industry often views P2 practices as a separate and perhaps small component of an EMS. This is wrong, because the goals of any EMS are to cost-effectively manage environmental affairs and to continually achieve greater degrees of environmental performance. P2 programs are critical to accomplishing these goals.

For those readers that are embracing the subject of EMS for the very first time, and who have skipped ahead to this part of the book, we recommend that you spend a few moments and read the first three to five pages of Chapter 1. These pages provide a succinct description of what an EMS is, and why enterprises need to do it. The reasons for doing an EMS are practically the same as those for doing P2 when enterprises face high costs for environmental compliance. But there are some important distinctions that should be made between P2 and EMS.

First, enterprises can clearly practice P2 without having a formal EMS, although the effectiveness of such programs may not be as great. In contrast, with perhaps only a few special cases in industry, having an EMS without a P2 program integrated throughout an organization makes little sense, simply because the underlying objective of EMS is to achieve a pollution-free operation.

The next question one might ask is what comes first — the EMS or P2? This is a "chicken and egg" question and, indeed, the authors decided the subject-presentation sequence over the toss of a coin. One can argue that P2 defines the financial incentives for an organization investing in developing and implementing a formal EMS, and obtaining certification for it. We may also equally argue that many enterprises can justify the investments required for a formalized EMS, recognizing that P2 is a component essential to effectively implementing it. Therefore, P2 programs evolve as a part of the development of an enterprise's EMS, and vice versa.

Finally, we should point out that, although the two subjects are intimately related, they do not necessarily carry equal interest to the senior managers of enterprises. An EMS, as the term implies, is a system that encompasses very specific guidelines on managing the complex and varied environmental components of a business operation. P2 is not a system. Rather, it is a collection of engineering and financial tools that help an enterprise achieve certain environmental performance goals. To management, P2 is but one of the means by which improved environmental performance can be achieved. EMS focuses on strategic management, and a systems approach; P2 focuses on technologies and best practices, which we refer to as the re-engineering of existing operations.

ONE MORE TIME

This chapter has provided the reader with a broad overview of the elements, terminology, and objectives of pollution prevention, or P2. Part of a sound environmental management system is pollution prevention. With few exceptions in industry, an EMS is simply not complete, or will not work properly, without a substantial P2 component. In fact, pollution prevention is a cornerstone of the ISO 14001 standard.

It is becoming more recognized that significant reductions in pollution can often be achieved with low-cost/no-cost investments. The efficient use of resources and the reduction in waste are obviously preferable to reliance on costly end-of-pipe treatment. In the United States, as well as in European Union countries, strong enforcement of environmental regulations was the initial driving force behind industry acceptance of pollution prevention. But there are many parts of the world today where regulatory enforcement is weak. In these situations, the true driving force for P2 programs are likely the non-environmental economic incentive — that is, the savings in raw materials and energy. Unfortunately, many enterprises do not apply dedicated P2 programs, while others force-fit P2 activities to their enterprises for the wrong reasons.

In the following chapters, we will explain the elements that go into establishing and maintaining a P2 program. Where possible, we've included examples from typical industrial settings. Our objective is to provide a step-by-step methodology for managers to save money by reducing pollution and waste. These savings will improve profitability, ensure future sustainability, and enable future investments into green technologies.

QUESTIONS FOR THINKING AND DISCUSSING

1) Does your enterprise have a deliberate organized program that focuses on P2 opportunities? If not, why not? If yes, what motivated top management to put it in place?

2) Does your enterprise have regular meetings or taskforces that focus on P2 opportunities? Have these meetings or taskforces ever been tried before, and what kind of success did they have?

3) What are the three essential elements of a successful P2 program? Describe the objectives of each.

4) What are the three sub-elements of the P2 audit?

5) What is the importance of LCA? How could LCA help your enterprise?

6) What are the Tier 2 and 3 costs that your enterprise absorbs because it does not have an effective P2 program?

7) Does your enterprise take corrective actions aimed at reducing pollution, or does it monitor and react to environmental problems? Which do you think is a better approach to managing environmental issues and why?

8) How many federal environmental statutes must your enterprise comply with? List them in accordance with the business operations they impact.

9) What does it cost your enterprise to comply with applicable environmental regulations? Which environmental regulations cost your enterprise the most?

10) Has your enterprise ever conducted an energy efficiency audit? If so, what corrective actions or programs did your enterprise put in place to save energy, and what results did they achieve? What environmental credits did your enterprise achieve from the energy efficiency activities it implemented?

11) Does your enterprise give priority to proposed P2 projects over normal pollution-control projects?

12) Has your enterprise been faced with a history of environmental litigation, fines, penalties, or corrective actions? How many of these have occurred over the last five years, and what were the reasons?

13) Is your enterprise a member of any of the EPA's voluntary programs on P2 or energy efficiency? If not, does management know about these programs and the possible benefits?

14) If your enterprise could reduce its overall emissions in all media categories (air, water, solid wastes) by, say, 5 percent, what would be the yearly savings? What would the savings be for a 10 percent reduction?

Chapter 6
INDUSTRY-SPECIFIC POLLUTION PREVENTION PRACTICES

DEFINING P2 ONE MORE TIME

In the simplest of terms, all outputs from a manufacturing facility fall into two categories — they are either product or waste. Anything that the customer pays for is product; everything else that leaves the facility is simply waste — whether it is regulated by environmental legislation or not. Ideally, enterprises should not produce any waste. In reality, enterprises must strive to reduce the waste from operations, because this represents an inefficient use of scarce resources. We can argue that *all* waste can be indirectly associated with pollution, because the management of waste consumes resources that would not otherwise be used, and pollution is often generated in these waste management activities.

Prevention is the act of taking advance measures against something possible or probable. Prevention is different than control or cure. For example, if an enterprise builds quality into its products, then it can *prevent* defects. Inspection, on the other hand, enables an enterprise to *control* defects. Hence, *quality*, *inspection*, and *prevention* are key terms and concepts in implementing P2 practices.

In general, the effort, time, and money associated with prevention is less than that associated with control or cure. This idea is captured in the maxim, "an ounce of prevention is worth a pound of cure." Thus, in many cases, it is worthwhile for an enterprise to prevent pollution rather than control it. In a nutshell, P2 is any practice that:

- Reduces the amount of any hazardous substance, pollutant, or contaminant reentering any waste stream or otherwise released into the environment prior to recycling, treatment, and disposal
- Reduces the hazards to public health and the environment associated with the release of such substances, pollutants, or contaminants
- Reduces or eliminates the creation of pollutants through increased efficiency in the use of raw materials; or through protection of natural resources by conservation

P2 is any action that reduces or eliminates the creation of pollutants or wastes at the source, achieved through activities that promote, encourage or require changes in the basic behavioral patterns of an enterprise. As noted in Chapter 5, some terms that are often used in connection and even sometimes interchangeably with P2 are cleaner production, clean technology, waste reduction, waste prevention, eco-efficiency, and waste minimization.

In other words, P2 focuses attention away from the treatment and disposal of wastes, and focuses toward the elimination or reduction of undesired by-products. In the long run, because it emphasizes waste minimization and cleaner production, P2 is more cost-effective and environmentally sound than traditional pollution-control methods. P2 techniques apply to any manufacturing process, and range from relatively easy operational changes and good housekeeping practices (also referred to as "best practices") to more extensive changes, such as the substitution of toxic substances with benign ones, the implementation of clean technology, and the installation of state-of-the-art recovery equipment. P2 can improve plant efficiency, enhance the quality and quantity of natural resources for production, and make it possible to invest more financial resources into the economic development of an enterprise.

HIERARCHY OF POLLUTION MANAGEMENT

Pollution management involves several strategies for dealing with wastes or pollution. Strategies that reduce or eliminate wastes before they are created are preferable to those that deal with treating or disposing wastes that are already generated. The hierarchy of these strategies, from high to low, are:

- *Prevention.* The best waste reduction strategy is one that keeps waste, or pollution, from being formed in the first place. Waste prevention may in some cases require significant changes to processes, but it can provide the greatest environmental and economic rewards.
- *Recycling.* If waste generation is unavoidable in a process, then an enterprise should pursue strategies that minimize the waste to the greatest extent possible, such as recycling and reuse.
- *Treatment.* When wastes cannot be prevented or minimized through reuse or recycling, enterprises should pursue strategies to reduce their volume or toxicity through treatment. Though end-of-pipe strategies can sometimes reduce waste, they are not as effective as preventing waste in the first place.

- *Disposal.* The last strategy to consider is alternative disposal methods. Proper waste disposal is an essential component of an overall environmental management program; however, it is the least effective technique.

Table 1 provides examples of practices in the hierarchy.

Table 1. Pollution Management Hierarchy Examples.

Priority	Method	Example	Applications
1	Prevention (Source reduction)	✓ Process changes ✓ Design of products that minimize environmental impacts ✓ Source elimination	✓ Modify process to avoid/reduce solvent use ✓ Modify product to extend coating life
2	Recycling	✓ Reuse ✓ Reclamation	✓ Solvent recycling ✓ Metal recovery from spent bath ✓ Volatile organic-compound recovery
3	Treatment	✓ Stabilization ✓ Neutralization ✓ Precipitation ✓ Evaporation ✓ Incineration ✓ Scrubbing	✓ Thermal destruction of organic solvents ✓ Precipitation of heavy metals from spent plating bath
4	Disposal	✓ Disposal at a facility	✓ Land disposal

WHY AND HOW IS P2 DONE?

Why Pollution Prevention Is Practiced

We have already explained why P2 should be done. But the general statements about the advantages of implementing P2 at an enterprise are not necessarily convincing. A few practical examples will help to illustrate why and how enterprises implement P2.

It is important to first define the objectives of a P2 activity. In reality, P2 does not necessarily pay in all situations. The audit process (covered in Chapter 7) helps enterprises identify and quantify a P2 opportunity. When an enterprise combines the results from an audit with the financial tools for project assessment, sometimes including life-cycle analysis (LCA) tools, then it can make a critical judgment as to whether or not project opportunities are justified from a business standpoint.

A very critical point to always bear in mind is that P2 is not a replacement for modernization investments or industry rationalization issues. For example, some heavily polluting industry sectors must make major investments not only into cleaner technologies, but into more efficient technologies, simply to sustain their operations. The iron and steel industries in Russia and Ukraine are is one example of this.

On the whole, P2 can be thought of as a means to developing incremental, yet important, savings. In some cases, a single P2 project will not produce enormous savings. But even then, when you add up incremental savings from a number of P2 projects, you can begin to see a long-term positive impact on business operations and profitability.

Therefore, pollution prevention needs to be thought of as a form of corporate religion, where the true believer practices the philosophy on a daily basis. If we are just occasional churchgoers, then we will be less likely to see a significant benefit to our lifestyle.

Pollution Prevention in Auto Painting

Let us start off by examining a simple example of a P2 practice. It is common for auto companies to change paint color with each car that goes through the paint process. As a result, old paint must be purged from the lines before painting each car. This results in excess paint sludge waste and fugitive emissions of toluene and xylene, which a company must use a scrubber — an end-of-pipe solution — to capture. In addition, the purging and refilling qualifies as a setup activity that adds time to the process.

Block painting — the process of painting batches of like colored cars — is a manufacturing-process change that reduces purged paint sludge and solvent emissions. Block painting not only decreases waste, it also decreases the setup time involved in the process.

There are other technological alternatives. Cars can be painted without toxic toluene and xylene solvents. Electrostatic painting can adhere paint to treated metal. While the scrubber represents treatment, and block painting represents waste reduction, shifting to the electrostatic painting process represents pollution prevention by design.

These are examples of potential P2 solutions that displace post treatment of pollution; however, a comparison of the economic ramifications of the two options, as well as of the status quo are needed to justify the investment in what is needed to change the existing process. The fact that alternative technologies are less polluting may not be enough by itself to convince management of the need for the investment.

Pollution Prevention in the Chemical Industry

The following are summaries of case studies for waste-minimization projects that were funded by the U.S. Agency for International Development, and largely implemented by the World Environmental Center through cooperative grant programs to foreign countries. The chemical industry in Poland is diverse, providing a multitude of chemical products to different domestic and export markets. Many plant operations are old — they lack modern, automated controls and instrumentation and, in general, have enormous spare capacity. The types of chemical companies included in the analysis are listed in Table 2.

To examine the collective benefits of a number of P2 projects, we introduce the use of a *pollution prevention matrix* (P2M). A P2M is a convenient way of tracking the performance of many pollution prevention projects within an organization. It can also serve as a useful management tool in developing P2 action plans for an enterprise (we discuss the *P2 Action Plan* in Chapter 7).

In this case, we apply the P2M to assessing the collective contributions of a number of P2 projects from different chemical companies. Table 3 provides the P2M constructed for this discussion. Company-specific P2 case studies are organized into the matrix. The matrix provides the following information for each company-specific P2 practice:

- The pollution emissions reductions achieved to air, water, and land, on a normalized tons-per-year basis,
- Raw materials savings, in tons per year,
- Dollar investments for each P2 measure,
- The total dollar savings achieved by each P2 practice, derived from reductions in pollution fees, waste processing costs, reduced raw materials consumption, improved product quality and yields, and, in some examples, energy savings, and
- The simple payback period for the investment, in months (the *Payback Period* is the cost of the P2 investment divided by the savings per month).

The matrix provides us not only with a sense of the overall cumulative benefits derived, but also identifies technology areas where P2 opportunities may exist in similar industry operations.

In other words, why reinvent the wheel? If someone else has applied a cost-effective solution to a pollution problem, why not use the same approach? That's one of the benefits to reviewing case studies.

Table 2. Polish Chemical Companies Included in the Analysis.

SPECIALTY CHEMICALS

Kedzierzyn Nitrogen Works, Kedzierzyn — nitrogen fertilizers, adhesive resins, technical gas components for plastics and intermediate products, specialty chemicals, high quality OXY-alcohols.

Boruta S A Dyestuff Industry Works, Zgierz — dyes and intermediate chemical by-products.

Organika Azot Chemical Works, Jaworzno — pesticides, herbicides, and fungicides. Principle chemicals synthesized are organochlorophosphates and copper oxychloride.

POLYMERS

Oswiecim Chemical Works, Oswiecim — synthetic rubbers and latex products, including styrene-butadiene, nitril-butadiene, and various monomers, vinyl plastics, styrene and related by-products, chlorinated polymers, chloroparaffins, esters, alkylbenzene, and detergents.

Boryszew S A Chemical and Plastics Works, Sochaszew — compounding ingredients used in the production of polyvinyl chloride, polypropylene, anti-freezing fluids, polyvinyl acetate, and pyrotechnic products.

Organika-Zachem Chemical Works, Bydgoszcz — plastics, dyes, and foam products that supply the following consumer markets: auto parts, furniture, electrical cable insulation, textiles, and chemical intermediates. Major products are chlorine, epichlorohydrin (EPI), toluenediisocyanate, polyesters, polyurethane foams, polyvinyl chloride insulation compounds, synthetic dyes, soda lye, sodium phosphate, hydrochloric acid, and dinitrotoluene.

COKE CHEMICALS

Blachownia Chemical Works, Kedzierzyn Kozle — chemical products derived from coal and coke.

INORGANIC CHEMICALS

POLCHEM Chemical Works, Torun — chlorosulfonic acids and sulfites.

Bonarka Chemical Works, Krakow — animal feed products, including dicalcium fodder phosphate, which is a granular phosphorous-based animal feed.

HEALTH CARE AND PHARMACEUTICALS

VISCOPLAST SA Chemical Works, Wroclaw — medical plasters, technical tapes and glues, medical adhesive bandages, and polypropylene fibers.

POLFA Pharmaceutical Works, Tarchomin — pharmaceutical dosage units of different drug forms. Main drug groups are antibiotics, insulin, and psychotropic drugs.

The results achieved by the companies sampled are significant. Overall, the total reductions in emissions to air, water, and land were 528,949 tons per year, with raw-materials savings amounting to 1,479,230 tons per year, both which translate into annual savings of $7,184,490. The total cost for these savings (i.e., total capital investment) was $1,479,230, which is a 400 percent simple return on investment (ROI) during the first year. Many of the P2 practices had payback periods ranging from immediate to fewer than three months.

Table 3. Pollution Prevention Matrix of Company-Specific Case Studies.

	Emissions Reductions (t/yr)			Mtls. Savings, t/yr	Invest-ment U.S. $	First Yr. Savings U.S. $	Pay-back months
	Air	Water	Solid				
COMPANY-SPECIFIC CASE STUDIES IN SPECIALTY CHEMICALS **Kedzierzyn Nitrogen Works Plant**							
Pollution Prevention Measure: Operational change to absorption towers resulted in collecting formaldehyde and methanol vapors that were normally released as air emissions. These wastes could be recycled, reducing purchase of raw materials.							
	27.5			27.5	0	19,600	-
Pollution Prevention Measure: Direct venting of methane was stopped and the gases sent to a fuel-gas collector during plant turnarounds. Methane losses of 177,000 cubic meters were stopped and reused as fuel.							
	31.2			31.2	0	19,800	-
Pollution Prevention Measure: Condensate discharges from cooler segments, normally sent to a sewer, were recycled. This also reduced freshwater feed purchases.							
		216,891		216,891	1,400	30,500	0.6
Pollution Prevention Measure: Replaced pressurized steam at 200 degrees C with steam condensate at 135 degrees C in venting-system applications. Decreasing the venting-system temperature caused vapor pressures and, hence, emissions of volatile organic compounds (VOCs) to decrease.							
	35.2				2,000	55,700	0.4
Pollution Prevention Measure: Evaluation of the post-reaction cooling process showed that installation of a liquid seal-type degasifier could enable condensate to be safely discharged to recirculation waters, and prevent post-reaction gases from being discharged to air. This also reduced freshwater feed to recirculation waters.							
		7,817		7,817	2,260	2,880	9.4
Pollution Prevention Measure: A new procedure was developed in the recovery of formalin from a urea adhesive plant, eliminating 350 cubic meters of hazardous wastes.							
			733		2,700	71,400	0.5
Pollution Prevention Measure: Wastewaters from the urea synthesis plant were found to be acceptable for recycling for process applications, thereby reducing wastes and raw-water use.							
		500		500	18,900	106,000	2.1
Pollution Prevention Measure: Operating procedures were changed in the urea production plant, reducing ammonia losses to the atmosphere. This was accomplished by installing a pump to transfer condensate, under pressure, from a cooler directly to the urea absorption tower. This enabled condensate to be used in place of ammonia water, which previously had to be produced.							
	1,000			1,831	40,100	89,000	5.4
Pollution Prevention Measure: A study of the heating system showed that combustion efficiency could be improved by installing a carbon monoxide/oxygen-monitoring instrument, resulting in a savings of 250,000 cubic meters per year of fuel gas and reductions in emissions of NOx and CO.							
				44.1	16,300	25,100	7.8

	Emissions Reductions (t/yr)			Mtls. Savings, t/yr	Invest-ment U.S. $	First Yr. Savings U.S. $	Pay-back months
	Air	Water	Solid				
Subtotals	1,094	225,208	733	227,142	83,660	419,980	
Pollution Prevention Measure: Application of a spectrophotometer enabled COD analysis on wastewaters to be made rapidly and reliably. This information was used by process personnel to adjust process conditions to maximize sodium-sulfate recovery from wastewater. This reduced non-mineral salts in wastewater discharges.							
		150			1,793	27,500	0.8
Pollution Prevention Measure: Application of a spectrophotometer enabled COD analysis on wastewaters to be made rapidly and reliably. This information was used by process personnel to adjust process conditions and run the oxidation process in wastewater treatment, reducing COD discharges.							
		40			1,794	31,200	0.7
Pollution Prevention Measure: Enhancements to a spectrophotometer enabled fast and reliable measurements of sulfites and sulfates in the post-absorption process. This enabled better process control and recovery of sodium dioxide, which is a raw material in the plant.							
		10			1,833	31,800	0.7
Subtotals		200			5,420	90,500	
Organika Azot Chemical Works							
Pollution Prevention Measure: Product recovery was done by redistillation of a part of organic waste during the raw production of DCAP. This also reduced solid waste.							
			1.7	5	800	11,900	0.8
Pollution Prevention Measure: Recycling of wastewaters in various parts of the plant resulted in reduced emissions and freshwater feeds.							
		165,146		165,145	1,050	21,600	0.6
Pollution Prevention Measure: Wastewaters from the second and third stages of product washing (DCAP) were reused in the hydrolysis process and replaced freshwater feed.							
		2			1,100	130,000	1.0
Pollution Prevention Measure: Water monitoring equipment was used to monitor and reduce consumption of hot and process waters at different locations in a plant							
				7,707	5,965	19,500	3.7
Subtotals		165,148	1.7	172,858	8,915	66,000	
COMPANY-SPECIFIC CASE STUDIES IN THE POLYMERS SEGMENT							
Oswiecim Chemical Works							
Pollution Prevention Measure: A new latex-degassing system and interstage latex-heating system by steam improved the efficiency of the process and reduced the content of vinyl chloride monomer (VCM) in air emissions.							
	54			54	31,400	2,100,000	0.2
Pollution Prevention Measure: Problems with latex density measurements were eliminated by use of a nuclear density gauge in the reactor. This improved polyvinyl chloride (PVC) product yield and quality, and reduced VCM emissions to air.							

	Emissions Reductions (t/yr)			Mtls. Savings, t/yr	Invest-ment U.S. $	First Yr. Savings U.S. $	Pay-back months
	Air	Water	Solid				
Subtotals	54.01			54	41,100	2,134,800	
Boryszew Chemical and Plastics Works							
Pollution Prevention Measure: By replacing acetic acid used in the production of vegetable oil with formic acid, wastewater discharges became more biodegradable, resulting in a reduction in pollution fees.							
		116			8,600	165,000	0.6
Pollution Prevention Measure: A series of low-cost measures were undertaken to improve process yields and distillation efficiency, resulting in reduced wastewater discharges.							
		11			20,000	48,100	5.0
Subtotals		127			28,600	213,100	
Organika-Zachem Chemical Works							
Pollution Prevention Measure: Allyl chloride present in chloroorganic wastes were found to be recoverable in an existing distillation column, and reused in the production of EPI							
	24				1,250	13,750	1.1
Pollution Prevention Measure: Introduction of a surface-active agent to a dye production process, along with process parameter changes, resulted in reduction in raw materials consumption, improved product quality, and reduced chemical oxygen demand (COD) discharges to wastewaters.							
		2.1			2,500	15,000	2.0
Pollution Prevention Measure: A cost-effective method for the recovery of calcium salts from process wastewaters and reprocessing into calcium hydroxide (lime) was found. Lime used as a raw material in hydrolysis of EPI.							
		208		83,349	7,200	109,200	0.8
Pollution Prevention Measure: This project was aimed at reducing steam and water use in toluenediisocyanate (TDI) production. By replacing a steam-water ejection system with a mechanical vacuum pump, a process solvent used as the pump's sealant absorbs product and reactants from the distillation columns, and is recycled back to the TDI synthesis reactors. Results include reductions in discharges to sewers and the recovery of valuable products.							
		135		341,542	99,800	258,000	4.6
Pollution Prevention Measure: The application of a plate-type heat exchanger at the EPI plant enabled recovery of heat energy from hot process wastewaters. The heat is now used to preheat raw materials sent to distillation units, achieving savings in steam and energy.							
	21.5			12,000	120,000	252,000	5.7
Pollution Prevention Measure: The application of a special evaporator enabled more-efficient TDI product recovery from after-distillation tars. This resulted in product yield improvements and decreases in waste-tar generation.							
			0.02		622,000	1,055,000	7.1
Pollution Prevention Measure: Application of a spectrophotometer enabled more-rapid measurements of COD in wastewaters. The more-rapid and reliable measurements enabled process changes to be made in TDI product production. This reduced wastewater discharges and improved product yield substantially.							

	Emissions Reductions (t/yr)			Mtls. Savings, t/yr	Invest-ment U.S. $	First Yr. Savings U.S. $	Pay-back months
	Air	Water	Solid				
		70		165,146	4,600	92,000	0.6
Pollution Prevention Measure: The production of dye is a pH-sensitive process. By automating pH measurement and control, the company reduced wastewater discharges and saved raw materials.							
		30		8,200	7,000	180,000	0.5
Pollution Prevention Measure: The application of steam-monitoring devices improved the control of steam consumption and accountability of users in different parts of the plant, reducing steam consumption.							
	7			3,900	12,100	84,000	1.7
Subtotals	52.5	441	0.02	614,137	876,450	2,058,950	
COMPANY-SPECIFIC CASE STUDIES IN COKE CHEMICALS **Blachownia Chemical Works**							
Pollution Prevention Measure: Compressed wheat substituted for coke as a filter media in the coke-tar recovery process, reducing operating costs and eliminating over-fired coke waste.							
			130		0	25,700	-
Pollution Prevention Measure: Reuse of washwater in the cooling-water loop of the carbochemical plant reduced water discharges and raw-water consumption.							
		12,551		12,551	400	53,000	0.1
Pollution Prevention Measure: The company made equipment modifications that decreased naphthalene losses from a naphthalene-oil distillation unit.							
	1.32			800	900	18,500	0.6
Pollution Prevention Measure: The installation of an aeration system in wastewater treatment enabled trace levels of organics to be removed from wastewater and be combusted in an existing catalytic burner.							
	600				11,500	255,000	0.5
Pollution Prevention Measure: Application of a gas chromatograph improved yields in the production of benzene, toluene, and light resin, and reduced the consumption of sulfuric acid.							
				0.01	25,200	157,000	1.9
Subtotals	601	12,551	130	126,311	38,000	509,200	
COMPANY-SPECIFIC CASE STUDIES IN INORGANIC CHEMICALS **POLYCHEM Chemical Works**							
Pollution Prevention Measure: The company found a solid waste containing high levels of lime. The waste was suitable as a raw material in a liquid-waste neutralization process. The company purchased an automatic pH meter to implement the recommendation.							
			204	10	2,700	3,600	9.0
Pollution Prevention Measure: The company made process modifications in which it could substitute selected liquid wastes for fresh water in a lime-milk preparation step for wastewater treatment. This reduced both direct wastewater discharges and freshwater feed consumption.							
		50,094		50,094	3,850	9,500	4.9

	Emissions Reductions (t/yr)			Mtls. Savings, t/yr	Invest-ment U.S. $	First Yr. Savings U.S. $	Pay-back months
	Air	Water	Solid				
Pollution Prevention Measure: The company redesigned the decarbonation system used to recirculate a portion of decarbonated water back into a reactor. This allowed the facility to reuse decarbonated water in place of raw water for process purposes.							
			2.5	220,234	10,350	14,500	8.6
Pollution Prevention Measure: Installation of automatic process controls for temperature, pressure, and flow improved production yields, reduced raw materials, and minimized wastewater flows.							
		44,500		40,800	250,500	1,100,000	2.7
Subtotals		50,094	44,707	311,139	267,400	1,127,600	
Bonarka Chemical Works							
Pollution Prevention Measure: The company was able to better control process changes through the use of a spectrophotometer. The faster control capability provided with a rapid measurement resulted in less phosphoric acid use, and a reduction in emissions from unreacted acid.							
	1,008			1,044	9,600	247,000	0.5
COMPANY-SPECIFIC CASE STUDIES IN COKE CHEMICALS							
VISCOPLAST SA Chemical Works							
Pollution Prevention Measure: Continuous monitoring of naphtha and steam during the absorption and desoprtion cycles in the production of acrylics and glues enabled the company to optimize solvent recovery operations and steam. The program also reduced fugitive air emissions.							
	88			88	18,500	19,360	11.5
POLFA Pharmaceutical Works							
Pollution Prevention Measure: The project improved operations in a butyl acid regeneration process. By changing operating procedures during the washing operations, washing steps could continue without process shutdowns. The new procedure minimized plugging of heat exchange equipment, and decreased steam and caustic consumption normally used for equipment cleanout purposes.							
		120		120	100	12,000	0.1
Pollution Prevention Measure: Repairs to leaking vacuum distillation equipment and cooling systems decreased solvent losses and improved process control. Additionally, major overhauls to vacuum pump equipment greatly improved yield and minimized losses. The use of an organic vapor analyzer was critical in identifying losses and leaks in the process operations.							
	248	26,203		26,203	51,000	198,000	3.1
Pollution Prevention Measure: The company implemented a plant-wide leak-detection and monitoring program to minimize solvent losses in the manufacture of various pharmaceuticals.							
			204	10	2,700	3,600	9.0
Subtotals	383	26,323		26,458	61,100	298,000	
TOTALS	3,283	480,094	45,571	1,479,230	1,438,745	7,184,490	

A closer look at the achievements of these projects reveals some interesting trends, which we can discern by first defining three *thresholds of investment* (i.e., the cost for a P2 practice). These levels are:

Level Designation	Investment Threshold	Dollar Range
A	Low-Cost/No-Cost	≤ 10,000
B	Moderate	> 10,000 to 50,000
C	Significant	> 50,000

The benefits derived for each investment threshold are summarized by the pie charts in Figure 1. The comparisons provided in Figure 1 show that:

- About 85 percent of the emissions reductions were achieved by low-cost/no-cost measures
- Nearly 56 percent of the materials savings were derived from low-cost/no-cost measures
- Up to 60 percent of the total dollar savings were derived from low-cost to moderate threshold levels of investments

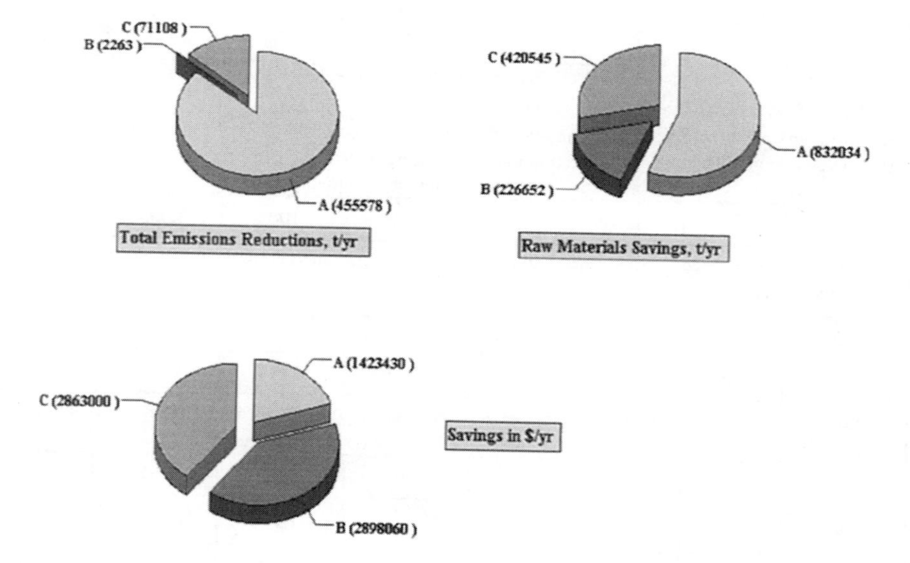

Figure 1. *Compares pollution reductions, raw materials savings, and dollar savings by investments.*

In addition, the simple returns on investment within the first year of implementation were the greatest for low-cost/no-cost investments, as shown in Figure 2. Many enterprises believe that the only way they can achieve more profitable operations, and to compete in international markets, is by making significant levels of investments. Though this may be true in some industry segments and in specific companies, the analysis presented here suggests that a number of small-scale investments in P2 practices can not only achieve significant reductions in emissions, but collectively add up to sizable savings — which, in turn, can be reinvested into modernizing a company's operations.

The reader should not fall into the belief that all P2 activities will result in huge returns on investments based upon small-scale projects or best practices alone. Indeed, in older plant operations such as those encountered in the Polish chemical industry, it is relatively easy to identify inefficiencies, which can be corrected with rather minor engineering or management practices. Many of the raw materials savings achieved in these examples were from water conservation and recycling opportunities, which tend to be overlooked by plant managers since historically the cost for water was low or government subsidized. In other cases, relatively low-cost monitoring instrumentation enabled emissions to be quantified, and hence potential savings to be readily identified.

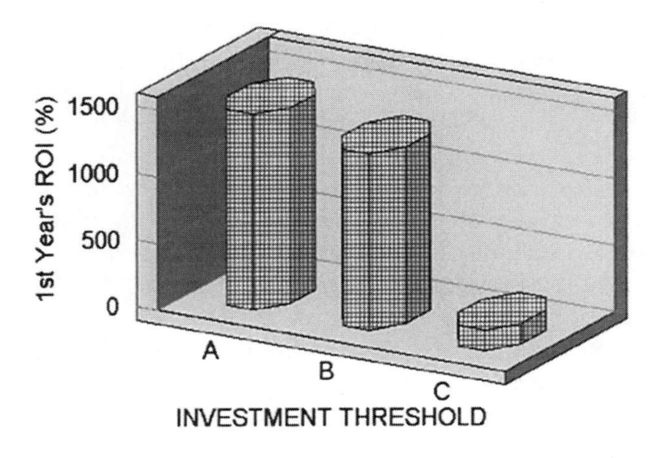

Figure 2. *The low-cost/no-cost P2 practices gave the highest returns.*

Elimination of a Toxic Solvent

Many industries use the solvent trichloroethylene (TCE) in their operations. This highly toxic chemical must be contained in a closed system, because large releases of

TCE can be fatal. Long-term exposure to low concentrations of TCE can lead to cancer; brain and central nervous system disorders; blood, liver and kidney diseases; and birth defects. Because of the toxicity of the chemical, such releases often require the evacuation of the facility.

Plant workers are the unwilling "internal customers" of TCE fumes. The external environment is also an unwilling customer. Effluents can affect rivers downstream. Aquatic life and people who depend on the river for drinking water are unwilling customers. Groundwater quality can be affected. In the United States, there is a landmark tort case in Redlands, Calif., involving thousands of residents from several communities that have undergone long-term chronic exposures to TCE in drinking water. The chemical made its way into the water as a result of dry-cleaning operations and jet fueling and blending operations practiced two decades ago.

TCE, like a number of other common industrial chemicals, served and continues to serve as an important industrial solvent. But, as we see in the Redlands case, uncontrolled releases to the environment — even those that are a result of innocent and standard handling practices on the part of industry — can result in hundreds of millions of dollars in potential liabilities to the generators of these wastes, not to mention the damage to quality of life and the environment generated by the handling of these chemicals.

It is important to recognize that quality can be built into, but not inspected into, a product. The producer must continuously identify and eliminate the *root cause* of any impediments to quality. Continuous improvement also is the key to reducing the environmental impacts of production processes.

The traditional approach to industrial waste has been to view it as a necessary by-product of manufacturing. Though production generates the waste, the responsibility to dispose of the waste in a safe and legal manner usually falls on the environmental engineering department. Because environmental engineers receive the waste after it has been created, they usually are not intimately familiar with the processes that created it. Further, because waste reduction is not a component of their performance review, environmental engineers do not have the institutional motivation to reduce the waste — only to dispose of it properly.

In all situations where toxic materials are handled, we run the risk of release, exposing workers, and endangering the public and the environment. This is a fact — no matter how well-designed an operation is, or how many process-safety reviews have been implemented to reduce the risks. As such, pollution *prevention* programs should *always* be considered as a serious option in the management of these materials in process operations.

The Ford Motor Company recognized this fact a number of years ago. Under its old technology, degreasing certain aluminum components with TCE required extensive safety mechanisms and procedures. Building better containment systems reduced the risk of exposure, but did not get to the root cause of the problem — namely, the use of TCE.

With this in mind, Ford looked for a TCE-free solution to degreasing radiator coils. The company formed a team that included a chemical engineer, an environmental engineer, a process engineer, an accountant, and a product engineer. The variety of backgrounds on the team ensured that the pertinent issues of cost, product quality, process feasibility, and environmental impact were all addressed. The team designed an aqueous degreasing system (i.e., soap and water) to replace the TCE. Now, not only is the toxic chemical removed from the plant, but the water in the new system is recycled as well. And, significantly, the aqueous degreaser exhibits better quality characteristics than the TCE degreaser.

This project is an example of: improved quality, reduced cost, and reduced environmental impact. Certainly, not all projects will prove so fruitful. Some clean alternatives may cost more than their polluting rivals, but that cost must be balanced with the benefits of the environmental improvement and the prospects of other long-term benefits. To justify this viewpoint, one needs only to look to the increasing expectations of external customers for environmentally friendly products.

Controlling Fugitive Emissions

OHIS is the abbreviated name for Organsko Hemiska Industrija Skopje, a joint stock company in the Republic of Macedonia. At its plant, which began operations in 1964, OHIS manufactures basic chemicals, including caustic soda and chlorine, herbicides, detergents, various synthetic fibers, PVA, and PVC.

Vinyl chloride monomer (VCM), the precursor to PVC, is one of the most widely used commodity plastics in the world. PVC resin is manufactured by a suspension process. The suspension polymerization is performed in reactors, with conversion rates typically between 89 to 91 percent of the monomer (VCM). The polymerization reaction is terminated by injecting an inhibitor, whereupon unconverted VCM is removed from the polymerization reactors by heating up the polymer slurry and applying vacuum technology to remove unreacted VCM vapors. The polymer slurry is removed from the reactors and sent to slurry tanks. From there, the slurry resin product goes on to polymer drying and finishing stages to produce a final resin product. The PVC plant comprises four single agitator reactors (CSTRs).

As noted, residual or unconverted VCM is removed from the polymer slurry before the resin is sent to the finishing stages of the operation. This is traditionally accomplished by the use of a series of stripper columns, where significant quantities of unreacted VCM that can potentially escape in the form of fugitive emissions. VCM is a highly toxic chemical that is a suspected human carcinogen (exposure has been linked to angiosarcoma, a liver cancer). In addition, long-term or chronic exposure to low levels of VCM has been linked to several other occupational diseases, including a bone condition (acro-osteolysisli); a condition leading to a narrowing of the blood vessels (Raynaud's phenomenon); and hardening of the skin (scleroderma).

The losses or fugitive emissions are not always at a steady rate; under transient process operating conditions, there are periods where there are greater losses than at others. This poses an additional problem — a serious threat of fire and explosion, because, under the proper conditions, VCM is a highly flammable gas. A catastrophic fire at this facility would represent a potential loss of lives for plant personnel, an extremely high health risk to the surrounding community because of exposure to noxious vapors resulting from the fire, and the possibility of hundreds of millions of dollars in property damage.

The company decided to perform a P2 audit, and determined that about 5 percent of the monomer was lost through agitator seals, various draw-off points, and pump seals in the vacuum system (refer to Figure 3). The audit revealed that recycling the VCM vapors lost to the atmosphere from the stripper column operations was feasible. The company also found out that product-quality demands in the international marketplace call for lower levels of residual VCM in the PVC resin product than what the company was allowing.

Existing fugitive VCM-monitoring practices involved slow and inaccurate methods based on random-grab samples of product and air samples emanating from the equipment and in the shop areas where technicians are exposed. These methods were incapable of accurately determining the sources and quantities of fugitive VCM emissions. Based on the P2 audit, OHIS invested in reliable, portable monitoring instruments and began using them as diagnostic tools to identify the sources of fugitive emissions, which are now being corrected. A thermoparamagnetic (O_2) oxygen analyzer and sample conditioning system cost the company $15,000, but it identified approximately 240 tons per year of VCM emissions that can be recycled through the process (to the polymerization reactors), with only minor modifications to existing process piping. Not only did this result in a projected annual material savings of $144,000, but it gave the company a more-consistent and higher-quality product.

Figure 3. *Monomer supply tank where fugitive emissions found at pump seals.*

Graphite Recovery in a Steel Mill

The Donetsk Industrial Hazardous Waste Management project was a major component of the Environmental Policy and Technology Programme (EPT) sponsored by the U.S. Agency for International Development (USAID) in Ukraine. This three-year project involved a full-time resident manager, who was responsible for establishing and managing the country's first office on pollution prevention and energy management, staffed by 20 full- and part-time local employees.

A major focus of the project was demonstration of sound environmental management practices to both industry and environmental regulators. To this end, the project performed training and helped develop local experts capable of performing P2 audits. This activity, performed in conjunction with numerous missions by the U.S. Environmental Protection Agency, the World Environmental Center, and the British Know-How Foundation, enabled hundreds of Ukrainians to be trained in pollution prevention, waste minimization, and energy efficiency practices for industry.

To strengthen these principles, the project implemented a major P2 demonstration at Azovstal, Ukraine's largest steel mill. This $250,000 project focused on graphite recovery from blast furnace sintering operations, and resulted in a payback period of less than four months. The technology made use of standard end-of-pipe treatment technologies (baghouses and cyclones) to recover fugitive PM emissions, which posed a serious threat to worker health and safety. In addition, the recovered graphite was of such high quality that it could be sold into a local off-take market.

The pilot demonstration was so successful that Azovstal repeated the project on a twin process within their plant, and the Illicha Iron and Steel Works (Ukraine's second largest iron and steel works) duplicated the technology.

With the help of the World Environmental Center, more than 50 pilot demonstrations were implemented.

THE P2 BENEFITS MATRIX

Most enterprises need to balance economic growth with environmental protection. It is increasingly being recognized that economic development, as well as the health and welfare of the workforce and public, are closely linked to proper management of an enterprise's natural resources and environment. P2 offers a government and industry a way to manage and mitigate the effects of industrial growth on the environment while enabling economic development. P2 addresses three important components of the environmental protection/economic development issue:

- *Environment.* P2 offers a better solution for environmental management than "end-of pipe" pollution solutions

• *Quality.* P2 encourages evaluation of production processes and product quality

• *Cost.* P2 Improves a facility's bottom-line by reducing treatment costs, saving on material and resource inputs, and reducing risk and liability insurance.

Dealing with environmental wastes through "end-of-pipe" measures (such as wastewater treatment systems, hazardous-waste incinerators and other treatment technologies, secure landfills, monitoring equipment, solid-waste hauling equipment, air-pollution control equipment, and catalytic converters) has proven to be very costly and does not address all environmental problems. P2 offers industry the advantages of:

1) Less need for costly pollution-control equipment
2) Getting ahead of environmental regulations
3) Reduced reporting and permitting requirements
4) Less operation and maintenance of pollution-control equipment

The process of identifying P2 opportunities also provides the opportunity to identify measures to improve product quality. The P2 audit requires the enterprise to examine its production process in-depth. After all, finding ways to reduce wastes also requires a study of the root causes for generating wastes and finding improvements to processes.

Some authors have coined the phrase *Total Quality Management* (TQM), which essentially refers to a management-system approach aimed at achieving high product and service quality. The management elements of TQM include:

- Customer focus
- Continuous improvement
- Teamwork
- Strong management commitment

At first glance, TQM seems unrelated to environmental concerns. Yet the inherent strengths of the TQM methodology can effectively address some of these issues. Professionals who apply TQM concepts to environmental issues have coined the term *Total Quality Environmental Management* (TQEM).

TQEM is a logical method for achieving P2. The first three elements of TQM — customer focus, continuous improvement, and a team approach — readily apply to environmental issues, too. For example, let's examine the area of customer focus. Customers fall into two categories — internal and external. The internal customer is the next person in the production chain, while the external customer is the end-user of the product. If the definition of the customer is expanded to include those people and environments that are affected by the production's process waste, TQEM requires us to understand the impact of this waste on those customers, and take steps to reduce it.

As in traditional TQM settings, the last element — strong management commitment — is perhaps the most important. No TQEM program will succeed without the commitment of top management. This group, made up of those who most likely have built their careers during a time when waste was seen as a necessary by-product, must come to understand that both internal and external customer expectations include products and processes that demonstrate environmental responsibility. Top management must learn to see the value of applying TQEM to get to the root causes of waste, and call on the cross-disciplinary teams to employ continuous improvement to implement ever "cleaner" solutions.

In many cases, P2 measures can have clear environmental benefits in terms of pollution that is *not* generated; reductions in the toxic materials used in the production process; and savings in energy use and other raw materials. Savings can accrue in five primary areas:

1) A company can save on raw materials and energy
2) A company can save on labor costs
3) A company can reduce or eliminate disposal costs
4) A facility can save on waste-handling and treatment costs, both in its own use of labor to collect, store, and process wastes, and in its external costs to transport wastes off-site
5) A facility can decrease the amount of toxic materials it uses, handles, and transports, reducing its future liability costs

P2 practices are driven by economic forces. This leads us to introducing the concept of a *Pollution Prevention Benefits Matrix*. The case studies presented in the previous section, and the case studies you can find out more about in the "Additional Resources" appendix at the end of the book, enable us to summarize the potential benefits of P2 programs into a number of benefit and cost-savings categories. Use the table below as a guide when identifying the potential cost-savings categories for P2 opportunities at your own enterprise.

The four cost-savings categories listed in Table 4 are:

- *Tier 0* — normal or direct costs;
- *Tier 1* — hidden costs;
- *Tier 2* — future liability costs; and
- *Tier 3* — less tangible costs.

We will return to these cost-savings categories in Chapter 8, but for now, we should bear the generalized definitions of these cost factors in mind as we review some the case histories.

Table 4. Generalized Pollution Prevention Benefits Matrix.

Benefit	**Cost Savings Categories**			
	Tier 0	**Tier 1**	**Tier 2**	**Tier 3**
Reduced labor from elimination of pollution	**X**			
Raw materials savings by recycling	**X**			
Raw materials savings by substitution of less costly, toxic, and regulated materials	**X**			
Reduced energy use	**X**			
Reduced O&M costs, by elimination of pollution-control equipment	**X**			
Elimination of off-site disposal practices for wastes	**X**	**X**	**X**	**X**
Elimination of off-site storage and disposal practices for wastes	**X**	**X**	**X**	**X**
Elimination of off-site treatment operations	**X**	**X**		
Elimination of pollution monitoring and analytical support		**X**		
Remedial actions for site operations minimized or eliminated			**X**	
Reduced risk of personal injury under OSHA			**X**	
Off-site property damages from environmental catastrophes minimized or eliminated			**X**	
Need for environmental-impact statements eliminated		**X**		
Need for full-time environmental		**X**		

Benefit	**Cost Savings Categories**			
	Tier 0	**Tier 1**	**Tier 2**	**Tier 3**
legal representation minimized				
Permitting requirements minimized		X		
Reporting and recordkeeping reduced		X		
Improved process efficiencies/productivity		X		
Improved product quality				X
Lower insurance risks under environmental policies		X		
Improved consumer confidence				X
Improved employee relations				X

INDUSTRY-SPECIFIC GUIDELINES AND PRACTICES

The authors have reviewed hundreds of case studies and implemented many dozens of P2 programs at enterprises. From this experience, we have compiled the following tables of industry-specific P2 practices. Use these as guides in generating ideas and in borrowing recommendations for addressing P2 opportunities in your own enterprise, paying special attention to the industry-specific practices in industries that are related to your own enterprise, because many pollution problems are often common.

Also, bear in mind the cost savings categories as you read over these tables. As an exercise, you may wish to list some of the P2 actions by savings categories and begin thinking of how some of these P2 applications could be adopted to your plant operations. Remember also that some industry specific practices may be generalized and could be applicable to other industry sectors. Therefore, it is worthwhile reading over some of the P2 practices that are in sectors different than your own enterprises operations.

Table 5. P2 Practices in the Chemical Processing Industries.

Area of Opportunity	RECOMMENDED POLLUTION PREVENTION PRACTICE
Reduction of air emissions	1. Minimize leakages of volatile organics, including benzene, vinyl chloride, and ethylene oxide, from valves, pump glands (through use of mechanical seals), flanges, and other process equipment by following good design practices and equipment maintenance procedures. 2. Use mechanical seals where appropriate. 3. Minimize losses from storage tanks, product transfer areas, and other process areas by adopting such methods as vapor-recovery systems and double seals (for floating roof tanks). 4. Recover catalysts and reduce particulate emissions. 5. Reduce nitrogen oxide (NO_x) emissions by using low-NO_x burners. Optimize fuel usage. 6. In some cases, organics that cannot be recovered are effectively destroyed by routing them to flares and other combustion devices.
Elimination or reduction of pollutants	7. Use nonchrome-based additives in cooling water. 8. Use long-life catalysts and regeneration to extend the cycle.
Recycling and reuse	9. Recycle cooling water and treated wastewater to the extent feasible. 10. Recover and reuse spent solvents and other chemicals to the extent feasible.
Improved operating procedures	11. Segregate process wastewaters from stormwater systems. 12. Optimize the frequency of tank and equipment cleaning. 13. Prevent solids and oily wastes from entering the drainage system. 14. Establish and maintain an emergency preparedness-and-response plan.

Table 6. P2 Practices in Fertilizer Manufacturing Plants.

Area of Opportunity	RECOMMENDED POLLUTION PREVENTION PRACTICE
Ammonia plant	1. Where possible, to minimize air emissions, use natural gas as the feedstock for the ammonia plant. 2. Use hot process gas from the secondary reformer to heat the primary reformer tubes (the exchanger-reformer concept), thus reducing the need for natural gas. 3. Direct hydrogen cyanide (HCN) gas in a fuel-oil gasification plant to a combustion unit, to prevent its release. 4. Consider using purge gases from the synthesis process to fire the reformer; strip condensates to reduce ammonia and methanol.

Table 6. Continued.

Area of Opportunity	RECOMMENDED POLLUTION PREVENTION PRACTICE
	5. Use carbon-dioxide removal processes that do not release toxic substances to the environment. When using monoethanolamine (MEA) or other processes, such as hot potassium carbonate, in carbon-dioxide removal, follow proper operation and maintenance procedures to minimize releases to the environment.
Urea plant	6. Use total-recycle processes in the synthesis process; reduce microprill formation and carryover of fines in prilling towers.
Ammonium nitrate plant	7. *Prill tower:* Reduce microprill formation and reduce carryover of fines through entrainment. 8. *Materials handling*: Where feasible, use covers and hoods on conveyors and transition points. Good cleanup practices must be in place to minimize contamination of stormwater runoff from the plant property. 9. *Granulators*: Reduce dust emissions from the disintegration of granules.

Table 7. P2 Practices in the Coking Industry.

Area of Opportunity	RECOMMENDED POLLUTION PREVENTION PRACTICE
General	1. Use cokeless iron- and steel-making processes, such as the direct reduction process, to eliminate the need to manufacture coke. 2. Use beneficiation (preferably at the coal mine) and blending processes that improve the quality of coal feed to produce coke of desired quality and reduce emissions of sulfur oxides and other pollutants. 3. Use enclosed conveyors and sieves for coal and coke handling. Use sprinklers and plastic emulsions to suppress dust formation. Provide windbreaks where feasible. Store materials in bunkers or warehouses. Reduce drop distances. 4. Use and preheat high-grade coal to reduce coking time, increase throughput, reduce fuel consumption, and minimize thermal shock to refractory bricks.
Coke oven emissions	5. *Charging:* Dust particles from coal charging should be evacuated by the use of jumper-pipe systems and steam injection into the ascension pipe, or controlled by fabric filters.

Table 7. Continued.

Area of Opportunity	RECOMMENDED POLLUTION PREVENTION PRACTICE
	6. *Coking:* Use large ovens to increase batch size and reduce the number of chargings and pushings, thereby reducing the associated emissions. Reduce fluctuations in coking conditions, including temperature. Clean and seal coke-oven openings, to minimize emissions. Use mechanical cleaning devices (preferably automatic) for cleaning doors, door frames, and hole lids. Seal lids, using a slurry. Use low-leakage door construction, preferably with gas sealings. 7. *Pushing:* Emissions from coke pushing can be reduced by maintaining a sufficient coking time, thus avoiding "green push." Use sheds and enclosed cars, or consider using traveling hoods. The gases released should be removed and passed through fabric filters. 8. *Quenching:* Where feasible, use dry instead of wet quenching. Filter all gases extracted from the dry quenching unit. If wet quenching is used, provide interceptors (baffles) to remove coarse dust. When wastewater is used for quenching, the process transfers pollutants from the wastewater to the air, requiring subsequent removal. Reuse quench water. 9. *Conveying and sieving:* Enclose potential dust sources, and filter evacuated gases.
By-product recovery	10. Use vapor recovery systems to prevent air emissions from light-oil processing, tar processing, naphthalene processing, and phenol- and ammonia-recovery processes. 11. Segregate process water from cooling water. 12. Reduce fixed-ammonia content in ammonia liquor by using caustic soda and steam stripping. 13. Recycle all process solid wastes, including tar decanter sludge to coke oven. 14. Recover sulfur from coke-oven gas. Recycle Claus-tail gas into coke-oven gas system.

Table 8. P2 Practices in Pharmaceuticals Manufacturing.

RECOMMENDED POLLUTION PREVENTION PRACTICE
1. Meter and control the quantities of active ingredients to minimize waste.
2. Reuse by-products from the process as raw materials or as raw-material substitutes in other processes.

Table 8. Continued.

RECOMMENDED POLLUTION PREVENTION PRACTICE
3. Recover solvents used in the process by distillation or other methods.
4. Give preference to the use of nonhalogenated solvents.
5. Use automated filling to minimize spillage.
6. Use "closed" feed systems into batch reactors. Use equipment washdown waters and other process waters (such as leakages from pump seals) as makeup solutions for subsequent batches.
7. Recirculate cooling water.
8. Use dedicated dust collectors to recycle recovered materials and prevent cross-contamination.
9. Vent equipment through a vapor-recovery system.
10. Use loss-free vacuum pumps.
11. Return toxic-materials packaging to the supplier for reuse, or incinerate/destroy it in an environmentally acceptable manner.
12. Minimize storage time of off-specification products through regular reprocessing.
13. Find productive uses for off-specification products to avoid disposal problems.
14. Use high-pressure hoses for equipment cleaning to reduce wastewater.
15. Provide stormwater drainage and avoid contamination of stormwater from process areas.
16. Label and store toxic and hazardous materials in secure, bunded areas. Spillage should be collected and reused.

Table 9. P2 Practices in Oil Refineries.

Area of Opportunity	RECOMMENDED POLLUTION PREVENTION PRACTICE
Reduction of air emissions	1. Minimize losses from storage tanks and product transfer areas by using such methods as vapor-recovery systems and double seals. 2. Minimize SO_x emissions either through desulfurization of fuels (to the extent feasible) or by directing the use of high-sulfur fuels to units equipped with SO_x-emissions controls.

Table 9. Continued.

Area of Opportunity	RECOMMENDED POLLUTION PREVENTION PRACTICE
	3. Recover sulfur from tail gases in high-efficiency sulfur recovery units.
	4 Recover nonsilica-based (i.e., metallic) catalysts and reduce particulate emissions.
	5. Use low-NO_x burners to reduce emissions of nitrogen oxides.
	6. Avoid and limit fugitive emissions by designing and maintaining processes properly.
	7. Keep fuel usage to a minimum.
Elimination or reduction of pollutants	8. For octane boosting, consider reformate and other octane boosters instead of tetraethyl lead and other organic lead compounds.
	9. Where inhibitors are needed, use nonchrome-based inhibitors in cooling water.
	10. Use long-life catalysts and regenerate to extend the catalysts' life cycle.
Recycling and reuse	11. Recycle cooling water and, where cost-effective, treated wastewater.
	12. Maximize recovery of oil from oily wastewaters and sludges. Minimize losses of oil to the effluent system.
	13. Recover and reuse phenols, caustics, and solvents from their spent solutions.
	14. Return oily sludges to coking units or crude distillation units.
Operating procedures	15. Segregate oily wastewaters from stormwater systems.
	16. Reduce oil losses during tank drainage carried out to remove water before product dispatch.
	17. Optimize frequency of tank and equipment cleaning to avoid accumulating residue at the bottom of the tanks.
	18. Prevent solids and oily wastes from entering the drainage system.
	19. Institute dry sweeping instead of washdown to reduce wastewater volumes.
	20. Establish and maintain an emergency preparedness-and-response plan and carry out frequent training.
	21. Practice corrosion monitoring, prevention, and control in underground piping and tank bottoms.
	22. Establish leak detection and repair programs.

Table 10. P2 Practices in Iron and Steel Making.

Area of Opportunity	RECOMMENDED POLLUTION PREVENTION PRACTICE
Eliminating coke with cokeless technologies	1. *The Japanese Direct Iron-Ore Smelting (DIOS) process:* This process produces molten iron directly with coal and sinter feed ore. A 500 ton-per-day pilot plant, started up in October 1993, attained the designed production rates as a short-term average. Data generated by this pilot plant is being used to determine economic feasibility on a commercial scale. 2. *HIsmelt process:* A plant using the HIsmelt process for molten iron production, developed by HIsmelt Corporation of Australia, was started up in late 1993. The process, using ore fines and coal, has achieved a production rate of 8 tons per hour using ore directly in the smelter. Developers anticipate reaching the production goal of 14 tons per hour. The data generated by the plant is being used to determine economic feasibility on commercial scale. If commercial feasibility is realized, Midrex is expected to become the U.S. engineering licensee of the HIsmelt process. 3. *Corex process:* The Corex or Cipcor process has integral coal desulfurizing, is amenable to a variety of coal types, and generates more electrical power than required by an iron and steel mill, allowing the excess to be sold to local power grids. A Corex plant is operating in South Africa, and other plants are expected to open soon in South Korea and India.
Reducing coke oven emissions with other technologies	4. *Pulverized coal injection:* This technology substitutes pulverized coal for a portion of the coke in the blast furnace. Use of pulverized-coal injection can replace about 25 to 40 percent of coke in the blast furnace, substantially reducing emissions associated with coke-making operations. This reduction ultimately depends on the fuel-injection rate applied to the blast furnaces — which, in turn, is dictated by the age of existing coking facilities, fuel costs, oxygen availability, capital requirements for fuel injection, and available hot blast temperature. 5. *Nonrecovery coke battery:* As opposed to the by-product recovery coke plant, the nonrecovery coke battery is designed to allow combustion of the gasses from the coking process, thus consuming the by-products that are typically recovered. The process results in lower air emissions and substantial reductions in coking-process wastewater discharges. 6. *The Davy Still Auto-process:* In this pre-combustion cleaning process for coke ovens, coke-oven battery process water is used to strip ammonia and hydrogen sulfide from coke-oven emissions.

Table 10. Continued.

Area of Opportunity	RECOMMENDED POLLUTION PREVENTION PRACTICE
	7. *Alternative fuels:* Steel producers can inject other fuels — such as natural gas, oil, and tar/pitch — instead of coke into the blast furnace, but these fuels can only replace coke in limited amounts.
Reducing wastewater	8. In Europe, some plants have implemented technology to shift from water quenching to dry quenching, to reduce energy costs. However, major construction changes are required for such a solution.
Recycling of coke by-products	9. Improvements are common in the in-process recycling of tar decanter sludge. Sludge can either be injected into the ovens to contribute to coke yield, or converted into a fuel that is suitable for the blast furnace.
Electric-arc furnace dust	10. Electric-arc furnace (EAF) dust is a hazardous waste because of the high concentrations of lead and cadmium it contains. With 550,000 tons of EAF dust generated annually in the United States, there is great potential to reduce the volume of this hazardous waste. U.S. steel companies typically pay a disposal fee of $150 to $200 per ton of dust. With an average zinc concentration of 19 percent, much of the EAF dust is sent off-site for zinc recovery. Most of the EAF dust-recovery options are only economically viable for dust with a zinc content of at least 15 to 20 percent. Facilities that manufacture specialty steels — such as stainless steel — with a lower zinc content still have opportunities to recover chromium and nickel from the EAF dust. In-process recycling of EAF dust involves pelletizing and then reusing the pellets in the furnace, however, recycling of EAF dust on-site has not proven to be technically or economically competitive for all mills. Improvements in technologies have made off-site recovery a cost-effective alternative to thermal treatment or secure landfill disposal.
Pig iron manufac-turing	11. Improve blast furnace efficiency by using coal and other fuels (such as oil or natural gas) for heating instead of coke, thereby minimizing air emissions. 12. Recover the thermal energy in the gas from the blast furnace before using it as a fuel. 13. Increase fuel efficiency and reduce emissions by improving blast furnace charge distribution. 14. Recover energy from sinter coolers and exhaust gases. 15. Use dry SO_x-removal systems, such as carbon absorption, for sinter plants or lime spraying in flue gases. 16. Recycle iron-rich materials — such as iron-ore fines, pollution-control dust, and scale — in a sinter plant. 17. Use low- NO_x burners to reduce NO_x emissions from fuel use by ancillary operations. 18. Improve productivity by screening the charge and using better taphole practices.

Table 10. Continued.

Area of Opportunity	RECOMMENDED POLLUTION PREVENTION PRACTICE
	19. Reduce dust emissions at furnaces by covering iron runners when tapping the blast furnace and by using nitrogen blankets during tapping. 20. Use pneumatic transport, enclosed conveyor belts, or self-closing conveyor belts, as well as wind barriers and other dust suppression measures, to reduce the formation of fugitive dust.
Steel manu-facturing	21. Use dry dust-collection and removal systems, to avoid the generation of wastewater. Recycle collected dust. 22. Use BOF gas as fuel. 23. Use enclosures for BOF. 24. Use a continuous process for casting steel to reduce energy consumption.
Finishing stages	25. *Pickling Acids:* The pickling process removes scale and cleans the surface of raw steel by dipping it into a tank of hydrochloric or sulfuric acid. If not recovered, the spent acid may be transported to deep injection wells for disposal, but as those wells continue to close, alternative disposal costs are rising. 26. Large-scale steel manufacturers recover HCl in finishing operations. However, the techniques used are not suitable for small- to medium-sized steel plants. An evolving recovery technique for smaller steel manufacturers and galvanizing plants removes iron chloride (a saleable product) from the HCl, reconcentrates the acid for reuse, and recondenses the water to be reused as a rinse water in the pickling process. Because the only by-product of the HCl-recovery process is a non-hazardous, marketable metal chloride, this technology generates no hazardous wastes. The manufacturer of the system projects industrywide HCl-waste reduction of 42,000 tons per year by 2010. This technology is less expensive than transporting and disposing waste acid; in addition, it eliminates the associated long-term liability associated with transport and disposal. The total savings for a small- to medium-sized galvanizer is projected to be $260,000 annually. 27. To simultaneously reduce spent pickling liquor and fluoride in plant effluents, one facility modified its existing treatment process to recover the fluoride ion from rinsewater and from spent pickling-acid raw-water waste streams. The fluoride is recovered as calcium fluoride (fluorspar), an input product for steelmaking. The melt shop had been purchasing 930 tons of fluorspar annually to use as a furnace flux material in the EAF, at a cost of $100 per ton. The recovered fluorspar is a better grade than the

Table 10. Continued.

Area of Opportunity	RECOMMENDED POLLUTION PREVENTION PRACTICE
	purchased material, which reduces the amount of flux used by approximately 10 percent. Not only does the facility reduce the amount of sludge it's producing (resulting in lower off-site sludge-disposal costs), but it saves money on its chemical purchases.
Process modifi-cations	28. Replace single-pass wastewater systems with closed-loop systems, to minimize chemical use in wastewater treatment and to reduce water use. 29. Continuous casting, now used for about 90 percent of crude steel cast in the United States, offers great improvements in process efficiency when compared with the traditional ingot-teeming method. This increased efficiency also results in a considerable savings in energy and some reduction in the volume of mill wastewater.
Materials substitution	30. If possible, use scrap steel with low lead and cadmium content as a raw material. 31. Eliminate the generation of reactive desulfurization slag generated in foundry work by replacing calcium carbide with a less-hazardous material.
Recycling miscella-neous materials	32. Recycle or reuse oils and greases. 33. Recover acids by removing dissolved iron salts from spent acids. 34. Use thermal decomposition for acid recovery from spent pickle liquor. 35. Use a bipolar membrane/electrodialytic process to separate acid from metal by-products in spent NO_3-HF pickle liquor. 36. Recover sulfuric acid using low-temperature separation of acid and metal crystals. 37. Use blast furnace slag in construction materials. Slag containing free lime can be used in iron making.

Table 11. P2 Practices in Mini Steel Mills.

RECOMMENDED POLLUTION PREVENTION PRACTICE
1. Locate EAFs in enclosed buildings.
2. Improve feed quality by using selected scrap to reduce the release of pollutants to the environment.
3. Use dry dust-collection methods, such as fabric filters.
4. Replace ingot teeming with continuous casting.
5. Wherever feasible, use continuous casting for semifinished and finished products. In some cases, continuous charging may be feasible and effective for controlling dust

Table 11. Continued.

RECOMMENDED POLLUTION PREVENTION PRACTICE
6. Use bottom tapping of EAFs to prevent dust emissions.
7. Control water consumption by proper design of spray nozzles and cooling-water systems.
8. Segregate wastewaters containing lubricating oils from other wastewater streams, and remove oil.
9. Recycle mill scale to the sinter plant in an integrated steel plant.
10. Where feasible, use acid-free methods (mechanical methods, such as blasting) for descaling.
11. In the pickling process, use countercurrent flow of rinsewater; use indirect methods for heating and pickling baths.
12. Use closed-loop systems for pickling; regenerate and recover acids from spent pickling liquor, using resin bed, retorting, or other regeneration methods, such as vacuum crystallization of sulfuric acid baths.
13. Use electrochemical methods in combination with pickling to lower acid consumption.
14. Reduce emissions of nitrogen oxides by using natural gas as a fuel; use low-NO_x burners; and use hydrogen peroxide and urea in stainless-steel pickling baths.
15. Recycle slags and other residuals from manufacturing operations, for use in construction and other industries.
16. Cooling requires a high amount of water use. Recycle wastewaters to reduce the discharge rate to less than 5 cubic meters per ton of steel produced, including indirect cooling waters.
17. Recover zinc from EAF dust containing more than 15% total zinc; recycle EAF dust to the extent feasible.

Table 12. P2 Practices in Zinc and Lead Smelting.

RECOMMENDED POLLUTION PREVENTION PRACTICE
1. Use doghouse enclosures where appropriate; use hoods to collect fugitive emissions.
2. Mix strong acidic gases with weak ones to facilitate production of sulfuric acid from sulfur oxides, thereby avoiding the release of weak acidic gases.
3. Maximize the recovery of sulfur by operating the furnaces to increase the SO_x content of the flue gas and by providing efficient sulfur conversion. Use a double-contact, double-absorption process.
4. Desulfurize paste with caustic soda or soda ash to reduce SO_2 emissions.

Table 12. Continued.

RECOMMENDED POLLUTION PREVENTION PRACTICE
5. Use energy-efficient measures, such as waste-heat recovery from process gases, to reduce fuel usage and associated emissions.
6. Recover acid, plastics, and other materials when handling battery scrap in secondary lead production.
7. Recycle condensates, rainwater, and excess process water for: washing; dust control; gas scrubbing; and other process applications where water quality is not of particular concern.
8. Give preference to natural gas over heavy fuel-oil for use as fuel, and to coke charge and breeze with lower sulfur contents.
9. Use low-NO_x burners.
10. Where appropriate, use suspension or fluidized bed roasters to achieve high SO_2 concentrations when roasting zinc sulfides.
11. Recover and reuse iron-bearing residues from zinc production for use in the steel or construction industries.
12. Give preference to fabric filters over wet scrubbers or wet electrostatic precipitators (ESPs) for dust control.
13. Good housekeeping practices are key to minimizing losses and preventing fugitive emissions. Losses and emissions are minimized by enclosed buildings, covered conveyors and transfer points, and dust-collection equipment. Yards should be paved and runoff water should be routed to settling ponds.

Table 13. P2 Practices in Copper Smelting.

RECOMMENDED POLLUTION PREVENTION PRACTICE
1. Use closed-loop electrolysis plants to prevent pollution.
2. Use continuous casting machines for cathode production, to avoid the need for mold-release agents.
3. Enclose furnaces to reduce fugitive emissions, and return dust from dust-control equipment to the process.
4. Apply energy-efficiency measures (such as waste-heat recovery from process gases) to reduce fuel use and associated emissions.
5. Recycle cooling water, condensates, rainwater, and excess process water used for washing, dust control, gas scrubbing, and other process applications where water quality is not a concern.

Table 13. Continued.

RECOMMENDED POLLUTION PREVENTION PRACTICE
6. Good housekeeping practices are key to minimizing losses and preventing fugitive emissions. Such losses and emissions are minimized by enclosed buildings, covered or enclosed conveyors and transfer points, and dust-collection equipment. Yards should be paved and runoff water routed to settling ponds. Regular sweeping of yards, and indoor storage or coverage of concentrates and other raw materials also reduces materials losses and emissions.

Table 14. P2 Practices in Pulp and Paper Manufacturing.

RECOMMENDED POLLUTION PREVENTION PRACTICE
1. Use energy-efficient pulping processes wherever feasible. Promote the acceptability of less-bright products. For less-bright products, such as newsprint, consider thermomechanical processes and recycled fiber.
2. Minimize the generation of effluents through process modifications and recycle wastewaters, aiming for total recycling.
3. Reduce effluent volume and treatment requirements by using dry instead of wet debarking; recovering pulping chemicals by concentrating black liquor and burning the concentrate in a recovery furnace; recovering cooking chemicals by recausticizing the smelt from the recovery furnace; and using high-efficiency washing and bleaching equipment.
4. To minimize unplanned or nonroutine discharges of wastewater and black liquor caused by equipment failures, human error, and faulty maintenance procedures, train operators, establish good operating practices, and provide sumps and other facilities to recover liquor losses from the process.
5. Minimize sulfur emissions to the atmosphere by using a low-odor design black-liquor recovery furnace.
6. Prevent and control spills of black liquor.
7. Where feasible, aim for zero-effluent discharge. Reduce wastewater discharges to the extent feasible. Incinerate liquid effluents from the pulping and bleaching processes.
8. Reduce the odor from reduced sulfur emissions by collection and incineration, and by using modem, low-odor recovery boilers fired at more than 75 percent concentration of black liquor.
9. Dewater and properly manage sludges.
10. Where wood is used as a raw material in the process, encourage tree plantations, to ensure the sustainability of forests.
11. Use energy-efficient processes for black-liquor chemical recovery, preferably aiming for a high solid content (for example, 70 percent).
12. Reduce bleaching requirements through process design and operation. Use the following measures to reduce emissions of chlorinated compounds to the environment: Before bleaching, reduce the lignin content in the pulp (Kappa number of 10) for hardwood by extended cooking and by oxygen delignification under elevated pressure; optimize pulp washing prior to bleaching; use TCF or, at a minimum, ECF bleaching systems; use oxygen, ozone, peroxides (hydrogen peroxide), peracetic acid, or enzymes (cellulose-free xylanase) as substitutes for chlorine-based bleaching chemicals; recover and incinerate maximum material removed from pulp bleach grade where chlorine bleaching is used; reduce the chlorine charge on the lignin by controlling pH and by splitting the addition of chlorine.

Table 15. P2 Practices in Sugar Manufacturing.

RECOMMENDED POLLUTION PREVENTION PRACTICE
1. Reduce product losses to less than 10 percent by better production control. Perform sugar auditing. 2. Discourage spraying of molasses on the ground for disposal. 3. Minimize storage time for juice and other intermediate products to reduce product losses and discharge of product into the wastewater stream. 4. Give preference to less polluting clarification processes, such as those using bentonite instead of sulfite, for the manufacture of white sugar. 5. Collect waste product for use in other industries — for example, bagasse for use in paper mills and as fuel. Cogeneration systems for large sugar mills can generate electricity for sale. Beet chips can be used as animal feed. 6. Optimize the use of water and cleaning chemicals. Procure cane washed in the field. Prefer the use of dry cleaning methods. 7. Recirculate cooling waters. 8. Continually sample and measure key production parameters to identify and reduce production losses, and thus reduce waste. Fermentation processes and juice handling are the main sources of leakage. 9. Odor problems can usually be prevented with good hygiene and storage practices.

Table 16. P2 Practices in Tanning and Leather Finishing.

Area of Opportunity	RECOMMENDED POLLUTION PREVENTION PRACTICE
General	1. Process fresh hides or skins to reduce the quantity of salt in wastewater, where feasible. 2. Reduce the quantities of salt used for preservation. When salted skins are used as raw material, pretreat the skins with salt elimination methods. 3. Use salt or chilling methods to preserve hides, instead of persistent insecticides and fungicides. 4. When antiseptics or biocides are necessary, avoid toxic and less degradable ones, especially those containing arsenic, mercury, lindane, or pentachlorophenol or other chlorinated substances. 5. Flesh green hides instead of limed hides. 6. Use sulfide and lime as a 20-50% solution to reduce sulfide levels in wastewater. 7. Split limed hides to reduce the amount of chrome needed for tanning. 8. Consider the use of carbon dioxide in deliming to reduce ammonia in wastewater. 9. Use only trivalent chrome when required for tanning.

Table 16. Continued.

Area of Opportunity	RECOMMENDED POLLUTION PREVENTION PRACTICE
	10. Inject tanning solution in the skin using high pressure nozzles; recover chrome from chrome-containing wastewaters, which should be kept segregated from other wastewaters. Recycle, chrome after precipitation and acidification. Improve fixation of chrome by addition of dicarboxylic acids. 11. Recycle spent chrome liquor to the tanning process or to the pickling vat. 12. Examine alternatives to chrome in tanning, such as titanium, aluminum, iron, zirconium, and vegetable tanning agents. 13. Use nonorganic solvents for dyeing and finishing. 14. Use photocell-assisted paint-spraying techniques to avoid overspraying. 15. Recover hair by using hair-saving methods to reduce pollution loads. For example, avoid dissolving hair in chemicals by making a proper choice of chemicals and using screens to remove hair from wastewater. 16. Precondition hides before vegetable tanning.
Water conservation	17. Monitor and control process waters; reductions of up to 50% can be achieved. 18. Use batch washing instead of continuous washing, for reductions of up to 50%. 19. Use low-float methods (for example, use 40-80% floats). Recycle liming, pickling, and tanning floats. Recycle sulfide in spent liming liquor after screening to reduce sulfide losses (by, say, 20-50%) and lime loss (by about 40 - 60%). 20. Use drums instead of pits for immersion of hides. 21. Reuse wastewaters for washing-for example, by recycling lime wash water to the soaking stage. Reuse treated wastewaters in the process to the extent feasible (for example, in soaking and pickling).
Waste reduction	22. Recover hide trimmings for use in the manufacture of glue, gelatin, and similar products. 23. Recover grease for rendering. Use aqueous degreasing methods. 24. Recycle wastes to the extent feasible in the manufacture of fertilizer, animal feed, and tallow, provided the quality of these products is not compromised. 25. Use tanned shavings in leather board manufacture. 26. Control odor problems by good housekeeping methods such as minimal storage of flesh trimmings and organic material. 27. Recover energy from the drying process to heat process water.

Table 17. P2 Practices for Breweries.

RECOMMENDED POLLUTION PREVENTION PRACTICE
1. Because breweries have a favorable steam-to-electricity ratio, they should consider cogeneration of electricity.
2. Reduce energy consumption through reuse of wort-cooling water as the process water for the next mash.
3. Collect broken glass, bottles that can be reused, and waste cardboard for recycling.
4. Consider the use of non-phosphate-containing cleaning agents.
5. Filter bottom sediments from final fermentation tanks, for use as animal feed.
6. Recover spilled beer, adding it to spent grain that is being dried through evaporation.
7. Dispose of crub by adding it to spent grain.
8. Use spent yeast that is not reused for livestock feed.
9. Dispose of wet hops by adding them to the spent grain.
10. Dispose of spent hop liquor by mixing it with spent grain.
11. Use spent grain as animal feed, either 80 percent wet, or dry after evaporation.
12. Use grit, weed seed, and discarded grain as chicken feed.
13. Install recirculating systems on cooling water circuits.
14. Use high-pressure, low-volume hoses for equipment cleaning.
15. Decontaminate equipment through clean-in-place methods.

Table 18. P2 Practices for Cement Factories.

RECOMMENDED POLLUTION PREVENTION PRACTICE
1. Install equipment covers and filters for crushing, grinding, and milling operations.
2. Operate control systems to achieve the required emissions levels.
3. Use low-sulfur fuels in the kiln.
4. Use low-NO_x burners with the optimum level of excess air.
5. Wet down intermediate- and finished-product storage piles.
6. Use enclosed, adjustable conveyors to minimize drop distances.

Table 19. P2 Practices for Vegetable Oil Processing.

RECOMMENDED POLLUTION PREVENTION PRACTICE
1. Where feasible, collect waste product for use in by-products, such as animal feed (while meeting animal-feed quality standards).
2. Identify and reduce production losses though continuous sampling and measuring of key production parameters.
3. Prevent odor problems through good hygiene and storage practices. Do not use chlorinated fluorocarbons refrigeration systems.
4. Recirculate cooling waters.
5. Optimize the use of water and cleaning chemicals.
6. Recover solvent vapors, to minimize losses.
7. Provide dust extractors, to maintain a clean workplace, recover product, and control air emissions.
8. Maintain volatile organic compounds (VOCs) well below explosive limits. Hexane should be below 150 mg per cubic meter of air (its explosive limit is 42,000 mg per cubic meter).
9. Reduce product losses through better production control.
10. Where appropriate, give preference to physical refining rather than chemical refining of crude oil, as active clay has a lower environmental impact than the chemicals generally used.
11. Where feasible, use citric acid instead of phosphoric acid in degumming operations.
12. Prevent the formation of molds on edible materials by controlling and monitoring air humidity.

Table 20. P2 Practices for Wood Preserving.

RECOMMENDED POLLUTION PREVENTION PRACTICE
1. Apply proper labels, and return used packaging to the supplier for reuse, or send it for other acceptable uses or destruction.
2. Store preservatives and other hazardous substances safely, preferably under a roof with a spill-collection system.
3. Select sites that are not prone to flooding, or adjacent to water intake points or valuable groundwater resources.
4. Cover process areas and collect surface runoff for recycling and treatment. Where water-based preservatives are used, prevent freshly treated wood from coming into contact with rainwater.
5. Minimize surface run-on by diversion of stormwater away from the process areas.
6. Use concrete pads for the wood treatment area and intermediate storage areas, to ensure proper collection of drippage. Treated wood should be sent for storage only after drippage has completely stopped.
7. Recycle collected drips after treatment, if necessary.
8. Minimize drippage by mechanically shaking extra preservative from the wood surface until no drippage is noticeable. Provide sufficient holding time after applying the preservative, to minimize free liquid.
9. Give preference to pressurized treatment processes, to minimize both wastage of raw materials and the release of toxic substances that may be present.
10. Minimize contamination of surface runoff and soil. Have a closed system for managing liquids, to avoid the discharge of liquid effluents.

Table 20. Continued.

RECOMMENDED POLLUTION PREVENTION PRACTICE
11. Exhaust streams should be treated, using carbon filters that allow the reuse of solvents, to reduce volatile organic compounds (VOCs) to acceptable levels before venting to the atmosphere. Where VOC recovery is not feasible, destruction should be carried out in combustion devices or bio-oxidation systems.
12. Do not use pentachlorophenol, lindane, tributyltin, or copper chrome arsenate (or its derivatives).

Table 21. P2 Practices for Electronics Manufacturing.

RECOMMENDED POLLUTION PREVENTION PRACTICE
1. Where liquid chemicals are employed, the plant, including loading and unloading areas, should be designed to minimize evaporation (other than water) and to eliminate all risk of chemicals entering the ground, any watercourse, or sewerage system, in the event of an accidental leak or spill.
2. Secure cylinders of toxic gases and fit them with leak-detection devices, as appropriate.
3. Require well-designed emergency preparedness programs. Note that fugitive emissions occurring when gas cylinders are changed do not normally require capture for treatment, but appropriate safety precautions should be in place.
4. No ozone-depleting chemicals should be used in the process unless no proven alternatives are available.
5. Equipment containing ozone-depleting chemicals, such as refrigeration equipment, should not be purchased unless no other option is available.
6. Toxic and hazardous sludges and waste materials must be treated and disposed of, or sent to approved waste-disposal or recycling operations.

Table 22. P2 Practices for Electroplating.

Area of Opportunity	POLLUTION PREVENTION RECOMMENDATION
Changes in process	1. Replace cadmium with high-quality, corrosion-resistant zinc plating. Use cyanide-free systems for zinc plating where appropriate. Where cadmium plating is necessary, use bright chloride, high-alkaline baths, or other alternatives. Note, however, that use of some alternatives to cyanides may lead to the release of heavy metals and cause problems in wastewater treatment.
	2. Use trivalent chrome instead of hexavalent chrome; work to promote acceptance of the change in finish.
	3. Where feasible, give preference to water-based surface-cleaning agents instead of organic cleaning agents, some of which are considered toxic.
	4. Whenever feasible, regenerate acids and other process ingredients.
Reduction in dragout and wastage	5. Minimize dragout through effective draining of bath solutions from the plated part — for example, by making drain holes in bucket-type pieces, if necessary.
	6. Allow dripping time of at least 10 to 20 seconds before rinsing.
	7. Use fog spraying of parts while dripping.

Table 22. Continued.

Area of Opportunity	POLLUTION PREVENTION RECOMMENDATION
	8. Maintain the density, viscosity, and temperature of the baths, to minimize dragout. 9. Place recovery tanks before the rinse tanks (also yielding make-up for the process tanks). The recovery tank provides for static rinsing with high dragout recovery.
Minimizing water consumption in rinsing systems	10. Agitate rinse water or work pieces to increase rinsing efficiency. 11. Apply multiple countercurrent rinses. 12. Spray rinses (especially for barrel loads).
Management of process solutions	13. Recycle process baths after concentration and filtration. Send spent bath solutions for recovery and regeneration of plating chemicals; do not discharge into wastewater-treatment units. 14. Recycle rinse waters (after filtration). 15. Regularly analyze and regenerate process solutions, to maximize useful life. 16. Clean racks between baths to minimize contamination. 17. Cover degreasing baths containing chlorinated solvents when not in operation, to reduce losses. Spent solvents should be sent to solvent recyclers and the residue from solvent recovery properly managed (that is, blended with fuel and burned in a combustion unit with proper controls for toxic metals).

Table 23. P2 Practices for Foundries.

RECOMMENDED POLLUTION PREVENTION PRACTICE
1. Reclaim sand after removing binders. 2. Reduce emissions of nitrogen oxides by using natural gas as a fuel; use low-NO_x burners. 3. Control water consumption by recirculating cooling water after treatment. 4. Store chemicals and other materials in such a way that spills, if any, can be collected. 5. Use closed-loop systems in scrubbers, in cases where scrubbers are necessary. 6. Wherever feasible, use continuous casting for semifinished and finished products. 7. Use dry dust-collection methods, such as fabric filters, instead of scrubbers. 8. Provide hoods for cupolas, or doghouse enclosures for EAFs and induction furnaces. 9. Improve feed quality — use selected and clean scrap to reduce the release of pollutants to the environment. Preheat scrap, with afterburriting of exhaust gases. Store scrap under cover to avoid contamination of stormwater. 10. Use induction furnaces, rather than cupola furnaces. 11. Where feasible, replace the cold-box method for core manufacture.

Table 24. P2 Practices for Fruit/Vegetable Processing.

RECOMMENDED POLLUTION PREVENTION PRACTICE
1. Reuse concentrated wastewaters and solid wastes for production of by-products.
2. Remove solid wastes without using water.
3. Minimize the use of water for cleaning floors and machines.
4. Use steam instead of hot water to reduce the quantity of wastewater going for treatment (taking into consideration, however, the tradeoff with increased use of energy).
5. Where washing is necessary, use countercurrent systems. Recirculation of process water from onion preparation, for example, reduces the organic load by 75 percent and water consumption by 95 percent. Similarly, the liquid-waste load (in terms of biochemical oxygen demand, or BOD) from apple juice and carrot processing can be reduced by 80 percent.
6. Separate and recirculate process wastewaters.
7. Use dry methods, such as vibration or air jets, to clean raw fruit and vegetables. Dry peeling methods reduce the effluent volume by up to 35 percent, and reduce pollutant concentration (organic load is reduced by up to 25 percent).
8. Use solid wastes, particularly from processes such as peeling and coring, as animal feed. They typically have a high nutritional value.
9. Procure clean raw fruit and vegetables, thus reducing the concentration of dirt and organics (including pesticides) in the effluent.

Table 25. P2 Practices for Meat Processing/Rendering.

RECOMMENDED POLLUTION PREVENTION PRACTICE
1. Recover and process blood into useful by-products. Allow enough time for blood draining (at least seven minutes).
2. Process paunches and intestines, and utilize fat and slime.
3. Minimize water consumed in production by, for example, using taps with automatic shutoff, using high water pressure, and improving the process layout.
4. Eliminate wet transport (pumping) of wastes (for example, intestines and feathers), to minimize water consumption.
5. Reduce the liquid-waste load by preventing any solid wastes or concentrated liquids from entering the wastewater stream.
6. Cover collection channels in the production area with grids, to reduce the amount of solids entering the wastewater.
7. Remove manure (from the stockyard and from intestine processing) in solid form.
8. Dispose of hair and bones to the rendering plants.
9. Reduce air emissions from ham processing through some degree of air recirculation, after filtering.
10. Isolate and ventilate all sources of odorous emissions. Oxidants, such as nitrates, can be added to wastes to reduce odor.
11. Equip the outlets of wastewater channels with screens and fat traps, to recover and reduce the concentration of coarse material and fat in the combined wastewater stream.
12. Implement dry precleaning of equipment and production areas prior to wet cleaning.
13. Separate cooling water from process water and wastewaters, and recirculate cooling water.
14. Optimize the use of detergents and disinfectants in washing water.

Table 25. Continued.

RECOMMENDED POLLUTION PREVENTION PRACTICE
In *rendering plants,* odor is the most important air-pollution issue. To reduce odor: 1. Minimize the stock of raw material and store it in a cold, closed, well-ventilated place 2. Pasteurize the raw material before processing it, to halt biological processes that generate odor 3. Install all equipment in closed spaces and operate under partial or total vacuum 4. Keep all working and storage areas clean

Table 26. P2 Practices for the Printing Industry.

RECOMMENDED POLLUTION PREVENTION PRACTICE
1. Treat metal-containing effluents from the manufacture of gravure cylinders and printing blocks by applying the established methods of chemical precipitation, sedimentation, and filtration. Collect fixing baths for recovery or destruction. Evaporate solvents from regeneration of active carbon filters. Perform closed-screen chase washing; recirculate solvents and separate sludge. Fit developing machines with counterflow fixing, or connect them to an organic ion exchanger. Collect film-developing agents for destruction. Perform high-pressure water-jet cleaning. Use ultrafiltration to treat washing water. 2. Where possible, replace chemicals used for form preparation and cleaning with more environmentally friendly alternatives. Maintain a record of chemicals and environmentally hazardous waste. Do not use halogenated solvents and degreasing agents in new plants. Replace them with nonhalogenated substances in existing facilities. 3. Estimate the quantity of developing bath and fixing bath used per year, and maintain these at acceptable levels. 4. Store chemicals and environmentally hazardous waste, such as dyes, inks, and solvents, so that the risk of spillage into the wastewater system is minimized. Examples of measures that should be considered including retaining dikes, or areas with no outlet. Minimize noise disturbance from fans and presses. 5. Return toxic materials packaging to the supplier for reuse. 6. Label and store toxic and hazardous materials in secure, bonded areas. 7. Recover plates by remelting. 8. Recover energy from combustion systems, when they are used. 9. Use equipment washdown waters as make-up solutions for subsequent batches. Use counter-current rinsing. 10. Use counter-current flow-fixing processes. Aim for a closed washing system. 11. Minimize the rinsewater flow in the developing machines by, for example, use of "stand-by." Collect fixing bath, developer, used film, photographic paper, and blackened ends of photosetting paper, and manage them properly. 12. Where feasible, enclose presses and ovens, to avoid diffuse evaporation of organic substances entering the general ventilation system. Use suction hoods to collect vapors and other fugitive emissions. 13. Engrave, rather than etch, gravure cylinders, to reduce the quantity of heavy metals used. 14. Give preference to the use of radiation-setting dyes. 15. Where feasible, replace solvent-based dyes and glues with solvent-free or water-based dyes and glues. Choose water-based dyes for flexographic printing on paper and plastic, and for screen printing and rotogravure.

Table 26. Continued.

RECOMMENDED POLLUTION PREVENTION PRACTICE
16. Control emissions of gases from web offset with heat-setting thermic or catalytic incineration. Recover toluene from rotogravure by absorption, using active carbon. Carry out adsorption of solvents, using zeolites, and recover organic solvents.
17. Evacuate air from printing presses and drying ovens into a ventilation system.
18. Estimate and control, typically on an annual basis, the quantities of volatile organic solvents used, including the amount used in dyes, inks, glues, and damping water. Estimate and control the proportion that is made up of chlorinated organic solvents.

Table 27. P2 Practices for the Textile Industry.

RECOMMENDED POLLUTION PREVENTION PRACTICE
1. Match process variables to type and weight of fabric. This reduces wastes by 10 to 20 percent.
2. Manage batches to minimize waste at the end of cycles.
3. Avoid nondegradable or less-degradable surfactants (for washing and scouring), and spinning oils.
4. Avoid the use, or at least the discharge, of alkyl-phenol ethoxylates. Ozone-depleting substances should not be used, and the use of organic solvents should be minimized.
5. Use transfer printing for synthetics. This reduces water consumption from 250 liters per kilogram to 2 liters per kilogram of material, and reduces dye consumption. When feasible, use water-based printing pastes.
6. Use pad batch dyeing. This saves up to 80 percent of energy requirements and 90 percent of water consumption, and reduces dye and salt use. For knitted goods, choose exhaust dyeing.
7. Where feasible, use jet rivers with a liquid-to-fabric ratio of 4:1 to 8:1, instead of winch rivers with a ratio of 15:1.
8. Avoid benzidine-based azo dyes and dyes containing cadmium and other heavy metals. Do not use chlorine-based dyes.
9. Use less-toxic dye carriers and finishing agents. Avoid carriers containing chlorine, such as chlorinated aromatics.
10. Replace dichromate oxidation of vat dyes and sulfur dyes with peroxide oxidation.
11. Reuse dye solution from dye baths.
Where feasible, use peroxide-based bleaches instead of sulfur and chlorine-based bleaches.
12. Use biodegradable textile-preservation chemicals. Do not use polybrominated diphenylethers, dieldrin, arsenic, mercury, or pentachlorophenol in mothproofing, carpet backing, and other finishing processes. Where feasible, use permethrin for mothproofing instead.
13. Control makeup chemicals.
14. Reuse and recover process chemicals, such as caustic (this reduces chemical costs by 30 percent) and size (up to 50 percent recovery is feasible).
15. Replace nondegradable spin-finish and size with degradable alternatives.
16. Use countercurrent rinsing.
17. Control the quantity and temperature of water used.
18. Improving your cleaning and housekeeping measures may reduce water usage to less than 150 cubic meters per ton of textiles produced.
19. Recover heat from wash water. This reduces steam consumption.

Table 28. P2 Practices for the Mining Industry.

Area of Opportunity	RECOMMENDED POLLUTION PREVENTION PRACTICE
Mining operations	1. Remove and store topsoil properly. 2. Restore worked-out areas and spoil heaps early, to minimize the extent of open areas. 3. Divert and manage surface and groundwater, to minimize water-pollution problems. Simple treatment to reduce the discharge of suspended solids may also be necessary (treatment of saline groundwater may be difficult). 4. Identify and manage areas with high potential for AMD generation. 5. Minimize AMD generation by reducing disturbed areas, and isolate drainage streams from contact with sulfur-bearing materials. 6. Prepare a water-management plan for operations and postclosure that includes measures to minimize liquid wastes by such methods as recycling water from tailings wash plant. 7. Minimize spillage losses by properly designing and operating coal-transport and transfer facilities. 8. Reduce dust by early revegetation, and by good maintenance of roads and work areas. Specific dust-suppression measures, such as minimizing drop distances, covering equipment, and wetting storage piles, may be required for coal-handling and loading facilities. Control the release of dust from crushing and other coal-processing and beneficiation operations. 9. Control the release of chemicals (including floatation chemicals) used in beneficiation processes. 10. Minimize the effects of subsidence by using careful extraction methods in relation to surface uses. 11. Control methane, a greenhouse gas, to levels of less than 1 percent by volume, to minimize the risk of explosion in closed mines; where feasible, recover the methane. When methane content is above 25 percent by volume, it should be recovered and used as a fuel. Co-generation or use in natural gas vehicles are possibilities (*see the case studies section earlier in this chapter*). 12. Develop restoration and revegetation methods appropriate to the specific site conditions. 13. Conduct proper storage and handling of fuel and chemicals used on-site, to avoid spills.
Mine closure and restoration	14. Return the land to conditions capable of supporting prior land use, equivalent uses, or other environmentally acceptable uses. 15. Use overburden for backfill, and use topsoil (or other plant-growth medium) for reclamation. 16. Contour slopes, to minimize erosion and runoff. 17. Plant native vegetation, to prevent erosion and encourage self-sustaining development of a productive ecosystem on the reclaimed land.

Table 29. U.S. Auto P2 Cases Studies.

Target Substance(s)	Project Description	Results	Savings $
General Motors, Inland Fisher Guide (Livonia, Michigan)			
Pollution Prevention Project: *Solvent-Free Spray Adhesives for Interior Trim*			
VOCs: methylene chloride, methyl ethyl ketone, hexane, and toluene	Substitution with a solvent-free adhesive for interior trim.	1. Substituted water-based adhesive, eliminating 20 tons of VOC emissions. 2. Converted solid-waste stream from hazardous to non-hazardous. 3. Achieved lower annual disposal-cost savings.	26,000
General Motors, Hamtramck Assembly			
Pollution Prevention Project: *Copper and Nickel Reclamation from Plating Waste*			
Copper and Nickel	Off-site reclamation of nickel and copper from nitric acid, by using a small tanker to ferry the acid from an inaccessible storage tank to the reclamation facility.	1. Reclaimed 68 tons of copper and 40 tons of nickel annually; normally a hazardous-waste sludge. 2. Achieved annual savings in wastewater-treatment costs.	23,000
General Motors, Hamtramck Assembly			
Pollution Prevention Project: *Adjusting Paint Equipment Reduces Emissions and Solid Waste*			
VOCs and paint sludge (toluene, xylene, methanol, and butyl cellosolve acetate)	Paint-spray equipment timing was fine-tuned; excess paint is no longer sprayed after the target body moves out of range.	1. Source reduction of 5.5 tons of VOCs. 2. Reduction of 4 tons of paint sludge from paint overspray.	85,000
Chrysler Newark Assembly Plant			
Pollution Prevention Project: *Glycol Ether Reduction in Surface Preparation Materials*			
Glycol ethers	Reformulation of a surfactant to remove glycol ether content.	1. Ninety-one percent source reduction of 71,920 kg of glycol ether. 2. Reduced employee exposure.	500,000
Ford Utica Plant			
Pollution Prevention Project: *Molded Fiberglass Headliner Offal Reduction Project*			
Chalet cloth offal	Changed the specs for chalet cloth, fiberglass matte, and foam/glue composite used to make headliners, to produce less offal.	1. Fifteen percent (45 ton) reduction of offal generated per year.	430,000

Table 29. Continued.

Target Substance(s)	Project Description	Results	Savings $
Ford Assembly Plant			
Pollution Prevention Project: *Basecoat and Solvent Reduction Project*			
Paint and solvent	Fine-tuned the timing of a paint-spray nozzle to improve its transfer efficiency.	1. Reduction of 12,862 kg of paint and solvent use per year. 2. Reduction in usage of paint-sludge treatment chemical.	63,224
Chrysler Warren Stamping Plant			
Pollution Prevention Project: *Adhesive Waste Reduction*			
Waste adhesives	Inefficiencies in the adhesive pumping process resulted in the disposal of usable adhesive material.	1. Elimination of 50 cubic meters of waste adhesive per year. 2. Improved quality.	500,000
Ford Norfolk Assembly Plant			
Pollution Prevention Project: *Paint Shop VOC Reduction Program*			
VOC Emissions from paint solvents	Paint-area management asked suppliers to review and improve the current method of solvent usage and to review the spray-booth cleaning procedure.	1. VOC solvent reduction of approximately 4 cubic meters per week. 2. VOC reduction of 155,000 kilograms per year	95,355
GM Fairfax Assembly Plant			
Pollution Prevention Project: *Turning off the Water Saves Millions of Gallons*			
Wastewater	Installation of a photoelectric cell connected to a timer to control the flow of rinse water.	1. Annual reduction of 25,000 cubic meters of water.	33,000
GM Technical Center			
Pollution Prevention Project: *Reducing the Volume and Cost of Plating Wastes*			
Nickel, copper, cyanide, and zinc	Reduced treated rinse water from the plating operations by extending the retention time in the settling tank.	1. Reduction of wastewater pump-out by 62.4 tons during the first year of implementation.	8,282

QUESTIONS FOR THINKING AND DISCUSSING

1) Select a portion of your plant and develop a P2 benefits matrix about the process. Which cost-savings categories are your enterprise most likely to benefit from?

2) Look at the appropriate *Pollution Prevention General Practices* tables starting on page 233, find the type of industry your enterprise is engaged in, and compare how many of these practices are followed by your enterprise. How many, and which others of these P2 practices might benefit your enterprise's operations?

3) If your enterprise practices pollution prevention, then develop a pollution prevention matrix of company-specific case studies to see exactly what the benefits are to your enterprise.

4) Try to gather information that enables your enterprise to compare its yearly average emissions and unit costs of production to your industry sector. Then see if you can make a comparison to the industry sector leader or leaders. How does your company compare to industry averages and to the leading companies?

Chapter 7
THE POLLUTION PREVENTION AUDIT

INTRODUCTION

What is an Audit?

Webster's dictionary defines an audit as "an examination or verification of records or accounts," most often associated with finances. However, audits or auditing practices are applied for many other reasons throughout industry, and do not just focus on records or accounts, but actual field conditions and events.

In other words, an audit is a physical examination of records *and* actual practices. Most audits have the same initial objective - that is, to gather information aimed at developing a baseline description of the subject under investigation.

Once an enterprise knows the baseline, then it can compare it to a known standard, to assess how well the subject under investigation performs. The standard by which comparisons are made to the baseline can be based on other companies within an industry sector, on industry-specific standards, on legal standards, or on internal goals set by the enterprise. This is known as *benchmarking*. By performing benchmarking, an enterprise can formulate a basis for recommending corrective actions aimed at improving the performance of the subject.

There are four essential elements of any audit:

- Examination
- Developing a baseline
- Benchmarking
- Developing recommendations for corrective actions

Types of Audits and How the P2 Audit Differs

Some of the more common types of non-financial audits applied throughout industry are listed in Table 1. There are subsets to a number of these audits.

Although it is beyond the scope of this book to discuss the objectives and functions of each type of audit, we should point out some of the characteristics that distinguish a P2 audit from all others.

First, many environmental engineers make the mistake of viewing a P2 audit as a subset of the environmental audit. This is simply not the case. Though environmental audits and their subsets have different objectives, their overall intent is to develop corrective actions that largely focus on *lowest achievable emission rate* (LAER) technologies, and only sometimes on *best available control technologies* (BACT). To understand the importance of this, the reader must have a clear understanding of LAER and BACT.

Table 1. Various Types of Audits and Their Subsets.

Type of Audit	Audit Subset
Environmental	Phase I Environmental
	Phase II Environmental
	Property Transaction (often a subset of Phase I and/or II Type Environmental Audits)
	Due Diligence
	Environmental Impact
	Air-Quality Impact
Health and Safety	Process Safety
	Indoor Air Quality
	Accident Statistic
	Accident Prevention
	Post Incident (Accident)
	Fire Prevention
ISO 14001	EMS Audit (see Chapter 2)
	Initial Environmental Review (see Chapter 4)
	Gap Analysis
ISO 9000 Quality Family	No subsets
Pollution Prevention	Waste Minimization
	Energy Efficiency
Waste Minimization	No subsets
Energy Efficiency	No subsets

The term LAER from a practical standpoint refers to corrective actions (i.e., technologies) capable of achieving the lowest emission rates. So, if a process is emitting nitrogen oxides (NO_x) and LAER is applied as the basis for reducing NO_x emissions, then we are forced to select only those technologies that will achieve the lowest emissions (based on a legal standard), without regard to the cost of the technology. In contrast, the term BACT (sometimes known as BAT, for best available technology) generally refers to those technologies that cost-effectively achieve the goals (reduced emissions) of the corrective action. In addition, LAER and BACT most often focus on end-of-pipe technologies. P2 technologies can be BACT, but there are far more industry examples of end-of-pipe treatment approaches.

The vast majority of environmental audits and subsets are driven by regulations, whether for the purpose of site clean-up (i.e., remediation) or for compliance purposes for operating facilities. This means that, in most situations where the driving force is purely regulatory, the cost for compliance is a secondary consideration or not a consideration at all.

A P2 audit is very different, simply because corrective actions are always based upon cost-sensitivity considerations. As we stated in Chapter 5, a P2 program only makes sense when it is financially attractive or, at a minimum, is break-even from a cost standpoint. In additions, P2 programs are not driven by regulations (though they often anticipate them) and, like environmental management systems (EMSs), they are totally voluntary.

A P2 audit differs from most other types of audits because it makes use of a dual benchmarking approach; namely, it uses both technical (environmental performance) and financial performance as its basis for making corrective actions. The status quo most often serves as the benchmark, though other standards certainly can be devised.

A P2 audit can also affect non-environmental issues. As shown from the case studies in Chapter 6, the types of corrective actions (i.e., P2 activities) can affect other types of wastes, energy, occupational safety, product quality, and worker productivity. It is worthwhile to look back at the *P2 Benefits Matrix* in Chapter 6 (page 232), because this clearly shows that benefits of a P2 program can be a part of the corrective actions derived from all other types of audits listed in Table 1. Consequently, when implementing a P2 audit, it makes a lot of sense to refer to other ongoing and past audits at a facility, and to coordinate efforts among different audits.

The P2 Audit: Its Objectives and a Three-Phase Approach

The audit protocol presented in this chapter is significantly different from other general approaches. The objectives of the audit are:

- To develop a baseline of the environmental performance
- To assess one or more corrective actions (P2 activities) with attractive economics which improve environmental performance.

The same four elements common to all audits are applied - examination, developing a baseline, benchmarking, and developing corrective actions. However, we make two distinctions here. First, environmental performance benchmarking will always be against the status quo of operations; and second, financial benchmarking will be used as the basis for selecting BATs. Financial benchmarking is treated in Chapter 8.

This book follows a step-by-step approach for the P2 audit that is distributed among three consecutive phases:

- Phase I: Pre-assessment for audit preparation
- Phase II: The in-plant assessment
- Phase III: Synthesis, benchmarking, and corrective actions

It is possible that not all of the steps will be relevant to every situation. Similarly, in some situations, additional steps may be required. The audit should therefore be considered flexible, not rigid, and the auditors should be prepared to adapt and modify the steps outlined in this chapter to the specific site conditions.

THE PRE-ASSESSMENT (PHASE I)

Step 1: Audit Focus and Preparation

Step 1.1: Getting Ready. A thorough preparation for a P2 audit is a prerequisite for an efficient and cost-effective evaluation. Gaining support for the assessment from top-level management, and for the implementation of results, is particularly important. Otherwise, there will be no real action on recommendations. In other words, early in the process, management needs to accept that, at a bare minimum, the audit is a worthwhile exercise and that resources - human and financial - should be diverted from other activities to the task of auditing.

In an enterprise built around an ISO 9000 environment, it goes without saying that employees are empowered to take on certain actions, bring to management's attention the need for change, or identify possible changes leading to improved performance. But you will find that some P2 audits can require significant effort, especially when there are complex unit processes involved in the manufacturing train. As such, members of local management (the group leader, the section head, the general manager) need to be involved at the very start of developing a P2 program.

Step 1.2: Assemble the Audit Team. The members of the P2 auditing team should be identified and assembled. The number of people required on a team will depend on the size and complexity of the processes to be investigated. A P2 audit of a small factory may be undertaken by one person with contributions from the employees. A more complicated process may require at least three or four people: for example, technical staff, production employees, and an environmental specialist.

Select team members on the basis of their expertise. Don't just limit the team members to environmental specialists. Environmental and environmental health and safety (EHS) engineers and specialists will understand what the current costs are for end-of-pipe treatment technologies and for disposal, and well- trained and experienced individuals will also have a good understanding of the subject facility's compliance issues. But these team members will generally not have a keen understanding of the process details and, more importantly, of the specific hardware associated with each unit operation. Further, we must recognize that because environmental engineers traditionally are trained in pollution *control* technologies, they are not always a source of P2 ideas. As such, the team needs to combine the brainstorming chemistry of different disciplines - for example, between a process engineer, an operator, possibly a chemist, laboratory or product quality personnel, and the EHS personnel.

Involving personnel in varying degrees or on an as-needed basis from each stage of the manufacturing operations will increase employee awareness of pollution and waste, and promote input to and support for the program.

Step 1.3: Identify and Allocate Additional Resources. The audit may require external resources, such as laboratory facilities and possibly equipment for air sampling, flow measurements, energy measurements, and product-quality testing, to name a few. A useful exercise is to review what the control and monitoring parameters are for different parts of the process. Develop a checklist that includes parameters into the following three categories:

1) *Control Variables.* This category includes those parameters that control a process - for example, the temperature, pressure, rate of catalyst feed to a reactor, agitator speed, residence time, etc.
2) *Dependent Variables.* This category includes parameters affected by the process controls - for example, production rate, product properties (viscosity, color, density, molecular weight, etc.), emissions, etc.
3) *Monitoring Parameters.* This category includes those parameters that are actually measured and used as a basis to control the process. Ideally, these are the control variables, but in many cases the corrective actions taken to control processes are derived from inferences drawn from product-quality measurements. One example of this is in the manufacturing of synthetic rubber from a solution-based process. The operator of the polymerization reactor controls pressure, temperature, residence time, catalyst feed rate, and other control variables, based upon a series of product-quality control tests. Quality control labs rely on "fast-response" tests – "quick and dirty" tests that determine whether or not the product is being made within spec. The results of these tests provide guidance to the process operator; hence, process control is performed by inference. This is widely practiced in many industry

sectors, including cosmetics manufacturing, paint and pigments, the paper industry, and oil refining, to name a few.

The reader may very well ask, "Why do I need to know this? After all, my focus is on pollution prevention. Therefore, shouldn't we only focus on the waste and pollution parameters?" The reason we need to know and understand what these parameters are and how they effect the production and product quality is that P2 activities only fall into two categories - namely, best practices (where only operational changes and/or minor design changes are made to a process) and re-engineering (where certain changes to the process are made, involving simple to complex process modifications, and even replacement of an entire process with an alternative technology). To assess as many options as possible to arrive at the optimum recommendation, we must have a technical understanding of the process, the unit operations, the hardware, and the parameters that are normally measured for control purposes.

From this list of three (*control variables*, *dependent variables*, and *monitoring parameters*), the team will gain an idea of what analytical support it may need once the audit gets underway. You should attempt to identify external-resource requirements at the outset. Analytical services and equipment may not be available to a small factory. If this is the case, investigate the possibility of forming a P2 association with other factories or industries. This will enable all enterprises involved to share the external resource costs.

Step 1.4:Select the Subject Facility. It is important to decide on the focus of your audit during the preparation stage. You may wish to audit a complete process or you may want to concentrate on a selection of unit operations within a unit process. The focus will depend on the overall objectives of the audit.

A good way to approach this is by conducting a brainstorming session with operators and environmental specialists. Such sessions often help to bring to the surface those parts of the plant where suspect waste issues or pollution problems are most critical. Such areas of operations are most likely to benefit from an audit and P2 program.

Step 1.5: Define the Audit Objectives. If the objectives are not entirely environmental, then you may wish to look at waste minimization as a whole or you may wish to concentrate on particular wastes - for example, raw material losses, wastes that cause processing problems, wastes considered to be hazardous or for which regulations exist, and/or wastes for which disposal costs are high. A good starting point for designing a P2 audit is to determine the major problems and wastes associated with your particular process. This can be accomplished by a thorough review of existing documentation, as described in the next sub-step.

Step 1.6: Review Documentation. Collate and review all existing documentation and information regarding the process. Regional or plant surveys may have been undertaken, environmental audits may have been conducted, safety audit findings may be available; these could yield useful information indicating the areas for concern, and will show gaps where no data are available. The following prompts give some guidelines on documentation worthy of review.

- Is a site plan available?
- Are process flow diagrams available?
- Have the process wastes ever been monitored and do you have access to the records?
- Do you have a map of the surrounding area, indicating watercourses, hydrology, and human settlements?
- Are there any other factories or plants in the area that may have similar processes?
- What are the obvious wastes associated with your process?
- Where is water used in greatest volume?
- Do you use chemicals that have special instructions for their use and handling?
- Do you have waste treatment and disposal costs - what are they?
- Where are your discharge points for liquid, solid, and gaseous emissions?

Step 1.7: Gain Employee Buy-in and Participation. The importance of top management support *and* employee participation cannot be overstated - for P2 or EMSs. Without either of these, a P2 or EMS program will not be successfully implemented.

The support of the staff is imperative for this type of interactive study. Inform plant employees that the assessment will be taking place, and encourage them to take part. It is important to undertake the assessment during normal working hours, so employees and operators can be consulted, the equipment can be observed in operation, and wastes can be quantified.

Step 2: Listing the Unit Operations

Manufacturing processes are made up of a number of *unit processes*, which are in turn made up of *unit operations*. Unit processes are distinct stages of a manufacturing operation. They each focus on one stage in a series of stages, successively bringing a product to its final form. For example, in our example of the synthetic-rubber manufacturing plant, the unit processes from start to finish in producing a final solid bale of rubber are:

- Catalyst preparation stage - a pre-preparation stage for monomers and catalyst additives

- Polymerization - where an intermediate stage of the product is synthesized in the form of a latex or polymer suspended as a dilute solution in a hydrocarbon diluent
- Finishing - where the rubber is dried, residual solvent is removed, and the rubber is compressed into a bale and packaged

As mentioned above, each unit process is made up of individual unit operations. A unit operation is an area where materials and energy are input, a function occurs, and materials are output, possibly in a different form, state, or composition. Unit operations, or "unit ops," are the individual process steps within each major unit process.

Quite often, the same unit ops are used in different unit processes within the same plant, but they have different functions, and hence do not use the same equipment. In the case of a synthetic-rubber producing plant, the unit operation of *filtration* can be found at several different points within the manufacturing train. In the polymerization process, the plant uses filtration equipment to remove contaminants resulting from the chemical reaction between monomers and catalyst. In this intermediate stage of the product, one can find cross-linked polymers, highly crystalline by-products, and catalyst residues that can damage product quality and performance. Therefore, filtration serves as a processing step in one of the unit processes. Filtration is also an important unit operation used in the finishing stages, which are a separate unit process. Here, filtration is used as a major step in treating wastewaters from the process, enabling process water for heat-exchange equipment and for rubber washing and cooling functions to be recycled.

Developing an accurate list of all the unit operations within a unit process is the starting point in identifying P2 and waste-minimization opportunities. Without such a list, the team will not be able to identify and, later on, quantify possible P2 and waste reductions. Furthermore, without the list of individual unit ops, the team cannot assess re-engineering options. Ultimately, each unit op can be related to a piece of process equipment. It is the process equipment that is ultimately modified, either physically or operationally, or that can be replaced with an alternative piece of equipment or newer technology.

It is best not to think of any one unit operation in terms of process equipment at this stage of the audit. Process-equipment options for each unit operation are considered much further into the P2 audit and, indeed, may not be necessary to consider at all if the P2 or waste minimization opportunity identified falls into the best practices category.

To develop a useful list, we recommend the following sub-steps.

Step 2.1: Refine the Initial Checklist. Table 2 is a checklist of unit operations. Apply this as an initial guide and checklist that can be used to identify each unit operation within the unit process that the audit will focus on. You may have to make additions to the list, because it is meant to be a general overview. Next to each unit operation that applies to the unit process of interest, provide a brief description of its function or purpose. This will prove useful in developing a process flow scheme in

Step 3. Much of this sub-step can be accomplished by sitting down with other team members and reviewing the plant or unit-process objectives.

Table 2. An Initial Checklist of Unit Operations.

Name of Unit Process: *Include the name, title or general statement of the unit process in this block. Then check and briefly describe the purpose of those unit ops that apply.*					
	Unit Op	**Objective/ Purpose**		**Unit Op**	**Objective/ Purpose**
□	Absorption		□	Filtration (gas-PM-vapor)	
□	Adsorption (liquid-liquid)		□	Filtration (liquid-PM)	
□	Adsorption (liquid-liquid)		□	Flotation - separation	
□	Blending and homogenization		□	Fluid transfer	
□	Centrifugation		□	Fluidization (gas-solid)	
□	Chemical reaction		□	Fluidization (gas-liquid)	
□	Classification (PM)		□	Gas compression	
□	Communition		□	Gas Scrubbing	
□	Crystallization		□	Homogenization	
□	Desalination		□	Incineration	
□	Distillation		□	Ion exchange	
□	Drying		□	Mixing	
□	Extraction (liquid-liquid)		□	Sedimentation (PM-gas)	
□	Extrusion		□	Sedimentation (solid-liquid)	

Step 2.2: An Initial Walk-through. The entire purpose of Step 2 is to develop an initial survey of the unit process. The step will set many of the ground rules for the audit. Using the modified checklist derived from Table 2, one or more of the team should walk around not only the unit process of interest, but neighboring unit processes that depend upon or affect the subject facility. If the plant is small enough, then take the time to walk around the entire plant, to gain a thorough understanding of all the processing operations and their interrelationships.

The walkthrough will help the auditors refine their list of unit operations, and enable them to decide how best to describe a process in terms of these unit ops.

During this initial overview, it is useful to record visual observations and discussions, and to make rough sketches of process layout, drainage systems, vents, plumbing, and other material-transfer areas. These help ensure that important factors are not overlooked.

The team should consult the production staff about normal operating conditions. The production or plant staff, are likely to know about waste-discharge points, unplanned waste-generating operations (such as spills and washouts), and give the team a good indication of actual operating procedures.

Investigations may reveal that night-shift procedures are different from day-shift procedures; also, plant personnel may disclose that actual material-handling practices are different from those set out in written procedures. A long-time employee could provide insight into recurring process problems. In the absence of any historical monitoring, this information can be very useful. Such employee participation must be free from the threat of disciplinary action. During this initial site survey, note imminent problems that need to be addressed before the assessment is complete. Set the notes generated from the initial walkthrough aside for use in Step 2.3.

Step 2.3: Organize the Collected Data. As previously noted, the audit team needs to understand the function and process variables associated with each unit operation. Similarly, the team needs to collate all the available information on the unit operations and the process in general that were collected in Steps 2.1 and 2.2. For complex unit processes and plant-wide audits, it's best to organize these materials into separate files. Always try to tabulate data, so team members can more easily review and cross-reference information.

A convenient way of organizing the files is on a unit-operation basis. Use the sample file index in Table 3 as a guide to developing these files. The sample shown in this table was developed around the polymerization unit-process example for a synthetic-rubber producing plant.

Identification of materials handling operations (manual, automatic, bulk, drums, etc.) covering raw materials, transfer practices, and products is also an important aspect that could usefully be included in the above organization and tabulation of information, as a prelude to developing material balances in Phase II.

Table 3. Sample Index to File Organization.

Unit Operation	Function	File
Blending	Mix tanks for catalyst, cocatalysts, and activation ingredients. Batch-mix formulation prepared in accordance with polymer-grade recipe for producing resin according to product specifications.	1
Chemical Reaction	Injection of catalyst pre-mix from blending to reactor vessel, to promote contact with monomer. Reaction takes place in a solution of hexane (diluent). Control variables include residence time, mixer speed (this is a constant stirred tank reactor, or CSTR), temperature, diene composition, and injection point.	2
Mixing	Better known as "quenching," the purpose of this process is to terminate the polymerization reaction by lowering the reaction temperature. The polymer solution (cement) is vigorously mixed with cold water to stop the reaction. Stabilizers are injected into the cement solution to prevent other post reactions and by-products of polymerization.	3
Sedimentation	Polymer cement is sent to the sedimentation basin to remove gels (cross-linked polymer), crab meat (crystalline polymer components, or PE), and solid catalyst residues.	4
Filtration	Polymer cement is filtered in a "polishing operation" to remove micro-size suspended PM (catalyst residue).	5
Flash Evaporation	Clean cement is sent to the "flash drum" to separate hexane solvent (diluent) for recovery.	6

Step 3.: Constructing Process Flow Sheets

The next step is to connect the individual unit operations in a block diagram, creating a *process flow sheet*. Figure 1 is an example of a simplified process flow diagram for a pattern-etch process for a printed wiring-board operation. The process flow sheet provides us with a description of the flow of all materials not only into and out of the entire unit process, but through each and every unit operation. Intermittent operations - such as cleaning, make-up, or tank dumping - may be distinguished by using broken lines to link the boxes.

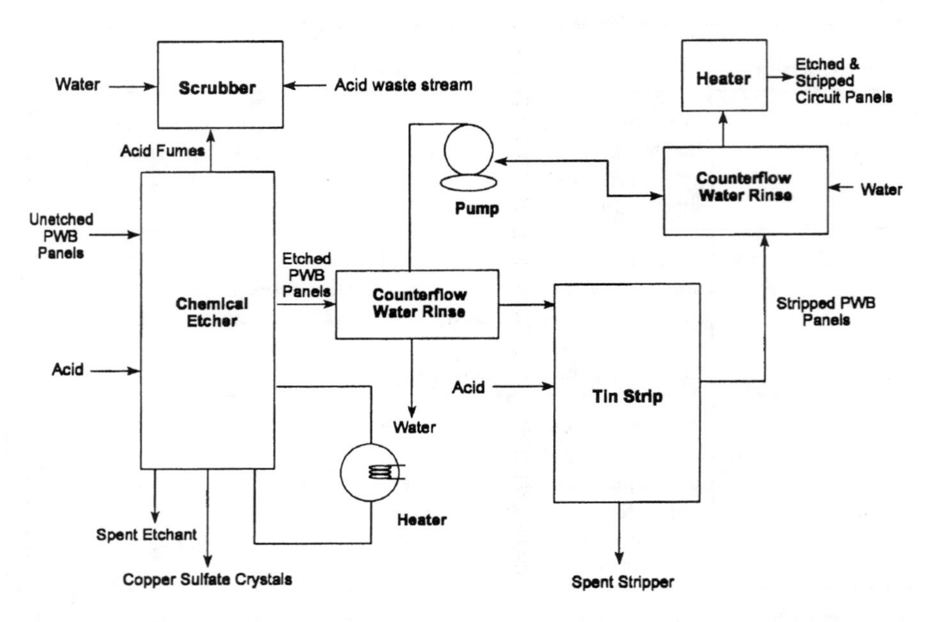

Figure 1. *Example of a process flow sheet of a pattern-etch process for a wiring-board operation.*

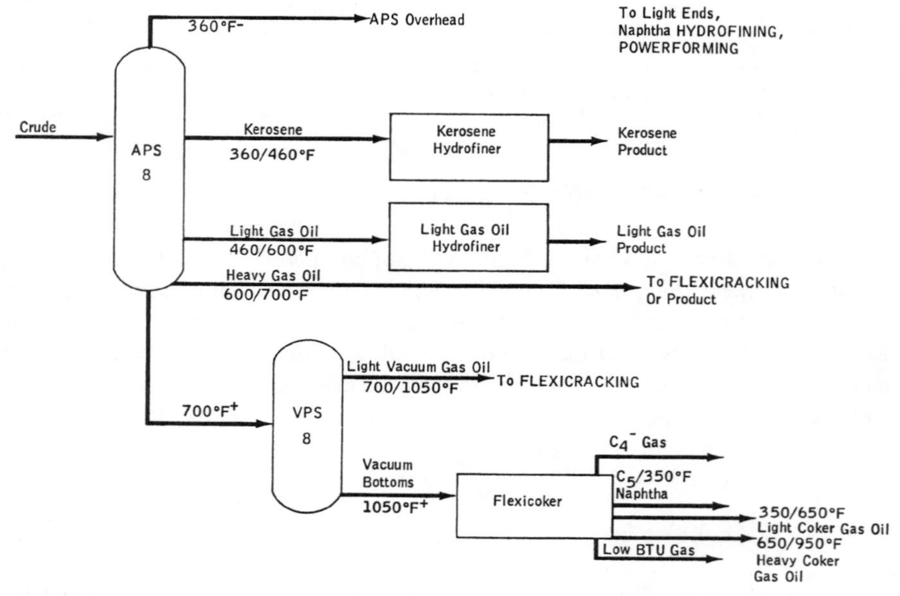

Figure 2. *Process unit flow scheme for a refinery pipestill operation.*

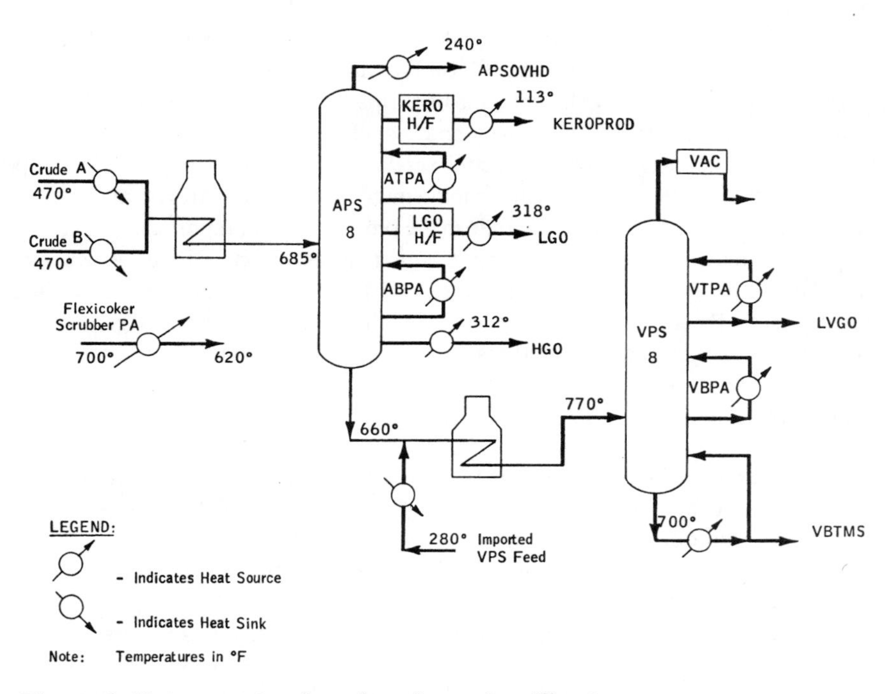

Figure 3. *Unit operation flow sheet for a pipestill unit process.*

For complex processes, prepare a general flow diagram illustrating the main process areas, and then separately prepare detailed flow diagrams for each main processing area. For examples, refer to Figures 2 and 3, which are the process flow sheets for a refinery pipestill train. Figure 2 illustrates the overall heat-recovery train for the process, whereas Figure 3 is a more detailed but simplified process flow diagram showing where the various streams enter or leave individual process units.

Decide on the level of detail that you require to achieve your objectives. It is important to realize that the less detailed or more far-reaching the assessment becomes, the more information is likely to be lost or masked by oversimplification. Establishing the correct level of detail and focusing on specific areas are very important at an early stage.

Step 4: Preliminary Assessment and Next Steps

Quite often, the walkthrough and analysis of the process flow sheets lead to an early identification of P2 and waste-reduction opportunities. By studying the process flow sheets and information gathered from the walkthrough, and by conducting brainstorming sessions with team members, the team will be able to identify sources of pollution and waste, and their causes. Pay particular attention to any obvious waste that can be reduced or prevented easily, and correct it before proceeding to the development of a material balance (Phase II). Simple changes at this early stage and the resulting benefits will help enlist the participation and stimulate the enthusiasm of employees for the total P2 assessment and reduction program.

Summarize the findings and recommendations in a brief pre-assessment report, and present and review this with management. This will reaffirm management's commitment to the next phase. Include in the pre-assessment report a list of actions or steps that the team intends to follow in the next phase. If some of these steps involve field measurements, then highlight these and their possible effects on production schedules and personnel assignments. Use the steps in Phase II as a guide to developing the recommended actions.

One More Time

At the end of Phase I, the team will have accomplished the following:

1) The team is organized and aware of the objectives of the pollution prevention assessment
2) Plant personnel are informed of the audit's purpose, to maximize cooperation between all parties concerned
3) Required financial and technical resources are secured and external facilities assessed for availability and capability in supporting the audit
4) The team is aware of the overall history and local surroundings of the plant and unit process
5) The scope and focus of the audit is established, and a rough timetable worked out, to fit in with production patterns
6) The audit team is familiar with the layout of the processes within the plant, and has developed a list of the unit operations associated with each unit process of interest
7) Sources of wastes and their causes have been identified
8) The team has identified obvious pollution and waste-saving measures that can be introduced and implemented immediately
9) The findings of the Phase I investigations are presented to top management, and commitment into the next phase obtained

To summarize, Phase I, the pre-assessment phase, is made up of the following steps:

Step 1 Audit Focus and Preparation

- Step 1.1 Get Ready
- Step 1.2 Assemble the Audit Team
- Step 1.3 Identify and Allocate Additional Resources
- Step 1.4 Select the Subject Facility
- Step 1.5 Define the Audit Objectives

Step 1.6 Review Documentation
Step 1.7 Gain Employee Buy-in and Participation

Step 2 Listing the Unit Operations
Step 2.1 Refine the Initial Checklist
Step 2.2 Conduct Initial Walkthrough
Step 2.3 Organize the Collected Data

Step 3 Constructing Process Flow Sheets

Step 4 Preliminary Assessment and Next Steps

THE IN-PLANT ASSESSMENT (PART 1 OF PHASE II)

Phase II is the heart of the P2 audit. There are a lot of steps in this phase, but they are implemented sequentially, and each one incrementally moves the team closer to a single objective - namely, developing a baseline.

An important engineering tool needed to implement Phase II is *stoichiometry*. Stoichiometry refers to the methods of *material balances,* which are defined as precise accounts of the inputs and outputs of an operation.

The first law of thermodynamics is the basis for material- and energy-balance calculations. Because there is no significant transformation of mass to energy in most manufacturing operations, for a material balance the first law can be reduced to the simplified form:

Mass in = mass out + accumulation

A similar statement, or equation, can be used to express the energy balance:
Energy in (above datum) + energy generated = energy out (above datum)

Energy balances differ from mass balances in that the total mass is known but the total energy of a component is difficult to express. Consequently, the heat energy of a material is usually expressed relative to its standard state at a given temperature. For example, the heat content, or enthalpy, of steam is expressed relative to liquid water at 273° K (0° C) at a pressure equal to its own vapor pressure.

Material-balance systems can be intimidating, depending upon the complexity of the unit process or the entire plant. And there is the additional complication of solving material-balance systems when dealing with chemical reactions and phase changes. Phase change situations require a combination of mass and energy balances. However, by relying on the detailed process-flow sheets developed under Phase I, and applying the systematic approach outlined in this chapter, material-balance systems can become a friendly tool in the P2 audit.

Step 5: Determining the Inputs

Step 5.1:Determine the Total Inputs. Inputs to a process or a unit operation may include raw materials, chemicals, water, compressed air, and energy. The inputs to the unit process and to each unit operation need to be quantified; as a first step toward quantifying raw material usage, examine purchasing records. This will rapidly give you an idea of the sort of quantities of raw materials involved.

In many situations, the unit operations where raw-material losses are greatest are raw-material storage and transfer. Look at these operations in conjunction with the purchasing records to determine the actual net input to the process.

Make notes regarding raw-material storage and handling practices. Consider evaporation losses, spillages, leaks from underground storage tanks, vapor losses through storage-tank pressure-relief vents, and contamination of raw materials. These can often be rectified very simply.

Record raw-material purchases and storage and handling losses in a table, to derive the net input to the process. Table 4 provides an example that your team can use as an initial guide in developing a tabulation of input parameters to the process.

Table 4. Example of Recording Raw Materials Storage and Handling Losses.

Raw Material	Quantity of Material	Yearly Quantity of Raw Mtl. Purchased	Type of Storage Used in Production	Average Length of Storage (months)	Est. Annual Raw Mtl. Losses
Reagent-grade sulfuric acid	2,500 liters	2,800 liters	vessels	1	100 liters
Nickel catalyst	600 kg	2,000 kg	sacks	3	50 kg
Surface treatment chemical	1,000 kg	7,000 kg	closed	2	50 kg

Step 5.2: Determine the Inputs to Unit Operations. Once you've determined the net input of raw materials to the process, quantify the raw material input to each unit operation. If accurate information about raw-material consumption rates for individual unit operations is not available, then take measurements to determine average quantities (or perform careful mass balances). Measurements should be taken for an appropriate length of time. For example, if a batch takes one week to run, then take measurements over a period of at least three weeks; these figures then can be extrapolated for monthly, quarterly, or annual figures. Some quantification is possible by observation and some simple accounting procedures. The idea is to develop a

database of inputs that reflect *steady-state* operations - that is, the input amounts that reflect average or normal consumption patterns. For solid raw materials, ask the warehouse operator how many sacks are stored at the beginning of the week or prior to using them in a unit operation; then ask him or her again at the end of the week or of the operation's production run. Weigh a selection of sacks to check compliance with specifications.

For liquid raw materials, such as water or solvents, check storage tank capacities and ask operators when a tank was last filled. *If automatic gauging is not used, tank volumes can be estimated from the tank diameter and depth.* Monitor the tank levels and the number of tankers arriving on site. While investigating the inputs, talking to staff, and observing the unit operations in action, the auditing team should think about how to improve the efficiency of the unit operations.

The audit relies on information gathered in the field, by interviewing operators and various shop personnel. These interviews help the team identify possible ways to save raw materials, reduce pollution, and conserve energy. These discussions should not be extemporaneous; they should be thought-out in advance, and initially formulated during the walkthrough. It's a good idea to have a list of questions and a checklist of issues for such meetings and encounters with operators. Table 5 provides a sample list of questions that your team can expand upon and modify, to make them more specific during the in-plant assessment.

Table 5. Sample Questions for Identifying Raw Materials Savings.

	Lead Question	Follow-on Question
1	Is the size of the raw material inventory appropriate to ensure that material-handling losses can be minimized?	How often is inventory checked?
2	Could transfer distances between storage and process, or between unit operations, be reduced to minimize potential wastage?	Are bins and silos a source of product losses?
3	Do the same tanks store different raw materials, depending on the batch product?	Is there a risk of cross contamination?
4	Are sacks of materials emptied or is some material wasted?	
5	Are viscous raw material used on site?	Is it possible to reduce residual wastage in drums?
6	Is the raw-material storage area secure?	Could a building be locked at night, or could an area be fenced off to restrict access?
7	How could the raw materials be protected from direct sunlight or from heavy downpours?	

Table 5. Continued.

	Lead Question	Follow-on Question
8	Is dust from stockpiles a problem?	
9	Is the equipment used to pump or transfer materials working efficiently?	Is it maintained regularly?
10	Could spillages be avoided?	Is there a formal spill-handling procedure?
11	Is the process adequately manned?	What is the experience level of operators?
12	How could the input of raw materials be monitored?	
13	Are there any obvious equipment items in need of repair?	Is there a regularly scheduled maintenance program? Describe.
14	Are pipelines self-draining?	Where does the residue go?
15	Is vacuum pump water recirculated?	

Step 5.3: Consider the Energy Inputs. The energy input to a unit operation should be considered at this stage; however, energy use deserves a full assessment in its own right. For our discussions, however, we will focus on energy only as it relates to evaluating a P2 opportunity. In other words, our primary focus is on waste- and pollution-reduction opportunities; if these could reduce energy consumption as well, all the better. If energy usage is a particularly prominent factor, then the team should recommend a separate energy audit.

Step 5.4. Record Information on Process Flow Sheets. The input data collected for the material balance can be recorded on the process flow diagram, and in tabular form on a spreadsheet. Table 6 provides a sample spreadsheet to follow as a guideline.

Table 6. Sample Spreadsheet for Input Materials

Unit Operation	Raw Material 1 (mm^3/yr)	Raw Material 2 (tons/yr)	Water (mm^3/yr)	Energy Source
Surface Treatment (A)				
Rinse (B)				
Painting (C)				
Totals				

Water is frequently used in the production process, for cooling, heat exchange, gas scrubbing, washouts, product rinsing, and steam cleaning. This water usage needs to be quantified as an input.

Some unit operations may receive recycled wastes from other unit operations. These also represent an input. Steps 6 and 7 describe how these two factors should be included in the audit.

Step 6: Accounting for Water Usage

The use of water, other than for a process reaction, is a factor that should be covered in all pollution prevention audits. The use of water to wash, rinse and cool is often overlooked, although it represents an area where waste reductions can frequently be achieved simply and cheaply. As we saw from many of the case studies in Chapter 2, water inputs to a process can represent both significant pollution and raw materials savings if opportunities for recycling are identified.

Consider these general points about the site water supply before assessing the water usage for individual unit operations or unit processes:

- Identify the water sources within the plant operations.
- Is water extracted directly from a borehole, river or reservoir; is water stored on site in tanks or in a lagoon?
- What is the storage capacity for water on site?
- How is water transferred C by pump, by gravity, manually?
- Is rainfall a significant factor on site?
- Is stormwater collected and used for any purposes?
- Does the facility have a formal stormwater management plant?
- Is water used for purposes other than continuous process operations; e.g., intermittent blowdowns, equipment washing, wetting down waste stockpiles?
- What is the unit cost of water to the enterprise?

For each unit operation, consider the following:

- What is water used for in each operation?
  Cooling
  Heat exchange
  Gas scrubbing
  Washing
  Product rinsing
  Dampening stockpiles
  General maintenance
  Safety quench
  Other

- How often does each action place?
- How much water is used for each action?

It is unlikely that the answers to these questions will be readily available - you may need to undertake a monitoring program to assess the use of water in each unit operation. Again, the measurements must cover a sufficient period of time to ensure that all actions are monitored.

Pay particular attention to intermittent actions, such as steam cleaning and tank washout. Water use is often indiscriminate during these operations. Find out when these actions take place, so you can make detailed measurements. Record water usage information in a tabular form - ensure that the units used to describe intermittent actions indicate a time period. Table 7 provides a sample spreadsheet to follow.

Table 7. Sample Spreadsheet for Water Balance.

Unit Operation	Cleaning	Steam	Cooling	Other
Latex batch mix				
Reactor washout				
Reactor feed				

Important *- Make sure that all measurements used for recording data in this table are standard units and consistent (e.g., m^3/day or m^3/yr, etc.). This will help avoid problems in later steps, when you tackle the material-balance system.*

Using less water can be a cost-saving exercise. In many older plant operations, water conservation programs are often overlooked. Even something as straightforward as a valve-maintenance program can result in significant reductions in water consumption.

Consider the following points while investigating water use. These could wind up being recommendations, with incremental financial savings for the plant:

- Tighter control of water use can reduce the volume of wastewater requiring treatment, and result in cost savings - it can sometimes reduce volumes and increase concentrations to the point of providing economic material recovery in place of costly wastewater treatment.
- Attention to good housekeeping practices often reduces water usage and, in turn, the amount of wastewater passing to drains.
- The cost of storing wastewater for subsequent reuse may be far less than the treatment and disposal costs.
- Counter-current rinsing and rinse-water reuse are useful tips for reducing usage. Counter-current contact systems are more efficient in promoting heat and mass exchanges, which are important to gas absorption, extraction, many types of chemical reactions, and absorption.

Step 7: Measuring Current Levels of Waste Reuse and Recycling

Some wastes lend themselves to direct reuse in production and may be transferred from one unit to another; others require some modifications before they are suitable for reuse in a process. These reused waste streams should be quantified.

The reuse or recycling of wastes can reduce the amount of freshwater and raw materials required for a process. While looking at the inputs to unit operations, think about the opportunities for reusing and recycling outputs from other operations.

Note that if reused wastes are not properly documented, double-counting may occur in the material balance, particularly at the process or complete plant level; that is, a waste will be quantified as an output from one process and as an input to another.

Let's Summarize Steps 5-7

Before moving on, let's quickly summarize the three steps covered thus far in Phase II. By the end of Step 7, the team will have quantified all of the process inputs. It will have established the net input of raw materials and water to the process, having taken into account any losses incurred at the storage and transfer stages. It will have documented any reused or recycled inputs. All notes regarding raw-material handling, process layout, water losses, obvious areas where problems exist should all be documented for consideration in Phase III.

So, steps 5 through 7 of Phase II are:

Step 5 Determining the Inputs
- Step 5.1 Determine the Total Inputs
- Step 5.2 Determine the Inputs to Unit Operations
- Step 5.3 Consider the Energy Inputs
- Step 5.4 Record Information on Process Flow Sheets

Step 6 Accounting for Water Usage

Step 7 Measuring Current Levels of Water Use and Recycling

Step 8: Quantifying Process Outputs

By now, the team has half the data needed for the material-balance system. To calculate the second half of the material balance, the outputs from unit operations and the process as a whole need to be quantified. Outputs include primary product, by-products, wastewater, gaseous wastes (emissions to atmosphere), liquid and solid wastes that need to be stored and/or sent off-site for disposal, and reusable or recyclable wastes. Your team may find that a spreadsheet along the lines of that shown in Table 8 will help organize the input information. It is important to identify units of measurement, and to keep them consistent.

The assessment of the amount of primary product or useful product is a key factor in process or unit-operation efficiency. If the product is sent off-site for sale, then the amount produced is likely to be documented in company records. However, if the product is intermediate, meant for input to another process or unit operation, then the output may not be so easy to quantify. *Production rates may have to be measured over a period of time. Similarly, the quantification of any by-products may require measurement.*

Table 8. Sample Spreadsheet for Tabulating Process Outputs.

Unit Operation	Product		Wastes/Pollutants				
	Main	By-product	Recycle	Water	Gas	Stored	Off-Site
A							
B							
C							
Totals							

Step 9: Accounting for Wastewater Flows

In many plant operations, significant quantities of both clean and contaminated water are discharged to a sewer or a watercourse. In many cases, this wastewater has environmental implications and incurs treatment costs. In addition, wastewater may wash out valuable unused raw materials from the process areas.

It is very important to know how much wastewater is going down the drain, and what that wastewater contains. The wastewater-flows from each unit operation, as well as from the process as a whole, need to be quantified, sampled, and analyzed.

Step 9.1: Identify the Effluent Discharge Points. Determine the points at which wastewater leaves the site. Wastewater may go to an effluent treatment plant or directly to a public sewer, or directly to a watercourse.

One factor that is often overlooked in audits is the use of several discharge points. It is important to identify the location, type, and size of *all* discharge flows. Identify where flows from different unit operations or process areas contribute to the overall flow. In this way, it is possible to piece together the drainage network for the site. This can lead to startling discoveries of what goes where!

Once the team understands the drainage system, the enterprise can design an appropriate sampling and flow measurement program to monitor the wastewater flows and strengths from each unit operation.

Step 9.2: Plan and Implement a Monitoring Program. Plan your monitoring program thoroughly and try to take samples over a range of operating conditions, such as full production, start up, shut down, and washing out.

In the case of combined stormwater and wastewater drainage systems, ensure that sampling and flow measurements are carried out in dry weather. For small or batch wastewater flows, it may be physically possible to collect all the flow for measurement using a pail and wristwatch. Larger or continuous wastewater flows can be assessed using flow measurement techniques. Refer to the list of suggested readings at the end of this book for references on field monitoring and measurement methods.

Step 9.3: Reconcile Wastewater Flows. The sum of the wastewater generated from each unit operation should be approximately the same as the input to the process. If its not, then there are two possible sources of error in your accounting:

1) As noted in our description of Step 7, double-counting can occur where wastewater is reused. This is one reason why it's important to understand your unit operation and the interrelationships between unit operations within a unit process.
2) The team has missed one or more discharges. These missing flows are most often associated with uncontrolled discharges - examples include large evaporative losses from cooling tower operations and other heat-exchange functions in the process; intermittent blowdowns and frequent unscheduled turnarounds for cleaning purposes; and excessive leaks from valves, fittings, pump, and reactor seals.

The inability to reconcile water flows is a good tip-off to P2 opportunities. To make sure it doesn't miss such an opportunity, the team should focus attention on accurate reconciliation of the inputs and outputs.

Step 9.4: Determine the Concentrations of Contaminants. Analyze wastewater to determine the concentration of contaminants. Common wastewater contaminants that may require measurement include:

- pH
- Chemical oxygen demand (COD)
- Biochemical oxygen demand (BODs)
- Suspended solids (SS)
- Turbidity
- VOCs (volatile organic compounds)
- Grease and oil

- Chemical-specific constituents, including mercury, TCE, such heavy metals as lead and copper, nickel, chromium, etc.

The last group of parameters depends on the raw-material inputs. For example, an electroplating process is likely to use nickel and chromium. The metal concentrations of the wastewater should be measured, to ensure that the concentrations do not exceed discharge regulations, but also to ensure that raw materials are not being lost to drain. *Any toxic substances used in the process should be measured.*

Take samples for laboratory analysis. Composite samples should be taken for continuously running wastewater. For example, a small volume - say, 100 ml - may be collected every hour through a production period of 10 hours, to gain a 1-liter composite sample. The composite sample represents the average wastewater conditions over that time.

Where significant flow variations occur during the discharge period, consider varying the size of individual samples in proportion to the flow rate, to ensure that you obtain a representative composite sample.

For batch tanks and periodic draindown, a single spot sample may be adequate (check for variations between batches before deciding on the appropriate sampling method).

Step 9.5: Tabulate Flows and Concentrations. Wastewater flows and concentrations should be tabulated. Table 9 provides a sample spreadsheet to follow.

Table 9: Example of Spreadsheet for Tabulating Wastewater Flows

	Discharge Receptacle				
Source	**Sewer**	**Storm Water**	**Reuse**	**Storage**	**Total**
Unit Ops. A					
Unit Ops. B					
Unit Ops. C					
Total					

Step 10: Accounting for Gaseous Emissions

Step 10.1: Quantify the Gaseous Emissions. It's necessary to arrive at an accurate material-balance quantification of gaseous emissions associated with the process. Consider the actual and potential gaseous emissions associated with each unit operation, from raw-material storage through product storage.

Gaseous emissions are not always obvious and can be difficult to measure. Also, continuous monitoring can be expensive and not always justified. Where quantification is impossible, estimations can be made using stoichiometric information. The following example illustrates the use of indirect estimation.

Consider coal burning in a boiler house. The assessor may not be able to measure the mass of sulfur dioxide (SO_2) leaving the boiler stack, because of access problems and the lack of suitable sampling ports on the stack. The only information available is that the coal is of soft quality, containing 3 percent sulfur by weight and, on average, 1,000 kg of coal is burned each day.

First calculate the amount of sulfur burned:

1,000 kg coal x 0.03 kg sulfur/kg coal = 30 kg sulfur/day

The combustion reaction is approximately:

$$S + O_2 = SO_2$$

The number of moles of sulfur burned equals the number of moles of sulfur dioxide produced. The atomic weight of sulfur is 32 and molecular weight of sulfur dioxide is 64. Therefore:

kg-moles S = 30 kg/32 kg per kg-mole = kg-moles of SO_2 formed

kg SO_2 formed = (64 kg SO_2/kg-mole) x kg-moles SO_2 = 64 x 30/32 = 60 kg

Thus, it may be estimated that an emission of 60 kg of SO_2 will take place each day from the boiler stack.

These types of stoichiometric calculations are commonplace and can provide reliable estimates for the material balance. As with any calculation method, one should list the assumptions, to qualify the accuracy of the estimate. Limited field measurements can always be done later, to verify the estimated emissions.

Step 10.2: Tabulate Flows and Concentrations. Record the quantified emission data in tabular form and indicate which figures are estimates and which are actual measurements. The team should consider qualitative characteristics when it quantifies gaseous wastes.

The following are some typical questions to address when developing the material balance around the gaseous emissions components.

- Are odors associated with a unit operation?

- Are there certain times when gaseous emissions are more prominent - are they linked to temperature?
- Is any pollution-control equipment in place?
- Are gaseous emissions from confined spaces (including fugitive emissions) vented to the outside?
- If gas scrubbing is practiced, what is done with the spent scrubber solution? Could it be converted to a useful product?
- Do employees complain of irritating vapors?
- Do employees wear protective clothing, such as masks?
- Do health and safety records indicate a high incidence of respiratory problems among workers?
- Are there visible fumes during any times during the operation?

Step 11: Accounting for Off-Site Wastes

Your process may produce wastes that cannot be treated on-site, and so must be transported off-site for treatment and disposal. Wastes of this type are usually non-aqueous liquids, sludges, or solids. Often, wastes for off-site disposal are costly to transport and to treat, and represent a third-party liability. Therefore, minimization of these wastes yields a direct cost benefit, both present and future.

Measure the quantity and note the composition of any wastes associated with your process that need to be sent for off-site disposal. Record your results in a table. Table 10 provides an example of a spreadsheet that can be used as an initial guide.

Table 10. Example of Spreadsheet for Tabulating Off-Site Disposal.

Unit Ops.	Liquids		Sludge		Solid Waste	
	Qty.[a]	Conc.[b]	Qty.	Conc.	Qty.	Conc.
A						
B						
C						

a - Quantities in cubic meters per year or tons per year
b - Concentrations in mass, per unit volume.

It is useful to ask the following questions during the data collection stage:

- Where does the waste originate from within our process operations?
- Could the manufacturing operations be optimized to produce less waste?

- Could alternative raw materials be used that would produce less waste?
- Is there a particular component that renders the waste hazardous, and can this component be isolated? *This can be a key question.* Under the U.S. Resource Conservation and Recovery Act (RCRA), for example, if an enterprise has a waste with only 1 percent of a carcinogenic material as a component, then the entire waste is classified as carcinogenic. By eliminating the hazardous, regulated component, the enterprise can potentially eliminate a much larger waste problem.
- Does the waste contain valuable materials that could be recovered or possibly sold offsite?
- Wastes for off-site disposal need to be stored on-site prior to dispatch. Does storage of these wastes cause additional emission problems? For example, are solvent wastes stored in closed tanks?
- How long are wastes stored on-site, and is the enterprise in compliance with RCRA storage requirements?
- Are stockpiles of solid waste secure, or are dust storms a regular occurrence? Also, do waste piles result in stormwater-runoff issues?

Let's Summarize Steps 8-11

At the end of Step 11, the P2 audit team should have collated all the information required for evaluating a material balance for each unit operation, and for a whole process. It should have quantified all actual and potential wastes. Where direct measurement is not possible, it should make estimates based on stoichiometric information. It should arrange the data in tables with consistent units. Throughout the data-collection phase, the auditors should make notes regarding actions, procedures, and operations that could be improved.

So, steps 8 through 11 of Phase II are:

Step 8 Quantifying Process Inputs

Step 9 Accounting for Wastewater Flows
- Step 9.1 Identify the Effluent Discharge Points
- Step 9.2 Plan and Implement a Monitoring Program
- Step 9.3 Reconcile Wastewater Flows
- Step 9.4 Determine the Concentration of Contaminants
- Step 9.5 Tabulate Flows and Concentrations

Step 10 Accounting for Gaseous Emissions
Step 10.1 Quantify the Gaseous Emissions
Step 10.2 Tabulate Flows and Concentrations

Step 11 Accounting for Off-site Wastes

This brings us to a break in our procedure. The next series of steps in Phase II focuses on the mathematical development, analysis, and interpretation of the material-balance system. The following chapter section provides a systematic approach to working with material-balance systems for common situations encountered in productions. However, if your team is well-versed in manipulating material-balance systems, then you may wish to skip over the next section and go directly to "The In-Plant Assessment (Part 2 of Phase II)," found on page 300.

WORKING WITH MATERIAL BALANCES

Systematic Approach to Working with Material-Balance Systems

Application of a systematic approach enables enterprises to resolve a material-balance system into a number of independent equations equal to the number of unknowns that it needs to solve for. The following steps should be followed with any material-balance system, regardless of complexity:

Step 1: State the Objective. More exactly, state the unknowns that you need to solve for. The team should actually have a good idea of these by the time Phase I is complete.

Step 2: Tabulate the Available Data. Later on you can refer to Table 11 for an example of how to organize your data.

Step 3: Refer to the Process Flow Sheets. All material balances are logical. The process flow sheets are the basis for rationalizing the mathematical statement form of material balances.

Step 4: Define the System Boundaries. This depends on the nature of the unit process and individual unit operations. For example, some processes involve only mass flow-through. An example is filtration. This unit operation involves only the physical separation of materials (e.g., solids from wastewater). Hence, we view the filtration equipment as a simple box on the process flow sheet, with one flow input (wastewater) and two flow outputs (filtrate and filter cake). This is an example of a system where no chemical reaction is involved. In contrast, if a chemical reaction *is* involved, then we must take into consideration the kinetics of the reaction, the stoichiometry of the reaction, and the by-products produced. An example is the combustion of coal in a boiler. On a process flow sheet, coal, water, and energy are the inputs to the box (the furnace), and the outputs are steam, ash, NO_x, SO_x, and CO_2.

Step 5: Establish the Basis for the System Parameters. Is the process steady-state, or do the parameters represent transient or intermittent conditions? These considerations qualify the nature of the operation, and ultimately qualify the conditions under which corrective actions are taken in any P2 activities the team recommends.

Step 6: Write the Component Material Balances. The Phase II auditing steps define the pollutants and wastes that are among the team's focus. Its objective has always been to identify specific wastes or pollutants that the enterprise can reduce; these are the components the team needs to assess in the material balances. It is important to note that once the material balance for each unit operation has been completed for raw-material inputs and waste outputs, it is necessary to repeat the procedure for each contaminant of concern. Also, it is highly desirable to carry out a water balance for all water inputs and outputs to and from unit operations, because water imbalances may indicate serious underlying process problems, such as leaks, frequent spills, and purging.

Step 7: Write an Overall Material Balance. Remember that the P2 audit focuses on a unit process, but that there are individual unit operations that make up this process. The team will need to develop a series of material balances for each unit operation, and an overall material balance about the entire unit process, to bring closure to a solution of parameters of interest. The individual or component material balances developed may be summed to give a balance for the whole process, production area, or factory.

Step 8: Solve the Equations. Many material balances can be stated in terms of simple algebraic expressions. For complex processes, matrix-theory techniques and extensive computer calculations will be needed, especially if there are a large number of equations and parameters, and/or chemical reactions and phase changes involved.

Step 9: Develop a Final Material Balance. You must develop a final material balance to check the accuracy of the calculations. You may find that it is necessary to go back onto the production floor and gather additional information and measurements, to improve the accuracy and reliability of the calculations. Note: *It is important to keep track of any assumptions made in developing the material balance. Oftentimes there are losses, emissions and, or throughputs that we cannot gain an accurate handle on. Assuming some of these to be small at one point in the analysis could lead to wrong conclusions. Therefore, assumptions or estimates will often require retesting in order to bring closure to a balance*. The material balance is really an evolving tool that the auditors may have to refine at several stages of the audit.

USE A STEP-WISE APPROACH TO MATERIAL BALANCES

Step 1 - State the objective
Step 2 - Tabulate the available data
Step 3 - Refer to the process flow sheets
Step 4 - Define the system boundaries
Step 5 - Establish the basis for the system parameters
Step 6 - Write the component material balances
Step 7 - Write an overall material balance
Step 8 - Solve the equations
Step 9 - Develop a final material balance

Common Forms of the Material Balance

Material Balances Where No Chemical Reaction Involved. In situations where no chemical reaction is involved, the balances are based on the masses of the individual components appearing in more than one incoming or outgoing stream. Components that appear in only one incoming stream and one outgoing stream can be lumped together as though they are one component, to simplify the calculations and increase precision. It is important at the start to select a convenient unit of mass (usually the kilogram or pound), and then express all components in that same unit.

To describe a continuous process on a standard basis, always choose a unit of time or a consistent flow rate per-unit-of-time basis. For batch processes, select a unit-weight basis on the basis of a single batch.

Illustration #1: Material Balance for a Filter. Consider a solid-liquid filtration unit in which a slurry containing 30 percent by weight of solids (total solids, or TS) is fed to the unit. The resultant filter cake contains 85 percent solids and the filtrate contains 2 percent TS. The slurry feed rate is 2,500 kilograms per hour (~5,500 pounds per hour). Determine the corresponding flow rates of the cake and the filtrate.

Solution. The first step is to develop a diagram, or process flow sheet, of the process. This example is straightforward, because there is only one piece of equipment or unit operation involved. A simplified flow scheme is given in Figure 4.

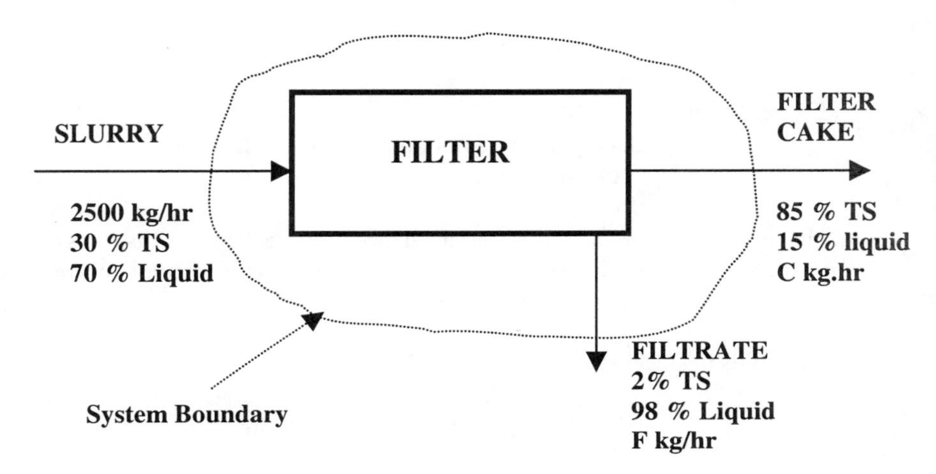

Figure 4. *Process flow sheet for filter material balance.*

Because this is a steady-steady operation, accumulation is zero, and hence the amount of mass in equals the amount of mass out on a per-unit-of-time basis. Note that in Figure 4 there are two unknowns - C and F. To solve for two unknowns, we need to develop two independent equations. One equation will be an overall balance, and the other can be either a liquid balance or a solid balance. The next step then is to write the equations.

For the overall balance:

Filtrate out + cake out = slurry in, or simply

$$F + C = 2500 \text{ kg/hr (or 5500 lb/hr)}$$

Choosing the liquid balance as the basis:

Liquid in filtrate + liquid in cake = liquid in slurry

A more rigorous way of stating this would be:

(wt. fraction of liquid in slurry)(mass of filtrate) + (wt. fraction of liquid in cake)(mass of cake) = (wt. fraction of liquid in slurry)(mass of slurry)

Mathematically, this may be stated as:

$$(1.0 - 0.02)F + (1.0 - 0.85)C = (1.0 - 0.30)(2500)$$

Thus, we have the following two equations that can be solved simultaneously:

$$F + C = 2500$$

$$0.98F + 0.15C = 1750$$

The results are:

C = 843.4 kg/hr (or 1855.6 lb/hr) of filter cake, and

F = 1656.6 kg/hr (or 3644.9 lb/hr) of filtrate

The final step to solving this material balance is to check the results. This can be done by substituting the calculated values into an independent equation, namely the one that we did not use (the solids balance).

The solids balance is:

Solid in filtrate + solid in cake = solid in slurry, or

$$0.02(1656.6) + 0.85(843.6) = 0.30(2500)$$

Because the right- and left-hand sides of this balance match, the material-balance system has been properly solved.

Material Balances Where Chemical Reactions Are Involved. Processes involving chemical reactions must be approached differently. It is best to express the compositions of flow streams entering the process or unit operation in terms of molar concentrations. Balances are developed in terms of the largest components that remain unchanged by the reactions.

In cases where the reactants involved are not present in the proper stoichiometric ratios, the limiting reactant will have to be determined and the excess amounts of the other reactants calculated. It is safe to assume that unconsumed reactants and inert components exit with the products in their original forms.

The example below follows that of Chopey and Hicks (see reference (11) in the *Additional Resources – Section II* section), and helps to illustrate a material balance when chemical reactions are involved.

Illustration # 2: Material Balance for Simple Combustion. Natural gas containing 98 percent methane and 2 percent nitrogen by volume is burned in a furnace with 15 percent excess air. The fuel consumption is 20 cubic meters per second, measured at 290° K and 101.3 kPa (or 14.7 psia). The problem is to determine how much air is required under these conditions. In addition, we want to determine the baseline environmental performance of the furnace by calculating the quantity and composition of the flue gas.

Solution. As with all material-balance systems, our first step is to start with a process flow sheet. Figure 5 provides a simplified flow scheme, along with a definition of the system boundary to be considered in the analysis. From this drawing we can establish the basis for the calculations.

To simplify calculations, but also by convention, the amount of excess reactant in a reaction is defined on the basis of the reaction going to completion for the limiting reactant. In the case of methane (CH_4) burned with excess air, the volume of air needed to combust the methane is calculated as though there is complete combustion of the methane, converting it entirely to carbon dioxide and water.

We assume one second as the basis for the calculations, and define A and F as the volumetric flow rates for air and flue gas, respectively. We will also adopt the consistent use of volumetric flow-rate units of cubic meters per second.

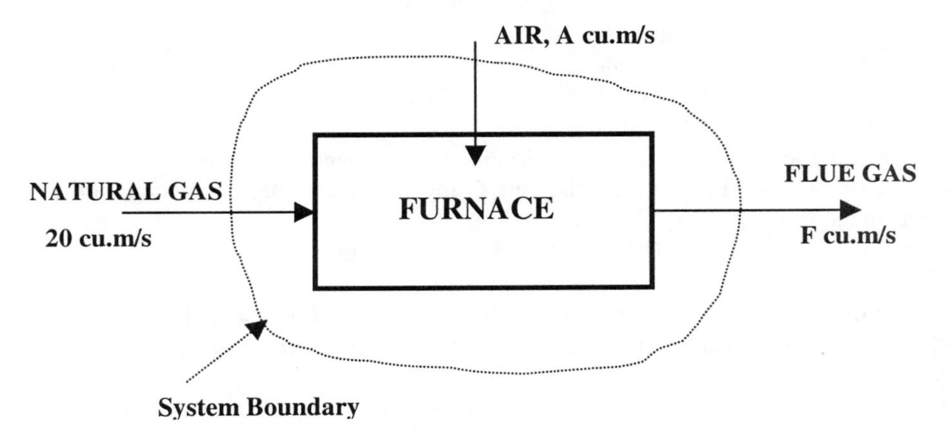

Figure 5. *Process flow sheet for furnace illustration.*

For our second step, we first summarize the information already known:

	Flow (m^3/s)	Component	Composition (%)	Mol. Weight
Natural gas	20	CH_4	98	16
		N_2	2	28
Air	A	O_2	21	32
		N_2	79	28

In our third step, we convert the compositions to a kg-moles-per-second basis. For this example (as well as many common industry cases), the ideal gas law can be used: n = PV/RT, where:

n = number of moles
R = ideal gas law constant, which is 8.314 kJ/(kg-mol)(K)
T = absolute temperature
P = pressure

For CH_4:

$$n = (101.3 \text{ kPa})[(20 \text{ m}^3/\text{s})(0.98)]/[8.314 \text{ kJ/(kg-mol)(K)}](298^\circ \text{ K})$$

$$n = 0.80 \text{ kg-mol/s}$$

For N_2, because the volumetric concentration is 98 % CH_4 and 2 % N_2, then:

$$n = (0.02/0.98)(0.80) = 0.016 \text{ kg-mol/s}$$

In our fourth step, we determine the amount of oxygen needed and the air-flow rate. To determine this, we must note the following stoichiometric reaction taking place in the furnace:

$$CH_4 + 2\ O_2 \rightarrow CO_2 + 2\ H_2O$$

From this, we know that 0.80 mol/s of CH_4 requires 2 H 0.80 = 1.60 mol/s O_2 for stoichiometric combustion. Because 15 percent excess air is used, the number of moles of O_2 in the air is:

$$(1.15)(1.60) = 1.84 \text{ kg-mol/s}$$

The amount of N_2 in with the air is:

$$[(0.79 \text{ mol } N_2/\text{mol air})/(0.21 \text{ mol } O_2/\text{mol air})](1.84 \text{ kg-mol/s } O_2)$$

$$= 6.92 \text{ kg-mol/s}$$

The total moles in the incoming air are:

$$1.84 + 6.92 = 8.76 \text{ kg-mol/s}$$

Now rearrange the ideal gas law to convert to volumetric flow rate:

$$V = nRT/P, \text{ or } (8.76)(8.314)(289)/101.3 = 207.8 \text{ m}^3/\text{s}$$

Our next step is to write the material balances and calculate the composition and quantity of flue gas. Unless there is a conversion between mass and energy, calculations such as these can be handled on a mass basis. We shall make this assumption, and also note that there is no accumulation. The flues gas includes N_2

from the air and from the natural gas, plus the 15 percent excess O_2, plus the reaction products (0.80 mol/s CO_2 and 2 H 0.80 = 1.60 mol/s H_2O). Summarizing the data:

Component	kg-mol x	kg per mole	= kg
	Inputs		
Natural gas:			
CH_4	0.80	16	12.8
N_2	0.016	28	0.448
Air:			
N_2	6.92	28	193.8
O_2	1.84	32	58.9
Total:			**265.9**
	Outputs		
Flue gas:			
N_2	(6.92 + .016)	28	194.2
O_2	(1.84 - 1.60)	32	7.68
CO_2	0.80	44	35.2
H_2O	1.60	18	28.8
Total:			**265.9**

The accumulation is zero. The overall material balance is 265.9 = 265.9 + 0. Thus the total quantity of flue gas is 265.9 kg/s.

Material Balances with Missing Composition and Flow Rate Data. It is quite possible that during the audit you'll run across situations where some streams entering or leaving a process may have incomplete data to express their compositions or flow rates. We can approach these types of situations by assigning letters to represent the unknown quantities in the material balances. To have a unique solution, there must be one independent material balance expressed for each unknown. The following illustration provides an example of such a case.

Illustration # 3: Material Balance for Batch Operation. In a pickling operation, metal parts are washed in a vat containing sulfuric acid. Concentrated sulfuric acid (60 percent by weight) is pumped into a vat, to which 1,200 kg of water is added. The resulting mixture in the vat contains 10 percent acid. Determine how much of this 10 percent acid solution is in the vat.

Solution: This is a batch operation. We therefore choose one batch as the basis for the calculation. There are two inputs: an unknown amount of acid solution having a known concentration (60 percent H_2SO_4), and a known amount of water (1,200 kg). There is one output: a final batch of unknown quantity but known composition (10 percent acid solution).

Let A represent the amount of input-acid solution, and V the amount of the final batch. Two independent material balances can be set up: one for the acid or for the water, and the other for the overall system.

Sulfuric acid: Input = Output, or $0.60A + 0 = 0.10V$
Water: Input = Output, or $(1 - 0.60)A + 1200 = (1 - 0.10)V$
Overall: Input = Output, or $A + 1200 = V$

Use the overall balance and one of the others; say, the one for sulfuric acid:

$$0.60A = 0.01V, \text{ or } A = (0.10/0.60)V = 0.167V$$

Substituting into the overall balance:

$$0.167V + 1200 = V, \text{ or } V = 1440 \text{ kg solution in the vat, and}$$
$$A = 0.167 \text{ H } 1440 = 240.5 \text{ kg sulfuric acid}$$

It is convenient to check the calculation by substituting into the balance equation not used (i.e., the water balance):

$$(1 - 0.60) \text{ H } 240.5 + 1200 = (1 - 0.10) \text{ H } 1440$$
$$1296 = 1296$$

Thus, the results are in agreement.

Combining Energy and Material Balances. There are many situations in industrial settings where one is faced with both energy and material flow issues. From a P2 standpoint, these offer the dual opportunity of material and energy savings. Consider the following simplified case.

Illustration # 4: Material Balance for an Evaporation Situation. The following problem is adapted from Badger and McCabe (*Elements of Chemical Engineering*, Walter L. Badger and Warren L. McCabe, 2nd Edition, McGraw-Hill Book Company, Inc, New York and London, 1936). A triple-effect evaporator is used as a means of concentrating a solution from 10 to 50 % solids. The steam available for this operation is 15 lb. Per sq. in gage (249° F), and a vacuum of 26 in. as referred to a 30-in. barometer, which is maintained in the vapor space of the last effect. This pressure corresponds to a vapor pressure of 125° F. The evaporator feed is 55,000 lb./hr at a temperature of 70° F. Condensate exits each effect at the steam temperature. The solution has a negligible elevation in boiling point. Its specific gravity is unity over the operational conditions of the evaporator. The evaporator is operated in the "feed-forward" mode. Each effect of the evaporator has the same heating surface area, and the heat transfer coefficients are 550 in the first, 350 in the second, and 200 in the third (units are Btu. Per sq. ft. per °F per hr.). We wish to evaluate the following: the heating surface, the steam consumption, the distribution of temperatures, and finally, the pounds of water evaporated per pound of steam.

Solution: This is a simple solution concentration case, where the evaporation is the same in all three effects, and may be calculated from an overall material balance. An assumption here is that the solids go through the evaporator without any losses. We may write:

	Mass (lbs.)		
	Solid	**Water**	**Total**
In feed liquor	5,500	49,500	55,000
In thick liquor	5,500	5,500	11,000
Evaporation		44,000	44,000

To solve this problem, we need some chemical engineering know-how. First we should recognize that the temperature drop across a given effect must be larger as the heat-transfer coefficient is smaller. In addition, if any effect has an extra load, then that effect requires a larger temperature drop. In this case the total drop is (249 – 125), or 124° F. The last effect will have the largest drop, and because of the large amount of heating done in the first effect, this effect must have a larger drop than the second effect. As such, we may state the following:

$$\Delta t_1 = 38^\circ \text{ F}; \Delta t_2 = 33^\circ \text{ F}; \Delta t_3 = 53^\circ \text{ F}$$

The steam to the first effect is at 249 ° F, and the Latent heat = 946 Btu. Per pound, and since $\Delta t_1 = 38^\circ$ F, we may write the following conditions:

Effect	Boiling Point, ° F	Δt, ° F	Latent heat, Btu. Per pound
1	211	33	971
2	178	53	991
3	125	--	1,022

It is now possible to write heat balance equations across each effect. The following terms are defined for the purpose of the calculations:

x = mass evaporated in the first effect
y = mass evaporated in the second effect
z = mass evaporated in the third effect
w = mass of steam supplied to the first effect

Reference datum temperatures must be chosen for the calculations – in this case it is convenient to choose the boiling points in each effect as the datums. In the feed from the first to the second effect, and from the second to the third, some amount of vapor is formed by flashing. Across the first effect:

(Heat in steam above 211 ° F) – (Heat to feed to 211 ° F) = (Latent heat of vapor at 211° F) + (Heat of condensate above 211 ° F)

In cases where the condensate exits at the steam temperature, the difference of the first and last terms is the latent heat of the steam entering at 249 ° F. We may then write:

$$946w - 55{,}000(211 - 70) = 971x$$

And for the second effect:

$$971x + (55{,}000 - x)(211 - 178) = 991y$$

For the third effect:

$$991y + (55{,}000 - x - y)(178 - 125) = 1{,}022x$$

For all three effects:

$$44{,}000 = x + y + z$$

The above algebraic expressions can be solved simultaneously to obtain the following results:

$x = 13{,}610$ lb.
$y = 14{,}720$ lb.
$z = 15{,}670$ lb.

And as a check, the Total = 44,000 lb.

By using the rate expressions for each effect, assigning A_1, A_2, and A_3 as the areas of the heating surfaces in the first, second and third effects, respectively, then the areas are solved for as follows (*note – it is assumed that the reader is familiar with standard heat transfer calculation methods using heat transfer coefficients*):

$A_1 = \{971x + (55{,}000)(211 - 70)\}/\{(38)(550)\} = 1{,}003$ sq.ft.
$A_2 = 971x/\{(33)(350)\} = 1{,}144$ sq.ft.
$A_3 = 991y/\{(53)(200)\} = 1{,}375$ sq.ft.
Average = 1,174 sq.ft.

Since we assumed that the heating surface is the same in all effects, the above solution is a contradiction. In fact, the above indicates that in order to maintain the assumed temperature distribution, widely different heating surfaces must be employed. To compensate, we must readjust the temperature distribution. Neglecting the small effects of latent heat, heating and flashing caused by a change in the temperature distribution, we can assume that the heating surfaces determined above change inversely with the temperature drops. The third effect is too large, and hence it must have a larger temperature drop to bring its surface down. The first two effects are small, and must have smaller drops. The tentative temperature drops can be corrected in proportion to the deviations from the mean surface as follows:

$\Delta t_1 = 38 \times (1{,}003/1{,}174) = 32^\circ$ F
$\Delta t_2 = 33 \times (1{,}144/1{,}174) = 32^\circ$ F
$\Delta t_3 = 53 \times (1{,}375/1{,}174) = 62^\circ$ F

In total these add up to 126° F, but the total average is only 124° F. Hence we can scale them down to: $\Delta t_1 = 32^\circ$ F, $\Delta t_2 = 31^\circ$ F, and $\Delta t_3 = 61^\circ$ F. This leads us to the revised set of conditions:

The steam to the first effect is 249° F, its latent heat is 946 Btu. Per pound, and $\Delta t_1 = 32^\circ$ F; and

Effect	Boiling Point, ° F	Δt, ° F	Latent heat, Btu. Per pound
1	217	31	967
2	186	61	986
3	125	--	1,022

The latent heat balances are restated as follows:

$$946w - 55{,}000(217\text{-}70) = 967x$$
$$967x + (55{,}000 - x)(217 - 186) = 986x$$
$$986y + (55{,}000 - x - y)(186 - 125) = 1{,}022x$$
$$x + y + z = 44{,}000$$

And solving, we obtain:

$x = 13{,}610$ lb.
$y = 14{,}660$ lb.
$z = 15{,}730$ lb.
$w = 22{,}460$ lb.
$A_1 = 1{,}027$ sq.ft.
$A_2 = 1{,}213$ sq.ft.
$A_3 = 1{,}185$ sq.ft.
Average = 1,202 sq. ft. per effect.
The evaporation per pound of steam is 44,000/22,460 = 1.96 lb.

In this example we have characterized the energy requirements by the steam consumption as well as the mass throughputs through the system. With an understanding of the base case conditions, we can now begin to search for improvements to the system that would result in savings. For examples, we might look at the steam economy by operating the system with backward feed, or with forward feed and some amount of extra steam withdrawn from the second effect.

Material Balances with Recycle Streams. The following example is adapted from Chopey and Hicks (see reference (11) in the *Additional Resources – Section II* section).

Illustration # 5: Material Balance for a Process with Recycling Stream and Chemical Reaction. Figure 6 provides the process flow sheet for the feed-preparation section of an ammonia plant, where hydrogen is produced from methane by the combination of steam reforming/partial-oxidation. Enough air is used in partial oxidation to provide a 3:1 hydrogen-nitrogen molar ratio in the feed to the ammonia plant section. The hydrogen-nitrogen mixture is heated to reaction temperature and then fed into a fixed-bed reactor. In the reactor 20 % conversion of reactants to ammonia is obtained per pass. Upon exiting the reactor the mixture is cooled and the ammonia is removed via condensation. Unreacted hydrogen-nitrogen mixture is recycled and mixed with the fresh feed. Determine the ammonia production and recycle rates.

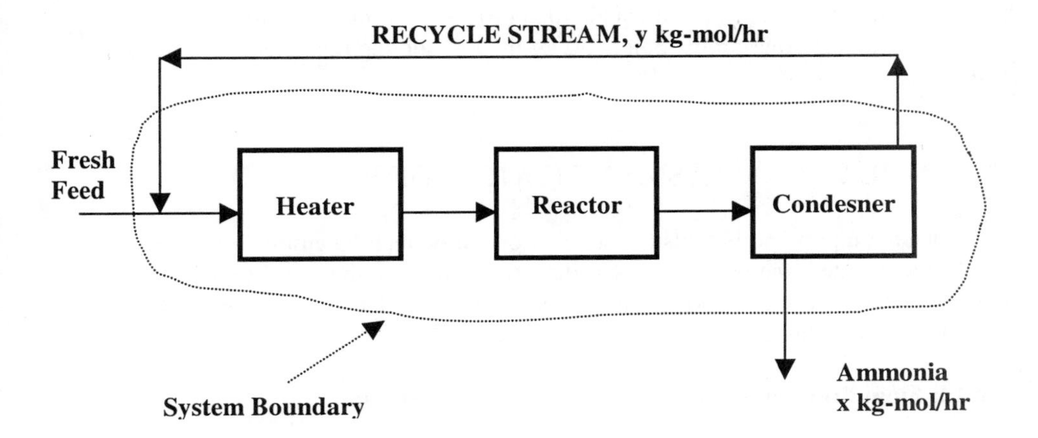

Figure 6. *Material balance system described by Chopey and Hicks.*

Solution. The first step in developing a material balance for this system is to select a basis for the calculations. We can assume a 100 kg-mol/hr feed to the process. Since the feed ratio is 3:1, then the feed will consist of 25 mol/hr of nitrogen and 75 mol/hr of hydrogen. Defining x = mol/hr ammonia produced, and y = mol/hr of recycle. To determine the amount of ammonia produced, we need to know the stoichiometry:

$$N_2 + 3H_2 \rightarrow 2NH_3$$

This means that for example, 4 mol hydrogen-nitrogen mixture in a 3:1 ratio will yield 2 mol ammonia. Noting the system boundary in Figure 6, the exiting and recycling reentering streams offset each other algebraically, and hence the net output from the system, consisting of liquid ammonia, must equal the net input, consisting of fresh feed. The amount of ammonia produced can thus be determined by:

$$x = [\ 100 \text{ kg-mol/hr}(H_2 + N_2)][(2 \text{ mol } NH_3)/4 \text{ mol } (H_2 + N_2)]$$

$$x = 50 \text{ kg-mol/hr } NH_3$$

To determine the recycle rate, we note that the total feed to the heater and the reactor consists of (100 + y) kg-mol/hr, and that 20 % of this feed is converted to ammonia. During the conversion 2 mol of ammonia is produced per 4 mol feed. Thus, the amount of ammonia produced equals $[0.20(100 + y)][2 \text{ mol } NH_3)/4 \text{ mol } (H_2 + N_2)]$. Since the ammonia production is 50 kg-mol/hr, solving this expression for y yields a recycle rate of 400 kg-mol/hr.

THE IN-PLANT ASSESSMENT (PART 2 OF PHASE II)

Once the inputs and outputs to each unit operation and the entire unit process have been tabulated, the team can begin developing material balances to solve for unknown concentrations and flows. The aim is to develop as detailed an understanding of the base-case environmental performance as needed.

Step 12: Final Preparation for the Material-Balance System

The Law of Conservation of Mass states that the total of what goes into a process must equal the total of what comes out. The team is now at the point where it can prepare a material balance at a scale appropriate to the level of detail required in the assessment. For example, you may require a material balance for each unit operation, or one for a whole process may be sufficient. Decide on various levels of detail that may be needed. Developing and analyzing material and energy balances usually winds up as an iterative process, one that continues until the appropriate or most revealing level of detail is achieved. Bear in mind that the idea behind performing a material balance is to gain a better understanding of the inputs and outputs (especially waste) of a unit operation, so areas where information is inaccurate or lacking can be identified. Imbalances require further investigation. Do not expect a perfect balance - your initial balance should be considered a rough assessment, to be refined and improved. Your first action in this step should be to assemble the input and output information for each unit operation, and then decide whether all the inputs and outputs need to be included in the material balance. For example, this is not essential where the cooling water input to a unit operation equals the cooling water output. Make sure to standardize units of measurement (liters, tons, or kilograms) on a per-day, per-year, or per-batch basis. Finally, summarize the measured values in standard units by referring to your process flow sheets (it may have been necessary to modify your process flow sheets following the in-plant assessment).

Step 13: Construct a Material Balance Information Sheet

In this step, the team organizes and makes a final review of the quantitative information on all material flows, and summarizes the data onto a spreadsheet. For each unit operation, the team should use data compiled from Steps 2 through 12 to construct the material-balance system. Organize the information clearly. Table 11 offers the team one way of presenting the material-balance information. This spreadsheet, of course, can be modified as necessary, but it does provide a general approach to tabulating the information in a concise manner.

Table 11: Sample Spreadsheet for a Preliminary Material Balance

INPUTS (AMOUNTS IN STANDARD UNITS, PER YEAR)	
Raw Material 1	
Raw Material 2	
Raw Material 3	
Waste Reuse	
Water	
Total	

Unit Process A

OUTPUTS (AMOUNTS IN STANDARD UNITS, PER YEAR)	
Product	
By-Product	
Raw Material Storage and Handling Losses	
Reused Wastes	
Wastewater	
Gaseous Emissions	
Stored Wastes	
Hazardous Liquid Waste Transported Off-Site	
Hazardous Solid Waste Transported Off-Site	
Non-Hazardous Liquid Waste Transported Off-Site	
Non-Hazardous Solid Waste Transported Off-Site	
Total	

Note that a material balance will often need to be carried out in units of weight, because volumes are not always conserved. Where volume measurements must be converted to weight units, take account of the density of the liquid, gas, or solids concerned.

Step 14: Evaluating Material Balances

Step 14.1: Classify the Material Balances. Though each unit process is different, leading to a unique material-balance system, there are general classes of material balances, regardless of the nature of the process. By recognizing these early on in the evaluation, the team will save time in setting up the family of equations needed to calculate key parameters, such as various mass and volumetric flow rates, and concentrations of individual chemical components.

Also, we can determine in this stage whether or not any portion of the material-balance system involves chemical reactions. It may be necessary to develop a kinetic model, either from experimental data or literature information, and incorporate this into the balances. Situations involving recycle streams, purge streams, and batch versus semi-batch operations can pose different constraints and boundary conditions when solving material balances. A few of the more common material balance cases are covered, along with sample illustrations, in the preceding section. What follows are general types of material-balance systems that you should look for in your analysis. Generalized solutions to these classes can be found in some of the references listed at the end of this book.

The common classes of material-balance systems you are likely to encounter include:

- No chemical reactions involved
- Chemical reactions involved
- Incomplete data on composition or flow
- Recycle streams
- Chemical reaction and recycle streams
- Chemical reactions, recycle streams, and purges
- Material balances with phase changes

Step 14.2: Determine the Gaps and Inaccuracies. The individual and sum totals making up the material balance should be reviewed to determine information gaps and inaccuracies. If you discover that there is a significant material imbalance, you should investigate it further. For example, if outputs are less than inputs, look for potential losses or waste discharges (such as evaporation, or fugitive emissions not accounted for, such as significant valve, pump, and reactor seals leakage). Outputs may appear to be greater than inputs if large measurement or estimating errors are made or some inputs have been overlooked.

At this stage you should take time to re-examine the unit operations to attempt to identify where unnoticed losses may be occurring. It may be necessary to repeat some data-collection activities. Remember that you need to be thorough and consistent to obtain a satisfactory material balance. The material balance not only reflects the adequacy of your data collection, but, by its very nature, ensures that you have a sound understanding of the processes involved.

Step 15: Refine the Material Balances

Now you can reconsider the material balance equations by adding those additional factors identified in the previous step. If necessary, estimates of unaccountable losses will have to be calculated. Note that, in the case of a relatively simple manufacturing plant, preparation of a preliminary material-balance system and its refinement (Steps 14 and 15) can usefully be combined. For more-complex P2 assessments, however, two separate steps are likely to be more appropriate.

An important rule to remember is that the *inputs should ideally equal the outputs* - but in practice this will rarely be the case. Some judgment will be required to determine what level of accuracy is acceptable, and we should have an idea as to what the unlikely sources of errors are (e.g., evaporative losses from outside holding ponds may be a materials loss we cannot accurately account for).

In the case of high concentrations of hazardous wastes, accurate measurements are needed to develop cost-effective waste-reduction options. It is possible that the material balance for a number of unit operations will need to be repeated. Again, continue to review, refine, and, where necessary, expand your database. The compilation of accurate and comprehensive data is essential for a successful P2 audit and subsequent waste-reduction action plan. Remember - *you can't reduce what you don't know is there*!

Let's Summarize

The final steps in Phase II are:

Step 12 Final Preparation for the Material Balance System

Step 13 Construct a Material Balance Information Sheet

Step 14 Evaluating Material Balances
- Step 14.1 Classify the Material Balances
- Step 14.2 Determine the Gaps and Inaccuracies

Step 15 Refine the Material Balances

By the end of Step 15, you should have assembled information covering process inputs and process outputs. These data should be organized and presented clearly in the form of material balances for each unit operation. Reflecting back on the 10 steps that constitute Phase II, the team's objective has been to arrive at a quantitative baseline that characterizes the performance of the operation.

By bringing closure to the material balances, we have developed an extensive database, based on field measurements and sampling, and on calculations. The database should contain information on essentially all the mass inputs and outputs to the unit process, as well as each unit operation. Hence, Phase II has not only defined for us the amounts and concentrations of liquid, gaseous, and solid forms of pollution, but also the products, the by-products, and all other wastes. The information compiled also provides us inference on wasted energy, and might very well divert our attention toward a separate energy efficiency study. We recommend, however, that this become part of the action plans for pollution-prevention and possible waste-minimization options we uncover in Phase III.

We now are ready to move onto the next phase, which involves developing P2 recommendations.

SYNTHESIS, BENCHMARKING, AND CORRECTIVE ACTIONS (PHASE III)

Introduction

Phases I and II have covered planning and undertaking a P2 audit, resulting in the preparation of a material balance for each unit operation. Phase III represents the interpretation of the material balance, to identify process areas or components of concern.

The material balance focuses the attention of the auditors on technical options for improving environmental performance and reducing wastes. The arrangement of the input and output data in the form of a material balance facilitates understanding of how materials flow through a production process. To interpret a material balance, it is necessary to have an understanding of normal operating performance C after all, how can you assess whether a unit operation is working efficiently if you do not know what is normal? Therefore, it is essential that you have a good working knowledge of the process. The results obtained from field measurements and material balance calculations will guide you to areas of concern and help to prioritize problems in the enterprise.

You should use the material balance information to identify the major sources of waste and pollution, to look for deviations from the norm in waste production, to identify areas of unexplained losses, and to pinpoint operations that contribute to flows that exceed national-, local-, or site-discharge regulations. Also, a good thing to remember is that, from a practical standpoint, *process efficiency* is synonymous with *waste minimization*.

Different pollution- and waste-reduction measures require varying degrees of effort, time, and financial resources. They can be categorized into two broad categories:

Group 1 includes obvious waste-reduction measures, including improvements in management techniques and housekeeping procedures that can be implemented cheaply and quickly. These are the best practices and minor (low-investment) re-engineering corrective actions, which can be referred to as low-cost/no-cost. They represent a category of corrective actions (P2 opportunities) where financial justifications are obvious, or trivial. For example, for a bottle-washing operation using once-through water, it is obvious that recycling would offer cost savings, especially if the re-piping required is trivial. Low-cost/no-cost P2 recommendations should be immediately reported to top management and implemented. Publicizing and monitoring the savings from such early success stories can help develop momentum for the P2 program.

Group 2 includes long-term reduction measures involving process modifications or process substitutions to eliminate problem wastes. This second group of corrective actions fall into two investment categories - moderate and high. Such projects involve re-engineering and equipment investments to the extent that a sufficiently detailed project-cost analysis will be needed to justify implementation. This involves financial benchmarking, which we discuss in Chapter 8.

The following steps will bring the team to the point of identifying options for both groups of P2 corrective actions.

Step 16: Low-Cost/No-Cost Recommendations

It may have been possible to implement very obvious waste-reduction measures already, before embarking on obtaining a material balance (even as early as at the end of Phase I). You should now consider the material balance information in conjunction with visual observations made during the data-collection period, to pinpoint areas or operations where simple adjustments in procedure could greatly improve the efficiency of the process by reducing unnecessary losses.

Use the information gathered for each unit operation to develop better operating practices for all units. Significant waste reductions can often be achieved by generally taking more care in performing operations and handling materials. Table 12 provides a list of waste-reduction hints that fall in the low cost/no-cost category of recommendations.

Table 12: Common Waste Reduction Hints.

Specifying and Ordering Materials

‘ Do not over-order materials, especially if the raw materials or components can spoil or are difficult to store.
‘ Try to purchase raw materials in a form that is easy to handle C for example, use pellets instead of powders.
‘ It is often more efficient, and certainly cheaper, to buy in bulk.

Receiving Materials

‘ Demand quality control from suppliers by refusing damaged, leaking, or unlabeled containers.
‘ Undertake a visual inspection of all materials coming on to the site.
‘ Check that a sack weighs what it should weigh, and that the volume ordered is the volume supplied.
‘ Check that composition and quality are correct.

Material Storage

‘ Install high-level controls on bulk tanks to avoid overflows.
‘ Bund tanks to contain spillages.
‘ Use tanks that can be pitched and elevated, with rounded edges for ease of draining and rinsing.
‘ Dedicated tanks (those that receive only one type of material) do not need to be washed out as often as tanks receiving a range of materials.
‘ Make sure that drums are stored in a stable arrangement to avoid damaging drums while in storage.
‘ Implement a tank-checking procedure C dip tanks regularly and document, to avoid discharging a material into the wrong tank.
‘ Use covered or closed tanks to reduce evaporation losses.

Material and Water Transfer and Handling

‘ Minimize the number of times materials are moved on-site.
‘ Check transfer lines for spills and leaks.
‘ Check your flexible pipework. Is it too long?
‘ Catch drainings from transfer hoses.
‘ Plug leaks and fit flow restrictions to reduce excess water consumption.

Table 12. Continued.

Process Control

' Design a monitoring program to check the emissions and wastes from key unit operations.

' Perform regular maintenance on all equipment, to reduce fugitive process losses.

' Inform employees about why actions are taken and what the enterprise hopes to achieve. Feedback on how waste reduction improves the process motivates operators, and is vital to the long-term viability of P2 programs.

Cleaning Procedures

' Minimize the amount of water used to wash out and rinse vessels. On many sites, indiscriminate water use contributes a large amount to wastewater flows. Ensure that hoses are not left running by fitting self-sealing valves.

' Investigate how washing water can be contained and used again before discharged to drains. Cleaning solvents also can often be used more than once.

Tightening up housekeeping procedures can reduce waste considerably. Simple, quick adjustments made to your process can achieve a rapid improvement in process efficiency. However, where such obvious reduction measures do not solve the entire waste-disposal problem, the enterprise needs to consider more-detailed consideration of waste reduction options. These are addressed in Steps 17 through 19.

Step 17: Targeting and Characterizing Problem Wastes

Use the material balance for each unit operation to pinpoint the problem areas associated with a process. The material-balance exercise may have brought to light the origin of wastes with high treatment costs, or may indicate which wastes are causing process problems in which operations. The material balance should be used to set priorities for long-term waste reduction.

At this stage, it may be worthwhile to consider the underlying causes of why wastes are generated and what factors lead to these causes; for example, poor or outdated technology, lack of maintenance, and non-compliance with company procedures may be contributing, or even underlying, factors. Additional sampling and characterization of wastes might be necessary, involving more in-depth analysis to ascertain the exact concentrations of contaminants. A worthwhile exercise is to list the wastes in order of priority for reduction actions. This will alert both the team and top management to the most costly waste issues and also help to better define what resources may be needed to address them.

Step 18: Segregation

Segregation of wastes can offer enhanced opportunities for recycling and reuse with resultant savings in raw-material costs. Concentrated simple wastes are more likely to be of value than dilute or complex wastes. Avoid the practice of mixing wastes, because this can aggravate pollution problems. If a highly concentrated waste is mixed with a large quantity of weak, relatively uncontaminated effluent, the result is a larger volume of waste that requires treatment. Isolating the concentrated waste from the weaker waste can reduce treatment costs. The concentrated waste could be recycled/reused or may require physical, chemical, and biological treatment to comply with discharge consent levels, whereas the weaker effluent could be reused or may only require settlement before discharge. Therefore, waste segregation can provide more scope for recycling and reuse, while reducing treatment costs. Step 18 then is to review your waste-collection and storage facilities to determine if waste segregation is possible. If so, then adjust your list of priority wastes accordingly.

Step 19: Developing Long-Term P2 and Waste-Reduction Options

Waste problems that cannot be solved by simple procedural adjustments or improvements in housekeeping practices will require more substantial long-term changes. It is necessary to develop possible prevention options for the waste problems. Process or production changes that may increase production efficiency and reduce waste generation include:

- Changes in production process - continuous versus batch
- Equipment and installation changes
- Changes in process control - automation
- Changes in process conditions, such as retention times, temperatures, agitation, pressure, and catalysts
- Use of dispersants in place of organic solvents, where appropriate
- Reduction in the quantity or type of raw materials used in production
- Raw-material substitution, through the use of wastes as raw materials, or the use of different raw materials that produce less waste or less hazardous waste
- Process substitution with cleaner technology

Waste reuse can often be implemented if materials of sufficient purity can be concentrated or purified. Technologies such as reverse osmosis, ultrafiltration, electrodialysis, distillation, electrolysis, and ion exchange may enable materials to be reused, and reduce or eliminate the need for waste treatment. Where waste treatment is necessary, a variety of technologies should be considered. These include physical, chemical, and biological treatment processes. In some cases, the treatment method can also recover valuable materials for reuse. For example, another industry or factory may be able to use or treat a waste that you cannot treat on-site. It also may be worth investigating the possibility of setting up a waste-exchange bureau as a structure for sharing treatment and reuse facilities.

A simple example of this kind of practice is the following. Figure 7 shows

Figure 7. *Cable producing plant in Samara, Russia.*

photographs of a wire and cable manufacturing operation that the authors visited in Russia. The operation is made up of a series of wire-drawing and cold-extrusion machines that draw copper wire down to various diameters and then spool the product. The spools (a closeup is shown in the lower photograph of Figure 7) are an intermediate product in the plant operation. These spools are sent to another unit operation, where the wire is coated with an insulator and multi-strand cables of various specifications are made.

Figure 8 shows a close-up photograph of the cable being drawn through a set of close tolerance rollers that are part of a cold-extrusion machine. From this stage in the operation, the wire is fed to a final stage of drawing and washing. The washing involves drawing the wire through a bath containing a solvent cleaning solution. In this manufacturing operation, a considerable amount of raw material (copper) is lost, because the abrasion of the soft wire surface occurs as it is drawn by the rollers. In addition, this fine dust is carried outside the shop area and contaminates the soil and becomes a contaminant in the stormwater runoff. The washing stage is where much of this loss accumulates. Figure 9 shows a photograph of one of the baths.

Figure 8. *Copper wire being drawn through rollers.*

The cleaning solution is recirculated, but because of the significant sludge buildup, the bath solution must be changed every 10 days or so. In the past, the sludge was bagged and stored on site. Because this not a huge accumulation of solid waste (up to 3 tons per year), the facility chose to stockpile the waste on site, and every quarter ship the waste, along with other solid wastes, to a local landfill. So, the normal practice is to mix the solid waste streams at the time of transport and to dispose of them in a common landfill operation off-site. Obviously, this would be totally unacceptable in the United States, but Russia has no solid-waste laws that are comparable to RCRA.

Figure 9. *Wire-cleaning operation produces sludge containing copper.*

The mixed waste comprised not only copper-contaminated sludge (copper is a heavy metal, and is toxic), but HDPE and other wire-coating insulating materials used in the downstream process. It never occurred to management to measure the concentration of copper in the waste sludge. Measurements showed the sludge to have copper levels as high as 60 percent, which made it attractive as a feedstock to a local smelter. The plant could capture benefits from the sale of the waste, while reducing transport and landfill-tipping fees by segregating this waste stream. Figure 10 shows one of the stockpiling stations for the sludge.

With virtually no investment on the part of the enterprise, the pollution problem was solved, resulting in a positive cash flow and savings of about $15,000 per year. The P2 opportunity in this example was relatively straightforward and seemingly very obvious - additional incremental cost savings are likely achievable from obvious housekeeping and better management of small-scale waste problems that have been overlooked.

Figure 10. *The final waste form being stockpiled.*

Let's Summarize Steps 16-19

At the end of Step 19 you and your team should have identified all the waste reduction options which could be implemented. Although we have not stated the following as a formal step, it is obvious that these options should be listed into the two general categories (Low-cost/No-cost and Longer-term/Higher Investments). There may

may be corrective actions among the first category that we may feel the need to justify the cost-savings further by considering the remaining steps to Phase III. The second category does require further actions before becoming formal recommendations.

So, the initial steps in Phase III are:

Step 16 Low-cost/No-cost Recommendations

Step 17 Targeting and Characterizing Problem Wastes

Step 18 Segregation

Step 19 Developing Long-term P2 and Waste-reduction Options

Step 20: Environmental and Economic Evaluation of P2 Options

When deciding which options to pursue as part of a waste-reduction action plan, consider each option in terms of environmental and economic benefits.

Environmental Evaluation. It is often taken for granted that reduction of a waste will have environmental benefits. Though this is generally true, there are exceptions to the rule. For example, reducing one waste may give rise to pH imbalances or may produce another waste which is more difficult to treat, resulting in a net environmental disadvantage. Hence, there may be environmental trade-offs between the status quo and the alternatives identified.

In many cases, the benefits may be obvious, such as the result of removing a toxic element from an aqueous effluent by segregating the polluted waste, or by changing the process so the waste is prevented from occurring. In other cases, the environmental benefits may be less tangible. Creating a cleaner, healthier workplace *will* increase production efficiency, but this may be difficult to quantify. These represent different cost-savings categories, which we discuss in Chapter 8. You should consider a series of questions or issues for each option:

- What is the effect of each option on the volume and degree of contamination of process wastes?
- Does the waste-reduction option have cross-media effects? For example, does the reduction of a gaseous waste produce a liquid waste?
- Does the option change the toxicity, degradability, or treatability of the wastes?
- Does the option use more, or fewer, non-renewable resources?
- Does the option use less energy?

In addition to these general questions, the team needs to begin focusing on the design aspects of the recommended options. At this stage of the audit process, detailed engineering practices begin to come into play. The possible recommendations will probably include equipment modifications, and changes involving newer and alternative pieces of equipment. At this point, the unit operations are no longer black boxes on a process flow sheet; instead, the team begins transforming these into preliminary shop drawings of actual equipment and layouts, piping changes, vendor quotes, and equipment supplier specs.

Economic Evaluation. The team should also undertake a comparative economic analysis of the P2 options and the existing situation. Where it cannot quantify benefits or changes (e.g., reduction in future liability, worker health and safety costs, etc.), it should make some form of qualitative assessment.

Economic evaluations of waste-reduction options should involve a comparison of operating costs, to illustrate where cost savings would accrue. For example, a waste-reduction measure that reduces the amount of raw material lost down the drain during the process will reduce raw-material costs. Raw-material substitution or process changes may reduce the amount of solid waste that must be transported off-site, reducing the transport costs for waste disposal.

In many cases, it is appropriate to compare the waste treatment costs under existing conditions with those associated with the waste-reduction option. The size of a treatment plant and the treatment processes required may be altered significantly by the implementation of waste-reduction options. This should be considered in an economic evaluation.

Calculate the annual operating costs for the existing process that needs waste treatment, and estimate how these costs would be altered by the introduction of waste-reduction options. Tabulate and compare the process and waste-treatment operating costs for both the existing and proposed waste-management options. The example given in Table 13 provides some of the typical cost components. In addition, if there are any monetary benefits (such as recycled or reused materials or wastes), then these should be subtracted from the total process or waste-treatment costs as appropriate. The expanded cost-analysis scheme discussed in Chapter 8 is appropriate to include at this point in the process.

Now that you have determined the likely savings in terms of annual process and waste-treatment operating costs associated with each option, consider the necessary investment required to implement each option. Investment can be assessed by looking at the payback period for each option C that is, the time taken for a project to recover its financial outlay. A more detailed investment analysis may involve an assessment of the internal rate of return (IRR) and net present value (NPV) of the investment based on discounted cash flows. An analysis of investment risk allows you to rank the options identified.

Consider the environmental benefits and the savings in process and waste-treatment operating costs, along with the payback period for an investment, to decide which options are viable candidates. Once this is done, the audit team can develop draft recommendations to include in the final report and presentation to top management.

Table 13. Summarizing Annual Process and Waste Treatment Operating Costs

Process Operating Costs	**Annual Cost**
Raw Material I	
Raw Material 2	
Water	
Energy	
Labor	
Maintenance	
Administration	
Other	
Total	

Waste Treatment Operating Costs	**Annual Cost**
Raw Material(e.g., Lime)	
Raw Material (e.g., Flocculent)	
Water	
Energy	
Trade Effluent Discharge Costs	
Transportation	
Off-Site Disposal	
Labor	
Maintenance	
Administration	
Other (e.g., violations, fines)	
Total	

Step 21: Developing and Implementing the Action Plan

The immediate reduction measures identified in Step 16, along with the long-term waste reduction measures evaluated in Steps 19 and 20, should form the basis of the P2 action plan. Prepare the ground for the waste-reduction action plan - precede its implementation by explaining its goals. It is necessary to convince those who must work with the new procedures that the change in philosophy from end-of-pipe treatment to pollution prevention makes sense and improves efficiency. Use posters around the site to emphasize the importance of waste reduction in minimizing production and waste-treatment and disposal costs, and (where appropriate) for improving the health and safety of company personnel.

Set out the intended action plan, within an appropriate schedule. Remember, it may take time for the staff to feel comfortable with a new way of thinking. It's a good idea to implement waste-reduction measures slowly but consistently, to allow everyone time to adapt to these changes.

Establish a monitoring program to run alongside the P2 action plan, so actual improvements in process efficiency, as well as in environmental performance, can be measured. For multiple P2 projects, the P2 matrix presented in Chapter 6 is a good way to track and report overall performance. Relay these results back to the workforce as evidence of the benefits of P2. Adopt an internal recordkeeping system for maintaining and managing data, to support material balances and waste-reduction assessments.

It is likely that you will have highlighted significant information gaps or inconsistencies during the P2 audit. You should concentrate on these gaps and explore ways of developing the additional data. Ask yourself repeatedly -- is outside help required?

A good way of providing P2 incentives is to set up an internal waste-charging system, so those processes that create pollution or waste in great volume, or that are difficult and expensive to handle, have to contribute to the treatment costs on a proportional basis. Another method of motivating staff is to offer financial rewards for individual waste-saving efforts, drawing on the savings gained from implementing waste-reduction measures.

P2 audits should be a regular event -- attempt to develop a specific P2-assessment approach for your own situation, keeping abreast of technological advances that could lead to waste reduction and the development of cleaner products and processes. Train process employees to undertake material-balance exercises. Training people who work on a process to undertake a P2 assessment of that or other processes will help to raise awareness in the workforce. Without the support of the operators, P2 actions will be ineffectual -- these are the people who can really make a difference to process performance.

Summarizing the Last Step

The key elements in the final step of the auditing process are:

1) Prepare the ground for the P2 action plan, ensuring that support for the audit, and implementation of the results, is gained from top management
2) Implement the plan slowly, to allow the workforce to adjust

3) Monitor process efficiency, and relay the results back to the workforce, to show them the direct benefits
4) Train process personnel to undertake P2 assessment for waste reduction

ONE MORE TIME

The systematic approach to conducting a P2 audit will enable your enterprise to develop a number of corrective actions that will fall into two general categories -- low-cost/no-cost, and those requiring medium to high capital investments. Just as you would for any other capital project, you must justify the P2 recommendations that require investment. Step 20 focuses on this cost evaluation; however, there are often indirect cost savings that tend to be missed in evaluations. These indirect cost savings often reveal that many P2 recommendations are actually more attractive from a financial standpoint than simple payback-period calculations indicate. We describe more in-depth project-cost estimating tools, meant to supplement Step 20, in Chapter 8.

Although there are variations of the three-phase auditing process described in this chapter, the general philosophy and protocol are similar. A key concept to bear in mind is that the audit itself is only a tool. For pollution prevention to work, there must be strong management support, and there must be a team approach to implementing the audit. Without this mix, the process simply will not yield positive results.

We close this chapter with three final notes. First, the following is a list of all the recommended steps to the audit. The reader can see all the steps below without having to flip back through this chapter. In reading over this list, decide what steps you might combine or even eliminate when implementing an audit at your enterprise. If you have been implementing the steps in an actual audit while reading the chapter, it is still worthwhile reading over the entire list of steps below. If your team has already gone through the actual exercise of an audit, reading the steps will help it streamline its approach when rolling out the audit to other parts of the plant.

Second, we've provided several exercises and questions for thinking and discussion, designed to help you sharpen your auditing skills.

Finally, we refer you to the "Additional Resources" appendix at the end of the book, where you'll find a list of recommended references organized by subject. Many of these references provide more in-depth reading materials, and calculation methods useful for material and energy balances, and for developing other analytical skills useful to the audit.

A SUMMARY OF ALL THE STEPS

Phase I -- The Pre-assessment

- **Step 1** Audit Focus and Preparation
  - Step 1.1 Get Ready
  - Step 1.2 Assemble the Audit Team
  - Step 1.3 Identify and Allocate Additional Resources
  - Step 1.4 Select the Subject Facility
  - Step 1.5 Define the Audit Objectives
  - Step 1.6 Review Documentation
  - Step 1.7 Gain Employee Buy-in and Participation
- **Step 2** Listing the Unit Operations
  - Step 2.1 Refine Our Initial Checklist
  - Step 2.2 Conduct an Initial Walk-through
  - Step 2.3 Organize the Collected Data
- **Step 3** Constructing Process Flow Sheets
- **Step 4** Preliminary Assessment and Next Steps

Phase II -- The In-plant Assessment

- **Step 5** Determining the Inputs
  - Step 5.1 Determine the Total Inputs
  - Step 5.2 Determine the Inputs to Unit Operations
  - Step 5.3 Consider the Energy Inputs
  - Step 5.4 Record Information on Process Flow Sheets
- **Step 6** Accounting for Water Usage
- **Step 7** Measuring Current Levels of Water Use and Recycling
- **Step 8** Quantifying Process Outputs
- **Step 9** Accounting for Wastewater Flows
  - Step 9.1 Identify the Effluent Discharge Points
  - Step 9.2 Plan and Implement a Monitoring Program
  - Step 9.3 Reconciling Wastewater Flows
  - Step 9.4 Determine the Concentrations of Contaminants
  - Step 9.5 Tabulate Flows and Concentrations
- **Step 10** Accounting for Gaseous Emissions
  - Step 10.1 Quantify the Gaseous Emissions
  - Step 10.2 Tabulate Flows and Concentrations
- **Step 11** Accounting for Off-Site Wastes
- **Step 12** Final Preparation for the Material-Balance System
- **Step 13** Construct a Material Balance Information Sheet
- **Step 14** Evaluating Material Balances
  - Step 14.1 Classify the Material Balances
  - Step 14.2 Determine the Gaps and Inaccuracies
- **Step 15** Refine the Material Balances

Phase III -- Synthesis, Benchmarking, and Corrective Actions

Step 16 Low-cost/No-cost Recommendations
Step 17 Targeting and Characterizing Problem Wastes
Step 18 Segregation
Step 19 Developing Long-term Waste-reduction Options
Step 20 Environmental and Economic Evaluation of P2 Options
Step 21 Developing and Implementing the Action Plan

QUESTIONS FOR THINKING AND DISCUSSING

1) Try the following problem to sharpen your skills in working with material balances. Hydrogen is produced from methane by a combination steam-reforming/partial-oxidation process. Sufficient air is used in partial oxidation to give a 3.5:1 hydrogen-to-nitrogen molar ratio in the feed section to an ammonia unit. The hydrogen-nitrogen mixture is heated to reaction temperature and then fed into a fixed-bed reactor, where 19 percent conversion of reactants to ammonia is obtained per pass. After exiting the reactor, the mixture is cooled and the ammonia is removed via condensation. The unreacted hydrogen-nitrogen mixture is recycled and mixed with fresh feed. Using a basis of 150 kg-mol/hr of fresh feed, develop a material balance, and calculate the ammonia-production and recycle rates.

2) Try the following problem to sharpen your skills in working with material and energy balances. Crude oil is heated to 525° K and then charged at a rate of 0.06 m^3/hr to the flash zone of a pilot-scale distillation tower. The flash zone is maintained at an absolute temperature of 115 kPa. Calculate the percent vaporized and the amounts of the overhead and bottoms streams. Assume that the vapor and liquid are in equilibrium.

3) Develop a list of the individual unit processes at your enterprise. Among these, which are the processes that are most likely to benefit from a P2 audit? Why?

4) For one of the unit processes identified in question 3, develop a list of all the unit operations involved, and prepare a simplified process-flow sheet. List all the material flows on the sheet, and highlight the pollution and waste streams.

5) From the process flow sheet developed in question 4, identify:
   a) those unit operations involving simple mass flows only
   b) those unit operations in which phase changes are occurring

c) those operations in which there are chemical reactions taking place
d) those operations that are batch
e) those operations that are continuous

5) From the previous set of questions, make a list of the different types of material balances that you could develop.

6) From your discussions about the previous series of questions, are there any best practices that you think might be recommended as possible low-cost/no-cost P2 actions? If so, develop a list of actions needed to verify their benefits, and how you intend to implement them.

7) Make a list of the important control variables, the dependent process variables, and the monitoring parameters important to the unit process that you have discussed in the previous series of questions.

9) What previous types of audits were conducted at your enterprise? Would any of these audits benefit you in a P2 audit?

Chapter 8
FINANCIAL PLANNING TOOLS

INTRODUCTION

Preparing a financial justification for a P2 project is often limited to declaring that if funding isn't made available, there could be an environmental incident resulting in fines or penalties, and perhaps litigation. Unfortunately, in the past, this approach has led to many poor decisions. Projects with limited benefit were often supported by management, and some projects that could have had large impacts on profit and cash flow were not. A P2 investment must be able to stand up to every other funding request and effectively compete for money on its own merits.

Unfortunately, P2 investments have often been among the first to be postponed in times of budget shortfalls. This has been due, in large part, to the inadequate support and defensive posture taken by enterprises toward environmental projects on an economic basis. Typically, when a production division requests money, all the necessary documentation, facts, and figures are ready for presentation. The production project is justified by showing how the project will increase revenues and how the added revenue will not only recover costs, but also substantially increase the earnings for the enterprise or enterprise's operating division. In fact, there is no difference between the two types of projects, because P2 project justification requires this same emphasis. To be competitive and to get management buy-in, proponents of a P2 program must understand the financial system.

Financial tools demonstrate the importance of the P2 investment on a life-cycle or total-cost basis; in terms of *revenues*, *expenses*, and *profits*. In this chapter, we provide principles and practices for cost accounting. The tools and principles discussed in this chapter are an integral part of Step 20 of the P2 audit, discussed in the previous chapter.

TOTAL COST AND COST ACCOUNTING

The term *total-cost accounting* (TCA) has come to be more commonly known as *life-cycle costing* (LCC). LCC is a method aimed at analyzing the costs and benefits associated with a piece of equipment or a practice over the entire time the equipment or practice is to be used. The idea actually originated in the federal government and was first applied in procuring weapons systems. Experience showed that the up-front purchase price was a poor measure of the total cost. Instead, costs such as those associated with maintainability, reliability, disposal and salvage value, as well as employee training and education, had to be given equal weight in making financial decisions. By the same token, justifying pollution prevention requires that all benefits and costs be clearly defined in the most concrete terms possible, and projected over the life of each option. How is this done? By applying the basic financial terms described below.

Present Worth or Present Value

The importance of *present worth*, also known as *present value*, lies in the fact that time is money. The preference between a dollar now and a dollar one year from now is driven by the fact that a dollar in-hand can earn interest. Present value can be expressed by a simple formula:

$$\mathbf{P = F/(1 + \mathit{i})^n}$$

where P is present worth or present value, F is future value, i is the interest or discount rate, and n is the number of periods. As a simple example, if we have or hold $1,000, in one year at 6 percent interest compounded annually, the $1,000 would have a computed present value of:

$$P = \$1{,}000/(1 + 0.06)^1 = \$943.40$$

Because our money can "work" at 6% interest, there is no difference between $943.40 now and $1,000 in one year because they both have the same value now. Economically, there is an additional factor at work in present value, and that factor is *pure time preference,* or *impatience* (see Pearce and Turner, *Economics of Natural Resources and the Environment*, 1977, pg. 213). However, this issue is generally ignored in business accounting, because the firm has no such emotions, and opportunities can be measured in terms of financial return.

But going back to our $1,000, if the money was received in three years, the present value would be:

$$P= \$1{,}000/(1 + 0.06)^3 = \$839.62$$

In considering either multiple payments or cash into and out of a company, the present values are additive. For example, at 6 percent interest, the present value of receiving both $1,000 in one year and $1,000 in three years would be $943.40 + $839.62 = $1,783.06. Similarly, if one was to receive $1,000 in one year, and pay $1,000 in 3 years the present value would be $943.40 - $839.62 = $103.78.

Present-value calculations allow both costs and benefits that are expended or earned in the future to be expressed as a single lump sum at their current or present value.

Financial Analysis Factors

It is common practice to compare investment options based on the present-value equation shown above. We may also apply one or all of the following four factors when comparing investment options:

- Payback period
- Internal rate of return
- Benefit-to-cost ratio
- Present value of net benefit

What is Payback Period?

The payback period of an investment is essentially a measure of how long it takes to break even on the cost of that investment. In other words, how many weeks, months, or years does it take to earn the investment capital that was laid out for a project or a piece of equipment?

Obviously, those projects with the fastest returns are highly attractive. The technique for determining payback period again lies within present value; however, instead of solving the present-value equation for the present value, the cost and benefit cash flows are kept separate over time.

First, the project's anticipated benefit and cost are tabulated for each year of the project's lifetime. Then, these values are converted to present values by using the present-value equation, with the firm's discount rate plugged in as the discount factor. Finally, the cumulative total of the benefits (at present value) and the cumulative total of the costs (at present value) are compared on a year-by-year basis. At the point in time when the cumulative present value of the benefits starts to exceed the cumulative present value of the costs, the project has reached the payback period. Ranking projects then becomes a matter of selecting those projects with the shortest payback period.

Although this approach is straightforward, there are dangers in selecting P2 projects based upon a minimum payback-time standard. For example, because the P2 benefit stream generally extends far into the future, discounting makes its payoff period very long. Another danger is that the highest costs and benefits associated with most environmental projects are generally due to catastrophic failure, also a far-future event. Because the payback period analysis stops when the benefits and costs are equal, the projects with the quickest positive cash flow will dominate. Hence, for a P2 project, with a high discount rate, the long-term costs and benefits may be so far into the

The minimum payback time standard is a good financial comparison, but it should not be the deciding basis for project selection.

future that they do not even enter into the analysis. In essence, the importance of life-cycle costing is lost in using the minimum payback-time standard, because it only considers costs and benefits to the point where they balance, instead of considering them over the entire life of the project.

What Do We Mean by the Internal Rate of Return?

Many readers are likely more familiar with the term *return on investment*, or ROI, than internal rate of return. The ROI is defined as the interest rate that would result in a return on the invested capital equivalent to the project's return. For illustration, if we had a P2 project with a ROI of 30 percent, that's financially equivalent to investing resources in the right stock and having its price go up 30 percent. As before, this method is based in the net present value of benefits and costs; however, it does not use a predetermined discount rate. Instead, the present-value equation is solved for the discount rate i. The discount rate that satisfies the zero benefit is the rate of return on the investment, and project selection is based on the highest rate. From a simple calculation standpoint, the present value equation is solved for i after setting the net present value equal to zero, and plugging in the future value, obtained by subtracting the future costs from the future benefits over the lifetime of the project. This approach is frequently used in business; however, the net benefits and costs must be determined for each time period, and brought back to present value separately. Computationally, this could mean dealing with a large number of simultaneous equations, which can complicate the analysis.

What is the Benefit-to-Cost Ratio?

The benefit to cost (B/C) ratio is a benchmark that is determined by taking the total present value of all of the financial benefits of a P2 project and dividing it by the total present value of all the costs of the project. If the ratio is greater than unity, then the benefits outweigh the costs, and we may conclude that the project is economically worthwhile.

The present values of the benefits and costs are kept separate, and expressed in one of two ways. First, as already explained, there is the pure B/C ratio, which implies that if the ratio is greater than unity, the benefits outweigh the costs and the project is viable. Second, there is the net B/C ratio, which is the net benefit (benefits minus costs) divided by the costs. In this latter case, the decision criteria is that the benefits must outweigh the costs, which means that the net ratio must be greater than zero

To better assess the financial attractiveness of a P2 project, identify and account for true costs and offsets to benefits.

(if the benefits exactly equaled the costs, the net B/C ratio would be zero). In both cases, the highest B/C ratios are considered as the best projects.

The B/C ratio can be misleading. For example, if the present value of a project's benefits were $10,000 and costs were $6,000, the B/C ratio would be $10,000/$6,000 or 1.67. But what if, upon further reassessment of the project, we find that some of the costs are not 'true' costs, but instead simply offsets to benefits? In this case, the ratio could be changed considerably. For argument sake, let's say that $4,500 of the $6,000 total cost is for waste disposal, and that $7,000 of the $10,000 in benefits is due to waste minimization; one could then use them to offset each other. Mathematically then, both the numerator and denominator of the ratio could be reduced by $4,500 with the following effect:

$$(\$10{,}000 - \$4{,}500) / (\$6{,}000 - \$4{,}500) = 3.7$$

Without changing the project, the recalculated B/C ratio would make the project seem to be considerably more attractive.

What is Present Value of Net Benefits?

The present value of net benefits (PVNB) shows the worth of a P2 project in terms of a present-value sum. The PVNB is determined by calculating the present value of all benefits; doing the same for all costs; and then subtracting the two totals. The result is an amount of money that would represent the tangible value of undertaking the project. This comparison evaluates all benefits and costs at their current or present values. If the net benefit (the benefits minus costs) is greater than zero, the project is worth undertaking; if the net is less than zero, the project should be abandoned on a financial basis. This technique is firmly grounded in microeconomic theory and is ideal for total-cost analysis (TCA) and P2 financial analysis.

Even though it requires a preselected discount rate, which can greatly discount long-term benefits, it assures that all benefits and costs over the entire life of the project are included in the analysis. Once an enterprise knows the present value of all options with positive net values, the actual ranking of projects using this method is straightforward; those with the highest PVNBs are funded first.

There are no hard and fast rules as to which factors one may apply in performing life-cycle costing or total-cost analysis; however, conceptually, the PVNB method is preferred. There are, however, many small-scale P2 projects where the benefits are so well defined and obvious that a comparative financial factor as simple as a ROI or the payback period will suffice.

ESTABLISHING BASELINE COSTS

To properly determine the cost of a project, we first need to establish a baseline for comparative purposes. If nothing else, a baseline defines for management the option of maintaining the status quo. Changes in material consumption, utility demands, staff time, etc., for options being considered can be measured as either more or less expensive than the baseline.

In this book, we follow the methodology of McHugh (McHugh, R.T., *The Economics of Waste Minimization*, McGraw-Hill Book Publishers, 1990). McHugh defines four tiers of potential costs that are integral to P2:

- *Tier 0:* Usual or normal costs, such as direct labor, raw materials, energy, equipment, etc.
- *Tier 1:* Hidden costs, such as monitoring expenses, reporting and record keeping, permitting requirements, environmental impact statements, legal, etc.
- *Tier 2:* Future liability costs, such as remedial actions, personal injury under the OSHA regulations, property damage, etc.
- *Tier 3:* Less tangible costs, such as consumer response and confidence, employee relations, corporate image, etc.

Tier 0 and Tier 1 costs are direct and indirect costs. They include the engineering, materials, labor, construction, contingency, etc., as well as waste-collection and transportation services, raw-material consumption (increase or decrease), and production costs. Tier 2 and Tier 3 represent intangible costs. They are much more difficult to define, and include potential corrective actions under the Resource Conservation and Recovery Act (RCRA); possible site remediation at third-party sites under Superfund; liabilities that could arise from third-party lawsuits for personal injury or property damages; and benefits of improved safety and work environments. Although these intangible costs often cannot be accurately predicted, they can be very important and should not be ignored when assessing a P2 project.

When it is not possible to analyze the intangible costs and benefits financially, they should be listed as additional factors to consider when making the pollution prevention investment decision.

A present-value analysis that contains such uncertain factors generally requires a little ingenuity in assessing the full merits of a P2 project.

Procurement versus Operating Costs

When analyzing the financial impact of projects, it is often useful to further categorize costs as either *procurement costs* or *operations costs*. This distinction better enables the projection of costs over time, because procurement costs are short-term, and refer to all costs required to bring a new piece of equipment or a new procedure on-line. Conversely, operations costs are long-term, and represent all costs of operating the equipment or performing the procedure in the post-procurement phase.

Establishing the Baseline

The baseline defines the current cost of doing business, and it gives management the option of doing nothing, of simply maintaining the status quo. To illustrate this, let's consider a beverage washing and bottling operation that one of the authors worked with in Samara, Russia. As photographs of the main production lines (shown in Figures 1 and 2) demonstrate, the plant cleans bottles in a series of automatic washing machines, dries the bottles, inspects them for defects, and then passes them on to bottle filling, labeling, and corking operations. Washing is done with fresh, hot, drinking-quality water, and a detergent solution. The washing stage is performed as once-through (meaning there is no recycling of wash water). Because the spent wash water contains a contaminant (organic matter and the detergent), the effluent is subject to a pollution fee, and it must be treated by a local municipal water-treatment plant.

Figure 1. *Automatic bottling line for vodka production in Russia.*

Figure 2. *Automatic bottle-washing machine.*

To establish the baseline, the current cost of doing business must first be determined. Once the present costs are known, all potential alternatives - such as substituting a more biodegradable cleaning fluid, or recycling a portion of the spent rinse waters - are then related to this baseline cost. There are three basic steps to establishing a baseline cost:

Step 1. Add up all the relevant input and output materials for the process, and compute their appropriate dollar value (note that we are only focusing on mass, not on energy or other issues). This makes use of our material-balance system.

Step 2. Check to make sure that the material balance makes sense. Specifically, is the volume of cleaning solvent purchased accounted-for in the losses, product, inventory, and/or the waste? Account for such losses as evaporative. In this example, the losses due to evaporation are about 5 percent. Once accomplished, determining the baseline costs becomes a matter of pricing each input and output, and then multiplying their volumes by the appropriate unit. The baseline costs for this example are tabulated in Table 1.

Step 3. Examine the expected business outlook and most likely changes, such as business expansions, new accounts, rising prices, cutbacks, etc. For simplicity's sake, costs and volumes in Table 1 are assumed as constant. In other words, the current annual costs will be the same in the out-years, with the exception of one very important aspect -- the time value of money.

Table 1. Current Yearly Costs for Bottle-Cleaning Operation.

Item	Cost/Unit	# Units	Cost/year
Cleaning solvent	\$1.75/gal.	3,500 gal.	\$6,125
Water	\$1.60/1,000 gal.	20,000 gal.	\$32
Water disposal	\$0.20/gal.	20,000 gal.	\$7,000
		Total Annual Cost	**\$13,157**

Because we are assuming in this example that the bottling plant's costs are constant, the \$13,157 annual cost shown in Table 1 will be repeated for each year. The present-value calculations shown earlier enable the annual expenditure to be expressed as a single sum that includes the effect of interest. Assuming that the bills are paid at the end of the year, the first year's cost would be the amount of money that would have to be banked starting today to pay a \$13,157 bill in one year.

Using a 10 percent interest rate, the computation is as follows:

$$P = \$13{,}157/(1 + 0.10)^1 = \$11{,}960.91$$

This means that if \$11,960.91 was banked at 10 percent interest at the beginning of the year, it would provide enough money to pay the \$13,157 bill at the end of the year. Similarly, the second, third, fourth, etc., years' expenditures can also be expressed in present value. This is shown in Table 2.

The simple analysis shows that the total cost of the bottle-washing system over the next 10 years, given a 10% interest rate, is \$80,844.07 in present-value terms. In other words, about \$81,000 invested today at 10 percent interest would be sufficient to pay the entire material and wastewater-discharge costs for the next 10 years.

Any changes to the operation of the firm can now be compared to this \$81,000 baseline. Any change that would result in a lower 10-year cost would be a benefit in that it would save money; any option with a higher cost will be more expensive and, from a financial or economic standpoint, should not be adopted.

We have, of course, not taken into account inflationary issues, which for Russian enterprises could be a serious unknown. In this example, the cost for raw water and for pollution fees are low by Western standards; the costs of these services could indeed rise significantly over a decade, and hence a decision to stay with the status quo could be a grave mistake.

Table 2. Present Value Calculations for the Vodka Bottling Plant.

Year	Expenditure, $	Present Value, $
1	13,157	11,960.91
2	13,157	10,873.55
3	13,157	9,885.05
4	13,157	8,986.41
5	13,157	8,169.46
6	13,157	7,426.78
7	13,157	6,751.62
8	13,157	6,137.84
9	13,157	5,579.85
10	13,157	5,072.59
	TOTAL	**$80,844.07**

Analyzing Present Value Under Uncertainty

Tier 2 and 3 costs are difficult to quantify or predict. While Tier 2 costs include potential liabilities, such as remedial actions, personal injuries covered under OSHA, property damage, etc., Tier 3 costs are even harder to predict -- for example, a typical Tier 3 cost could be the cost of sales lost due to adverse public reaction to a pollution incident, such as a leaking underground storage tank, a PCB spill, or a fire and explosion incident. Tier 2 and Tier 3 variables include the types of incidents that could occur, the severity of each incident, the ability of the firm to control or respond to the emergency, the public's reaction to the incident, or the company's ability to address the public's concerns -- a complex situation, to say the least.

In many cases, there is a probability that can be connected with a particular event. This enters into the calculation of *expected value*. The expected value of an event is the probability of an event occurring, multiplied by the cost or benefit of the event. Once all expected values are determined, they are totaled and brought back to present value as done with any other benefit or expense. Hence, the expected value measures the central tendency, or the average value that an outcome would have.

For example, there are a number of games at county fairs that involve betting on numbers or colors, much like roulette. If the required bet is $1, and the prize is worth $5, and there are 10 selections, the expected value of the game can be computed as:

(Benefit of Success) x (Probability of Success) - (Cost of Failure) x (Probability of Failure), or

($5) x (0.1) - ($l) x (0.9) = -$0.40

On the average, the player will lose -- meaning the game operator will win -- 40 cents on every dollar wagered. For tier 2 and 3 expenses, the analysis is the same. For example, there is a great deal of data available from Occupational Safety and Health Administration (OSHA) studies regarding employee injury in the workplace. In justifying a material-substitution P2 project, if the probability of injury and a cost can be found, the benefit of the project can be computed.

The concept of expected value is not complicated, though the calculations can be cumbersome. For example, even though each individual's chance of injury may be small, the number of employees, their individual opportunity costs, the various probabilities for each task, etc., could mean a large number of calculations. However, if one considers the effect of the sum of these small costs, or the large potential costs of environmental lawsuits or site remediation under either RCRA or the Comprehensive Environmental Response, Compensation, and Liability Act (CERCLA), the expected value computations can be quite important in the financial analysis.

REVENUES, EXPENSES, AND CASH FLOW

At first, P2 projects tend to require more financial justification than other programs that generate savings related to Tier 0 or possibly even Tier 1 costs. However, as an organization gets more sophisticated in its subsequent efforts, the less-tangible Tier 2 and Tier 3 costs become more important and more easily incorporated into its analysis. Even if these costs cannot be accurately predicted, in cases where two investment options - one a P2 project, the other not - appear to be financially equivalent, the Tier 2 and Tier 3 considerations can favor the P2 option. Generally, all that is needed is one or two success stories, and enterprises quickly become hooked on the *pollution prevention practices,* or P3, concept.

Because it is the goal of any business to make a profit, the costs-and-benefits cash flows for each option can be related to the basic profit equation:

REVENUES - EXPENSES = PROFITS

The most important aspect of this is that profits can be increased by either an increase in revenues or a decrease in expenses. P2 often lowers expenditures and increases profit. There are different categories of P2 revenues and expenses, and it is important to distinguish between them.

What Is Revenue?

Obviously, revenue is money coming into the firm; from the sale of goods or services, from rental fees, from interest income, etc. The profit equation shows that an increase in revenue leads to a direct increase in profit, and vice versa if all other revenues and expenses are held constant. Note that we are going to assume that the condition of other expenses/revenues are held constant in the discussions below.

Revenue impacts must be closely examined. For example, firms often can cut wastewater treatment costs if water use (and, in turn, the resulting wastewater flow) is limited to nonpeak times at the wastewater treatment facility. However, this limitation on water use could hamper production. Consequently, even though the company's actions to regulate water use could reduce wastewater charges, revenue could also be decreased, unless alternative methods could be found to maintain total production. Conversely, a change in a production procedure as a result of a P2 project could increase revenue. For example, moving from liquid to dry paint stripping can not only reduce water consumption, but also affect production output. Because clean-up time from dry paint-stripping operations (such as bead blasting) is generally much shorter than from using a hazardous, liquid based stripper, it could mean not only the elimination of the liquid waste stream (the direct objective of the P2 project), but also less employee time spent in the cleanup operation. In this case, production is enhanced and revenues are increased by the P2 practice. Another potential revenue effect is the generation of marketable by-products from the P2 practice. In Chapter 6, we discussed several case studies where certain waste streams were identified as having value in secondary markets. Such opportunities bring new, incremental revenues to the overall operation of the plant. The point to remember is that P2 has the potential to either increase or decrease revenues and profits - and that's the reason for doing a financial analysis.

A P2 project has the potential to either increase or decrease production rates – hence, overall cost savings should be assessed in terms of the relative importance of each benefit.

What Are Expenses?

Expenses are monies that leave the firm to cover the costs of operations, maintenance, insurance, etc. There are several major cost categories that P2 can have an effect on:

- Insurance expenses
- Depreciation expenses
- Interest expenses
- Labor expenses
- Training expenses
- Auditing and demo expenses
- Floor-space expenses

Insurance Expenses. Depending upon the P2 project, insurance expenses could either increase or decrease. For example, OSHA sets limits on workers' exposure to a number of chlorinated solvents. If one P2 option eliminated a hazardous, chlorinated solvent from production operations, the enterprise could realize a savings in employee health coverage, liability insurance, etc. Likewise, if an enterprise used a nonflammable solvent instead of a flammable one, it could decrease its fire insurance premium. Conversely, P2 projects can increase insurance expenses. For example, if an enterprise added a heat-recovery still to a process operation, it could increase its fire insurance premiums.

Depreciation Expenses. If a P2 project involves the purchase of capital equipment with a limited life (such as storage tanks, recycle or recovery equipment, new solvent-bath systems, new fabric dyeing baths, etc.), the entire cost is not charged against the current year. Instead, depreciation expense calculations spread the equipment=s procurement costs (including delivery charges, installation, start-up expenses, etc.) over a period of time by taking a percentage of the cost each year over the life of the equipment. For example, if the expect life of a piece of equipment is 10 years, each year the enterprise would charge an accounting expense of 10 percent of the procurement cost of the equipment. This is known as the *method of straight-line depreciation*. Although there are other methods available, all investment projects under consideration at any given time should use a single depreciation method to accurately compare alternative projects' expense and revenue effects. Because straight-line depreciation is easy to compute, it is the preferred method. Note that even though a firm must use a different depreciation system for tax purposes (e.g., the Accelerated Cost Recovery System, or ACRS), it is acceptable to use other methods for bookkeeping and analysis. In any event, any P2-related capital equipment must be expensed through depreciation.

Interest Expense. Investment in P2 implies that one of two things must occur: Either a company must pay for the project out of its own cash, or it must finance the cost by borrowing money from a bank, by issuing bonds, or by some other means. When a firm pays for a P2 project out of its own cash reserves, the action is sometimes called an *opportunity cost*. If the enterprise must borrow the cash, there is an interest charge associated with using someone else's money. It is important to recognize that interest is a true expense and must be treated, like insurance expense, as an offset to the project's benefits. The magnitude of the expense will vary with bank lending rates, the interest rate offered on the corporate notes issued, etc. In any case, there will be an expense.

The reason enterprises account for P2-equipment purchases as a cost is this: If cash is used for the purpose of pollution prevention, it is unavailable to use for other opportunities or investments. Revenues that *could* have been generated by the cash (for example, interest from a certificate of deposit at a bank) are treated as an expense and thus reduce the value of the P2 project.

Although the reasoning seems sound, opportunity costs are not really expenses. Though it is true that the cash will be unavailable for other investments, opportunity cost should be thought of as a comparison criteria and not an expense. The opportunity forgone by using the cash is considered when the P2 project competes for funds and is expressed by one of the financial analysis factors discussed earlier (net value of present worth, pay back period, etc.). It is this competition for company funds that encompasses opportunity cost, so opportunity cost should not be accounted directly against the project's benefits.

Many enterprises apply a minimum rate of return, or *hurdle rate*, to express the opportunity-cost competition between investments. For example, if a firm can draw 10 percent interest on cash in the bank, then 10 percent would be a valid choice for the hurdle rate as it represents the company's cash opportunity cost. Then, in analyzing investment options under a return-on-investment criteria, not only would the highest returns be selected, but any project that pays the firm a return of less than the 10 percent hurdle rate would not be considered.

It's important to remember that pollution prevention has good investment potential. In reducing or eliminating waste generation and the related disposal and treatment expenses, P2 can have a significant positive impact on the company=s bottom line. Even in cases where revenues are not generated, reducing the expenses and liabilities associated with generating hazardous waste or an occupational health risk represents a substantial reduction in overall expenses and, thus, an increase in profit.

Labor Expenses. In the majority of situations, the P2 project will cause a company's labor requirements to change. This change could be a positive effect that increases available productive time, or there could be a decrease in employees' production time depending upon the P2 practice.

When computing labor expenses, the Tier 1 costs could be significant. For example, if a material-substitution project eliminates a hazardous input material that eliminates a hazardous waste, there could be a significant decrease in the labor required to complete and track manifests, a decrease in the costs of labeling, a decrease in transportation costs, elimination of handling and storing hazardous waste drums, etc. Hence, both direct (Tier 0) expenses (for example, five hours per week of preventive maintenance on the P2 equipment) and secondary (Tier 1) expenses can have an effect on manpower costs.

Labor expense calculations can be simplistic or comprehensive. The most direct and basic approach is to multiply the wage rate by the hours of labor. More comprehensive calculations include the associated costs of payroll taxes, administration, and benefits. Many enterprises routinely track these costs and establish an internal 'burdened' labor rate to use in financial analysis.

Training Expenses. P2 practices may also involve the purchase of equipment or new, nonhazardous input materials that require additional operator training. In computing the total training costs, the enterprise must consider as an expense both the direct costs and the staff time spent in training. In addition, any other costs for refresher training, or for training for new employees, that is above the level currently needed must be included in the analysis. Computing direct costs is simply a matter of adding the costs of tuition, travel, per diem, etc., for the employees. Similarly, to compute the labor costs, simply multiply the employees' wage rates by the number of hours spent away from the job in training.

Auditing and Demo Expenses. Labor and other expenses associated with defining a P2 project are often overlooked. Although these tend to be small for low-investment projects, some P2 practices may require extensive auditing; pilot or plant trials can incur significant up-front costs from production down times, personnel, monitoring equipment, and laboratory measurements, as well as engineering design time and consultant-time charges. Some enterprises may prefer to absorb these costs as part of their R&D budget -- for organizations that practice pollution prevention on a regular basis, many of these expenses simply are a part of the baseline cost of operations, and are more than paid for by a few successful projects.

Floor Space Expenses. As with any opportunity costs, the floor-space costs must be based on the value of alternative uses. For example, multiple rinse tanks have long been used to reduce water use in electroplating operations. If a single-dip rinse tank of 50 square feet is replaced with a cascade-rinse system of 65 square feet, then the floor-space expense is the financial worth of the extra 15 square feet; it must be included as an expense in the financial analysis for the P2 project.

Unfortunately, computing floor-space opportunity cost is not always straightforward, as it is in the case of training costs. In instances where little square footage is required, there may be no other use for the floor space, which implies a zero cost. In other cases, where the area is currently being used only for storage of extra parts, bench stock, feed materials, etc., the costs may involve determining the value of having a drum of chemical or an extra part closer to the operator. Alternatively, as the square footage required increases, calculating floor-space costs becomes more straightforward. For example, if a new building is needed to house the pollution prevention equipment, it's easy to compute a cost. Similarly, if installing the equipment at the production site displaces enough storage room to require additional sheds be built, the cost is again easy to compute. As a default, the cost of floor space can be estimated from information available from realtors. The average-square-foot cost for a new or used warehouse (or administrative or production space) that would be charged to procure the space on the local market is the average market worth of a square foot of floor space. Unless there is a specific alternative proposal for the floor space, this market analysis should work as a proxy.

Cash Flow Considerations

Though cash flow does not have a direct effect on a company's revenues or expenses, the concept must be considered with any P2 project. If the project involves procurement costs, they often must be paid upon delivery of the equipment - yet cash recovery could take many months or even years.

Three things about a P2 project can affect a firm's available cash. First, cash is used at the time of purchase. Second, it takes time to realize financial returns from the project, through either enhanced revenues or decreased expenses. Finally, depreciation expense is calculated at a much slower rate than the cash was spent. As a result of the investment, a company could find itself cash-poor. Conversely, P2 efforts can have a very positive effect on cash flow. For example, eliminating a hazardous waste via an input-material substitution could result in an increase in cash available, because the enterprise would not have to pay for hazardous-waste disposal every three months or so. Hence, even though cash flow does not directly affect revenues and expenses, it may be necessary to consider when analyzing P2 projects.

INTEREST AND DISCOUNT RATES

In determining the value of a pollution prevention project investment, the discount rate used becomes very important. If the potential project benefits are accrued far into the future, or if a larger discount rate is used, the effect on the present value (and hence the apparent value of the project) could be dramatic.

Figure 3 illustrates the relationship between percent of future worth regained over time at varying interest rates. On the average, enterprises prefer a return on investment (ROI), or hurdle rate, in the range of 10 to 15 percent. At 10 percent, more than half of a future benefit stream can be lost, due to the time-value of money within the first 10 years. This factor works against the acceptability of projects that provide benefits far in the future. Hence, to justify P2 project investments with long-term-benefit cash flows, it is often necessary to move to Tier 2 or Tier 3 criteria. We will again discuss the categories of cost savings so that theses terms become second nature in performing such analyses.

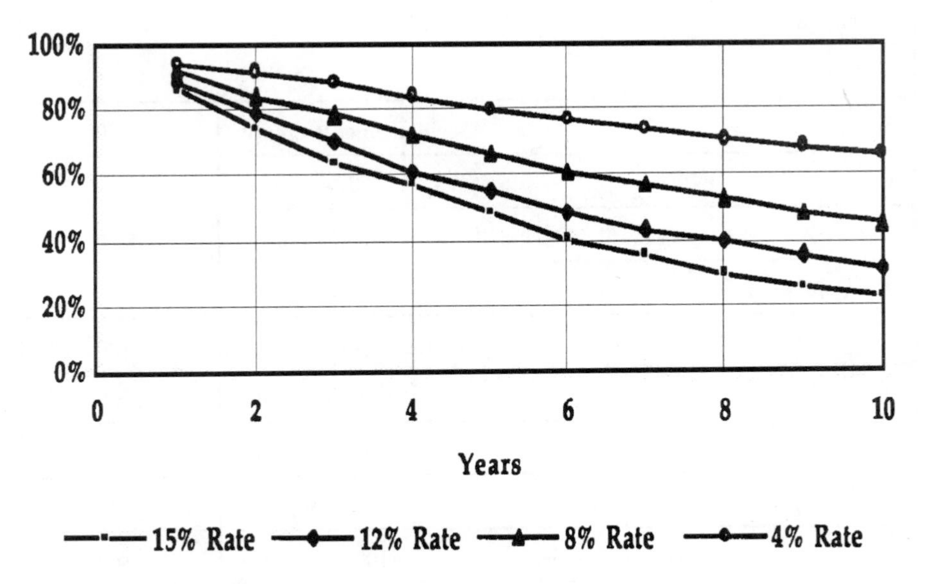

Figure 3. *Comparison of present value to future value.*

INCOME TAXES

Though most enterprises use only revenue and expense figures when comparing investment projects, income-tax effects can enter into each calculation if either revenues or expenses change from the baseline values. More expenses mean lower profits and less taxes, and vice versa. If an enterprise needs to know the effect of income taxes on profit, the computations are simple and can be done during or after the analysis.

As with expenses and revenues, the enterprise does not need to computer the total tax liability for each option. Instead, it only needs to look at the options' effect on revenues and/or expenses, and the difference in tax liability resulting from deviations from the baseline. The profit equation reflects gross or pretax profits. Income tax is based on the gross profit figure from this equation and cannot be computed until the enterprise knows what effect the options will have on its revenues and/or expenses.

Table 3: Effect of Changes in Revenues and Expenses on Pre-tax and Net Profits.

Revenue Increase	
Initial Condition:	
Beginning pre-tax profit:	$100
Tax liability:	$40
Net Profit without pollution prevention project:	**$60**
Post Pollution Prevention:	
Revenue increase subsequent to project:	$100
New pre-tax profit:	$200
New tax liability:	$80
New net profit:	**$120**
Increase in net profit due to +100 in revenues	**+$60**
Expense Increase	
Initial Condition:	
Beginning pre-tax profit:	$100
Tax liability:	$40
Net Profit without pollution prevention project:	**$60**
Post Pollution Prevention:	
Expense increase subsequent to project:	$100
New pre-tax profit:	$00
New tax liability:	$00
New net profit:	$00
Decrease in net profit due to +100 in expenses:	-$60

Refer to Table 3 for the following discussion. For the purposes of illustration, the income tax rate shall be taken as constant, at 40 percent of gross profit. Taxes act to soften the impact of P2 projects on net profit, due to changes in revenue/expenses as follows. If revenues increase by $100 with no other changes, pre-tax profits would also increase $100. Because income taxes take $40 of this increase, the effect on net profit would be to soften the $100 revenue increase to a $60 net profit increase. Similarly, if expenses increase $100, pretax income would decrease $100. The tax liability would be $40 less, so in this latter case, the $100 pretax impact would be softened to a $60 net-profit decrease.

Table 3, which illustrates this, was taken from Stephen, D. G., *A Primer for Financial Analysis of Pollution Prevention Projects*, EPA Cooperative Agreement No. CR-815932, April, 1993. The table shows that the profit impact of an increase or decrease in revenues or expenses is limited by 1 minus the tax rate (1 - t). If the tax rate is different from 40 percent, it can be inserted into (1 - t) and used to calculate the impact. For example, for a 33 percent tax rate, a $100 increase in revenue would increase profit by (1 - 0.33), or $67.

Tax credits are a special case allowed by the IRS at various times. For example, during the energy crunch of the 1970s, certain capital expenses that reduced energy consumption (such as solar energy projects) were given special treatment as tax credits. Unlike personal tax deductions, tax credits could be deducted directly from the tax obligation of a firm. As a result, in this special tax-credit case, capital expenses that would otherwise lower pretax income can be subtracted directly from the tax liability and increase profit. Unfortunately, there are currently no tax-credit P2 projects available in the United States, but this could be an option for countries with economies in transition.

APPLICATION OF TOTAL-COST ASSESSMENT

As explained, the economic analysis for a P2 project involves tabulating the financial costs, the revenues, and the savings that an enterprise expects a project to generate. Such estimates provide necessary information and data needed to evaluate the economic advantages of projects that are competing for a fixed amount of funds.

For a variety of reasons, many facilities often fail to apply the same financial tests to a P2 project that they would to other capital-intensive projects, such as those associated with operations, product quality, debottlenecking, and other more-conventional project categories that have direct relationships to revenues and profits. One reason for this is that there are still many firms that simply don=t believe or understand that P2 practices can indeed have the same positive and direct benefits as other projects. Another reason simply boils down to incomplete analyses of factors that fall into Tier 2 and Tier 3 cost parameters. This is particularly true when attempting to assess the economic impacts of reduced risks from occupational-safety issues, or from future and sometimes present environmental liabilities.

We have introduced some of the concepts above as they relate to *total-cost accounting* and the *total-cost assessment*. Total-cost accounting, also referred to as *full-cost environmental accounting*, is applied in management accounting to represent the allocation of all direct and indirect costs to specific products, the lives of products, or to operations. Total-cost assessment has come to represent the process of integrating environmental costs into the capital-budgeting analysis. It is generally defined as a long-term, comprehensive analysis of the entire range of costs and savings associated with the investment by the enterprise making the investment.

An additional term that has only been mentioned in passing to this point is the *life-cycle cost assessment*. Life-cycle cost assessment represents a methodical process of evaluating the life-cycle costs of a product, product line, process, system, or facility - starting with raw-material acquisition and going all the way to disposal - by identifying the environmental consequences and assigning monetary value. We shall expand on this important subject shortly.

For P2 projects to compete fairly with pollution control and other project alternatives, more potential costs and savings must be considered. The way to accomplish this is by expanding the cost and savings inventories in the analysis. Tables 4 and 5 provide a list of capital and operating costs that environmental managers can use to determine the financial costs and savings associated with a particular project opportunity.

The challenge in applying an expanded cost/savings inventory for investment analysis is that some of the cost data associated with a particular piece of equipment or process may be difficult to obtain. Many costs can be a challenge, because they may be grouped with other cost items in existing overhead accounts - for example, waste-disposal costs for existing processes are often placed in a facility overhead account. An expanded cost inventory would call for these costs to be directly allocated to the process that produces them. Consequently, it's likely that information for all the cost categories will not be identified during analyses. Analysts can use the list of categories provided in Tables 4 and 5 to incrementally expand their existing financial analyses whenever possible.

One approach to uncovering more of the true economic benefits of P2 projects is to expand the evaluation of costs and savings over a longer time period, usually five or more years. This is because many of the costs and savings can take years to materialize, and because the savings from P2 projects often occur every year for an extended period of time - for example, some projects may result in recurrent savings as a result of less waste requiring less management and disposal every year. Conventional project analysis, however, often confines costs and savings to a three- to five-year time frame, a time period shorter than the useful life of the equipment being evaluated. Using this traditional time frame in project evaluation will exclude some of the areas of savings generated by P2 projects.

Table 4. Partial Inventory of Potential Capital-Cost Items.

PURCHASED EQUIPMENT	SITE PREPARATION (Labor, Supervision, Materials)	SHAKE-DOWN AND START-UP (Labor, Supervision, Materials)
Equipment Delivery Sales and VAT tax Insurance Price for initial spare parts	Site studies (EIS, other) Demolition and cleaning Old equipment/garbage disposal Grading/landscaping Equipment rental Ties-ins to existing utilities & infrastructure	In-house Contractor/vendor/ consultant fees Trials/manufacturing variance Training
MATERIALS	**CONSTRUCTION/ INSTALLATION (Labor, Supervision, Materials)**	**REGULATORY AND PERMITTING (Labor, Supervision, Materials)**
Piping Electrical Instrumentation Structural Insulation Other	In-house Contractor/vendor/ consultant fees Equipment rentals	In-house Contractor/vendor/ consultant fees Permit fees
UTILITY SYSTEMS AND CONNECTIONS	**PLANNING AND ENGINEERING (Labor, Supervision, Materials)**	**WORKING CAPITAL**
General plumbing Electricity Steam Water (e.g., cooling, process) Fuel (e.g., gas, oil) Plant air Inert gas supplies Refrigeration Sewerage	In-house planning and engineering (e.g, design, shop drawings, cost estimating, etc.) Contractor/vendor/ consultant fees Procurement	Raw materials Other materials and supplies Product inventory Protective equipment
CONTINGENCY	**BACK-END**	
Future compliance costs Remediation	Closure and decommissioning Inventory disposal Site survey	

Table 5. Partial Inventory of Potential Operating Costs.

DIRECT MATERIALS	WASTE MANAGEMENT (Labor, Supervision, Materials)	INSURANCE, FUTURE LIABILITY, FINES AND PENALTIES, COST OF LEGAL PROCEEDINGS (e.g., transaction costs), PERSONAL INJURY
Raw materials (e.g., wasted raw-material costs/savings) Solvents Catalysts Transport Storage	Pre-treatment On-site handling Storage Treatment Hauling Insurance Disposal	Property damage Natural resource damage Superfund
DIRECT LABOR	**UTILITIES**	**REVENUES**
Operating (e.g., worker productivity changes) Supervision Manufacturing clerical Inspection/QA/QC	Electricity Steam Water (e.g., cooling, process) Fuel Plant air Inert gas Refrigeration Sewerage	Sale of product (e.g., from changes in manufacturing throughput, market share, corporate image) Marketable by-products Sale of recyclables
REGULATORY COMPLIANCE (labor, Supervision, Materials)	Permitting Training (e.g., Right-to-Know training, Hazmat, etc.) Monitoring and inspections Notifications Testing Labeling and packaging Manifesting Recordkeeping	Reporting General fees and taxes Closure and postclosure care Financial assurance Value of marketable pollution permits (e.g., SO_x) Avoided future regulation (e.g., CAAA)

Though expanding cost inventories and time horizons can greatly enhance the ability to accurately portray the economic consequences of a single P2 project, the financial-performance indicators - payback period, net present value, and internal rate of return - are needed to allow comparisons to be made between competing project alternatives. To review, the payback-period analysis focuses on determining the length of time it will take before the costs of a new project are recouped. A useful formula used to calculate payback period is:

Payback period (in years) = start up costs / (annual benefits - annual costs)

So, for example:

Payback period = $800 / ($600/yr - $400/yr) = 4 years

Those investments that recoup their costs before a pre-defined 'threshold' period of time are determined to be projects worthy of funding. Readers will remember from earlier discussions that the payback-period analysis does not discount costs and savings over future years. Furthermore, costs and savings are not considered if they occur in years later than the threshold time in which a project must pay back to be justified.

A more thorough analysis is based upon net present value, or NPV. The NPV method is particularly useful when comparing P2 projects against alternatives that result in higher annual waste-management and disposal costs. The increased costs of status-quo operations (or of the investment options that do not reduce wastes) will tend to lower their NPV. The method also easily accommodates the expanded cost inventory when analyzing all costs and benefits.

An additional financial analysis term to consider is the internal rate of return, or IRR (also known as return on investment, or ROI). The purpose of IRR calculations is to determine the interest rate at which NPV is equal to zero. If the rate exceeds the hurdle rate (the minimum acceptable rate of return on a project), the investment is deemed worthwhile. The following formula may be used:

$$\textbf{Initial Cost} + \textbf{Cash Flow in Year 1}/(1 + r)^1 + \textbf{Cash Flow in Year 2}/(1 + r)^2 + \textbf{Cash Flow in Year 3}/(1 + r)^3 + \ldots + \textbf{Cash Flow Year n}/(1 + r)^n = 0$$

where 'r' is the discount rate (let's assume that it's the IRR), and 'n' is the number of years of the investment. A trial-and-error procedure can be applied to solve the above equation for r.

We present Table 6 to help put the theory presented thus far into practice. We have modified Table 6, originally devised by the U.S. Environmental Protection Agency (*Federal Facility Pollution Prevention Project Analysis: A Primer for Life Cycle and Total Cost Assessment Concepts*, EPA 300-B-95-008, July 1995), for our discussions; it constitutes a *pollution prevention project analysis worksheet* (P3AW). Organizing such a worksheet in a spreadsheet program can help the reader analyze the costs and benefits associated with current operations, potential P2 projects, and alternative project opportunities.

Table 6. Pollution Prevention Project (P3) Analysis Worksheet.

	Section		ESTIMATED CASH FLOW IN EACH							
			Start-Up	1	2	3	4	5	6	7
CASH OUTFLOW	1	**CAPITAL COSTS**								
		Equipment								
		Utility Connections								
		Construction								
		Engineering								
		Training								
		Other								
		Subtotal								
	2	**OPERATING COSTS**								
		Materials								
		Labor								
		Utilities								
		Waste Management								
		Compliance								
		Liability								
		Other								
		Subtotal								
CASH INFLOWS	3	**REVENUES**								
		Sale of Products								
		Sale of By-products								
		Sale of Recyclables								
		Other								
		Subtotal								
	4	**PAYBACK**	years	*Equals Sec. 1 divided by (Sec. 2 - Sec. 3)*						
	5	**CASH FLOW (CF)**								
			Cash flow is calculated by subtracting cash outflows from cash inflows during each year of the investment (i.e., Sec. 3 minus Sec. 2 minus Sec. 1 subtotals)							
	6	**PV FACTORS**								
			For present values (PVs) for different investment durations, refer to the text for discussions							
	7	**CF x PV**								
	8	**NPV**	*Equals the sum of all values from Sec. 7*							

The use of worksheets such as Table 6 helps to demonstrate ways of capturing more cost categories by better allocating costs to specific activities, expanding the cost areas included in the analysis, and expanding the time horizon over which the project is analyzed. We have abbreviated the potential costs and revenues in Table 6 for ease of use. The reader can readily modify the worksheet to adapt it to his or her needs.

The P3AW enables its user to calculate two measures of financial performance: a simple payback analysis, and a net present-value calculation (which incorporates the time-value of money). Both of these calculations can help an enterprise compare competing project options, or a proposed project against the status quo. IRR calculations are not included on the worksheet, but they may be readily added by the reader.

It is important to note that when completing the worksheet, some data might not be available to complete all the sections. However, completing even only a few of the sections of the worksheet with data that would not have normally been collected will significantly enhance the accuracy of evaluating project opportunities.

Before attempting to organize information and data needed for the P3AW, enterprises should define the objective of the analysis. In addition, the analysis ultimately will be used by decision makers C that is, top management, which will decide whether or not a proposed P2 project will be funded. Like any other sales pitch, presentation means a lot. Therefore, knowing the audience, the decision making criteria for company projects, and the format to best present the analysis are important areas to consider. With some planning, you can make certain that the scope of the analysis is appropriate, and that the completed analysis will be presented in a readily understood and accepted manner. When using P3AWs to compare project alternatives, or to compare a potential project to current operations, you should use a separate worksheet for each option under consideration. The following guidelines will help the reader use the P3AW.

Sections 1-3. First, identify the economic consequences associated with the activity under review. Specific items (such as categories of cash outflows) noted in the P3AW may not necessarily represent a complete list of costs incurred for the facility under review. As such, Tables 4 and 5 should help identify additional capital- and operating-cost categories. If the focus is on a payback analysis, completing information for only the initial year is acceptable, provided that the data are available to describe annual costs and annual savings. If the focus is on analyzing the financial performance using a NPV calculation, then you will need to obtain estimates of future costs and benefits. It is not necessary to make adjustments for inflation if the calculations are addressed through a nominal discount factor. This option is described further on and some nominal discount factors are given in Table 7 for different periods of investments. Note also that, to allow comparisons with other project options, there are two measures of economic performance included on the P3AW (*payback analysis* and *net present-value analysis*). To conduct a payback analysis, refer to the Section 4 discussions, below. To conduct a NPV analysis, refer to sections 5 through 8 in the discussions below.

Table 7. Present Value Factors for Nominal Discount Rates.

Year	7.3%	7.6%	7.7%	7.9%	8.1%
1	0.93197	0.92937	0.92851	0.92678	0.92507
2	0.86856	0.86372	0.86212	0.85893	0.85575
3	0.80947	0.80272	0.80048	0.79604	0.79163
4		0.74602	0.74325	0.73776	0.73231
5		0.69333	0.69012	0.68374	0.67744
6			0.64078	0.63368	0.62668
7			0.59496	0.58729	0.57972
8				0.54429	0.53628
9				0.50444	0.49610
10				0.46750	0.45893
11					0.42454
12					0.39273
13					0.36330
14					0.33608
15					0.31090
16					0.28760
17					0.26605
18					0.24611
19					0.22767
20					0.21061
21					0.19483
22					0.18023
23					0.16673
24					0.15424
25					0.14268
26					0.13199
27					0.12210
28					0.11295
29					0.10449
30					0.09666

Section 4. This section focuses on calculating the number of years that it will take to recoup the initial capital expenditure. This value is obtained by dividing the initial investment needed to establish the project by the net annual benefits (which

are obtained by subtracting the annual cash outflows from the expected annual cash inflows). If only the payback analysis is important, then the reader may skip the following discussions.

Section 5. For each year included in the evaluation, calculate the annual net cash flow by subtracting the capital expenditures subtotal (Section 1) and the annual cash outflows (subtotals from sections 3, 4, 5) from the annual cash inflows (Section 2).

Section 6. To calculate the NPV, you need to determine the value of future cash flows, starting from today. To accomplish this, you can use present-value factors to discount future cash flows. The discount will be specific to your local region, but for a comparative basis, we can use the discount rates used by the U.S. government. Table 7 provides present value factors for nominal discount rates (based on 1995 figures).

Section 7. Multiply the net cash flows (Section 5) by the PV factors (Section 6) to determine the present value today of the cash flow in each year.

Section 8. Sum up all the annual discounted cash to determine the NPV of the project. If the value is positive, the project is cost-beneficial. If more than one investment is being studied, then the project with the greatest NPV is the most attractive.

Once you've completed the analysis, you many prepare a report explaining the results. The report should highlight the economic benefits of the proposed P2 projects. Also discuss the non-economic benefits, as these may tip the scales in favor of the P2 opportunity if the financial analysis is marginal or too close to call.

THE LIFE-CYCLE ANALYSIS

With pollution control based on end-of-pipe treatment technologies, as well as with many early P2 programs, enterprises approached the project-review process by considering only those environmental impacts that could be easily translated into financial terms (permitting costs, and pollution-control equipment costs, etc.). Consequently, the financial budgeting tools discussed here often did not fully capture the benefits of P2 opportunities, particularly those that reduce environmental concerns for the present and future.

Without the tools to completely document environmental benefits, P2 opportunities have often been difficult to support when competing against more-easily quantified environmental projects, such as end-of-pipe controls, and against non-environmental investments, such as retooling or plant expansion. Decision-makers require analytical tools that accurately and comprehensively account for the environmental consequences and benefits of competing projects. These environmentally based project-review tools must be flexible, easy-to-use, and require limited staff and funding so that they can be easily incorporated into the review process.

This is where we can begin to apply the principles of life-cycle assessment (LCA), which provides a means to evaluate environmental consequences and impacts. LCA identifies and evaluates "cradle-to-grave" natural resource requirements and environmental releases associated with processes, products, packaging, and services. LCA concepts also can be particularly useful in ensuring that identified P2 opportunities are not causing unwanted secondary impacts by shifting burdens to other places within the life-cycle of a product or process.

LCA is an evolving tool undergoing continued development. Nevertheless, LCA concepts can be useful in acquiring a broader appreciation of the true environmental impacts of current manufacturing practices and of proposed P2 opportunities. It has taken a good two decades for many environmental professionals to become more aware that the consumption of manufactured goods and services can have adverse impacts on the supplies of natural resources as well as the quality of the environment. These effects occur at virtually all stages of the life cycle of a product, starting with raw-materials harvesting, continuing through materials manufacturing and product fabrication, and concluding with product consumption and disposal. LCA is essentially a tool that enables us to evaluate the environmental consequences of a product or activity across its entire life.

LCA is made up of the components listed in Figure 4. These components or stages are defined as follows:

- *Goal Definition and Scoping.* This is a screening process that involves defining and describing the product, process, or activity; establishing the context in which the assessment is to be made; and identifying the life-cycle stages to be reviewed for the assessment.
- *Inventory Analysis.* This process involves identifying and quantifying energy, water, and materials usage, and the environmental releases (e.g., air emissions, solid waste, wastewater discharges) during each life-cycle stage.
- *Impact Assessment.* This process is used to assess the human and ecological effects of material consumption and environmental releases identified during the inventory analysis.
- *Improvement Assessment.* This process involves evaluating and implementing opportunities to reduce the environmental burdens as well as the energy and material consumption associated with a product or process.

LCA can be used in process analysis, materials selection, product evaluation, product comparison, and even in policymaking. LCA can be used by acquisitions staff, new-product design staff, and staff involved in investment evaluations.

What makes this type of assessment unique is its focus on the entire life cycle, rather than a single manufacturing step or environmental emission. The theory behind this approach is that operations occurring within a facility can also cause impacts outside the facility's gates that need to be considered when evaluating project alternatives. Examining these upstream and downstream impacts can identify benefits or drawbacks to a particular opportunity that otherwise may have been overlooked. For example, examining whether to invest in plastic bottle cartons for the beverage bottling facility described earlier, or to use wooden crates for staging and storing incoming bottles, should include a comparison of all major impacts, both inside the facility (e.g., disposing of the wooden crates) and outside the gate (e.g., additional wastewater discharges from off-site washing of the reusable plastic cartons).

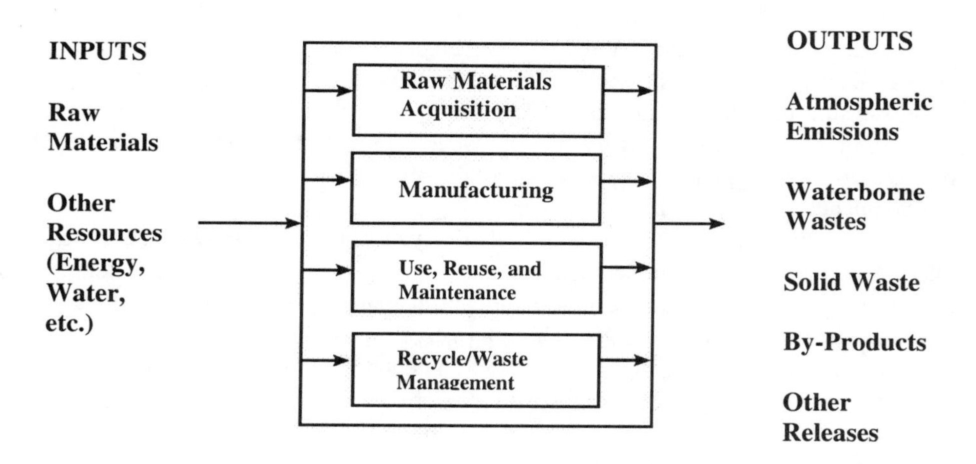

Figure 4. *Life-cycle stages.*

Gaining complete understanding of a proposed project's environmental effects requires identifying and analyzing inputs and releases from every life-cycle stage. However, securing and analyzing this data can be frustrating and, perhaps, an endless task. Process engineers in plants often are faced with immediate priorities and may not have the time or resources to examine each life-cycle stage or to collect all pertinent data. Despite this shortcoming, it is worthwhile to discuss the steps required to begin applying LCA concepts and principles to project analysis, and to give examples that demonstrate steps within selected life-cycle stages.

Before beginning to apply LCA concepts to projects under review, it is important to first determine the purpose and the scope of the study. In determining the purpose, facility managers should consider the type of information needed from the environmental review (for example, does the study require quantitative data or will qualitative information satisfy the requirements?). Once the purpose has been defined, the boundaries or the scope of the study should then be determined. What stages of the life cycle are to be examined? Are data available to study the inputs and outputs for each stage of the life cycle to be reviewed? Are the available data of an acceptable type and quality to meet the objectives of the study? Are adequate staff and resources available to conduct a detailed study?

This definition and scoping activity links the purpose and scope of the assessment with available resources and time, and allows reviewers to outline what will and will not be included in the study. In some cases, the assessment may be conducted for all stages of the life cycle (i.e., raw-materials acquisition, manufacturing, use/reuse/maintenance, and recycling/waste management). In many cases, the analysis may begin at the point where equipment and/or materials enter the facility. In other cases, primary emphasis may be placed on a single life-cycle stage, such as identifying and quantifying waste and emissions data. In all cases, managers should ensure that the boundaries of the LCA address the purpose for which the assessment is conducted and the realities of resource constraints. Whenever possible, managers should include in the analysis all life-cycle stages in which significant environmental impacts are likely to occur.

Conducting an LCA that includes all life-cycle stages will provide decision makers with the most complete understanding of environmental consequences. However, if resources are limited and an in-depth, quantitative analysis is not practical, a simplified approach may be taken. This alternative approach makes use of a simple checklist to identify and highlight certain environmental implications associated with competing projects. A checklist that uses qualitative data instead of quantitative inputs can be very useful when available information is limited or when used as a first step in conducting a more thorough LCA. In addition, a *life-cycle checklist* (LCC) should include questions regarding the environmental effects of current operations and/or potential projects that cover materials and resources consumed and wastes/emissions generated.

Table 8 provides a sample checklist. Of course, this example is very general, but the reader can expand upon it, making it specific to the facility and nature of the operations under evaluation. As an exercise, take the process flow sheet of your plant, or a small shop area, and trace the flow of materials through the system to the final product of the unit process (or plant). Construct a Life-Cycle Checklist that is specific to the process, using as a guide the generalized questions in Table 8. Now take your answers and see if you can quantify some of the conclusions in terms of potential cost savings. Many readers may be surprised to learn that this simple exercise helps to identify some substantial savings that are worthwhile going after.

Table 8. Example of a Life-Cycle Checklist.

Issue	Question	Yes	No
Material Usage	*Does the project minimize the use of raw materials?*		
	Is there a potential to reduce the number of suppliers for certain raw materials?		
	Is there any significant waste/discard/off-spec of raw materials due to inventory and shelf-life issues?		
Resource Conservation	*Does the project minimize energy use?*		
	Does the project minimize water use?		
	Does the project involve fuel switches or alternative energy technologies?		
Local Environmental Impacts	*Does the project eliminate or minimize impacts to the local environment (i.e., air, water, land)?*		
	Does the project have possible impacts on future environmental legislation?		
	Does the project reduce or eliminate dependency on external resources that are polluting (e.g., reduce dependence on electricity purchased from a utility that generates power at coal-fired plants)?		
Global Environmental Impacts	*Does the project eliminate or minimize impacts known to cause global environmental concerns (e.g., global warming, ozone depletion, acid rain)?*		
Toxicity Reduction	*Does the project improve the management of toxic and regulated hazardous materials and/or processes/operations that could result in human/ecological exposure?*		
Off-site Environmental Liabilities	*Does the project eliminate potential third-party liabilities (e.g., eliminating the need for regulated wastes to be disposed of off-site by contractors)?*		
OSHA and Health Risk	*Does the project eliminate the need for personal protective equipment and confined-space operations?*		
Liabilities	*Does the project have the potential to lower insurance premiums by reducing risks of fire, explosion, or occupational exposure?*		

Application of an LCC has specific advantages and disadvantages when compared to other life-cycle assessments. The primary advantage of an LCC is that completion of a checklist is relatively straightforward and requires only limited resources. However, a LCC does not provide a detailed or complete assessment of the environmental consequences associated with the activity under review. Regardless of how specific or detailed the checklist is, the method only provides general, qualitative data.

Detailed Project Reviews

Conducting a more-thorough review of the environmental consequences of P2 projects will require more and, most likely, dedicated resources. A more in-depth analysis would be aimed at identifying and evaluating the resource and material inputs, and the environmental releases, associated with each life-cycle stage. This is a resource-intensive operation and much more comprehensive than applying an LCC. As first steps, we recommend that you define and scope out the analysis to fit available resources while including all significant areas of environmental impact.

The first step in identifying and evaluating the inputs and outputs associated with the life-cycle stages under review is to describe and understand each step in the process. One common method to do this is to construct a system-flow diagram for the product, process, or activity being studied. Each step within the relevant life-cycle stages is represented by a box. Each box is connected to other boxes that represent the preceding and succeeding step. A simple example of a process-flow diagram is below. In this example, the life-cycle stages covered within the diagram begin at the point where a solvent is purchased for use and enters a facility property. Each of these boxes can be further divided into detailed process-flow steps.

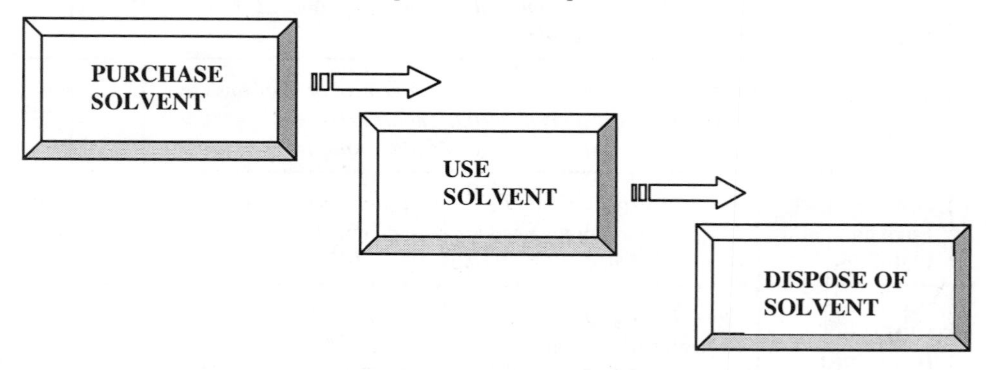

Once all of the relevant steps for each stage of the product, process, or activity under review have been identified, the flow diagram can be expanded to identify specific material and energy inputs, and the specific environmental releases and wastes associated with each box on the diagram. The process-flow diagram serves as a roadmap for us to follow the flows of all materials and energy into and out of each stage in the operation. Data on each of these identified inputs and releases will be collected later in the assessment, and will form the basis for our findings, conclusions, and recommendations. In this simple example, then, we can note the following:

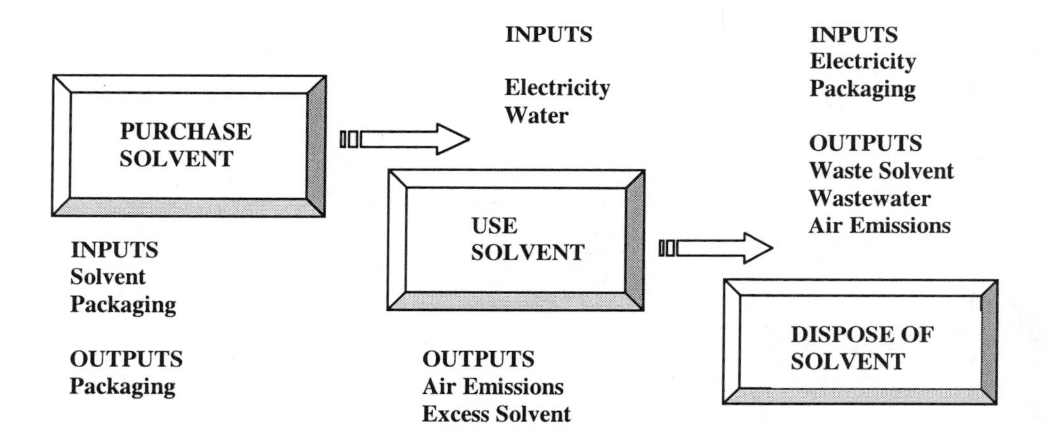

Although quantitative data are preferable (and necessary to accurately and completely conduct an impact assessment), qualitative data may be acceptable in cases where quantitative data are lacking. A sample LCA worksheet and instructions are provided in Table 9 to help readers complete a sample process-flow diagram. The intent of this worksheet is to acquaint the reader with both a suggested form and systematic approach to assessing project options or process changes under consideration.

This life cycle-based worksheet is organized into three sections. The first section asks for a flowchart of the process steps/activities to be included in the analysis. The second section asks for inputs (i.e., raw materials, energy, and water), and the third section asks for outputs (i.e., products, air, water, and land releases). The worksheet provides space for four process steps; however, the reader can readily expand this to as many steps as necessary by incorporating the worksheet into a spreadsheet on a computer.

Table 9. Example of a Life-Cycle Worksheet.

I. PROCESS STEPS					
Provide process-step name or unit operation in each box (Step 1... Step 4)	*1*	*2*	*3*	*4*	
II. INPUTS					
2a Raw materials(units)					
______	____	____	____	____	____
______	____	____	____	____	____
	____	____	____	____	____
2b Energy usage					
Electricity (kW-hr)	___ +	___ +	___ +	___ =	___
Natural gas (cu. ft)	___ +	___ +	___ +	___ =	___
Fuel oil (gal.)	___ +	___ +	___ +	___ =	___
Other	___ +	___ +	___ +	___ =	___
2c Water use (gal.)	___ +	___ +	___ +	___ =	___
2d Other inputs (units)					
______	___ +	___ +	___ +	___ =	___
______	___ +	___ +	___ +	___ =	___
III. OUTPUTS					
3a Products, useful by-products (item and amount)					
	___	___	___	___	
	___	___	___	___	
	___	___	___	___	
3b Releases to the air (including nonhazardous gaseous wastes)					
______	___ +	___ +	___ +	___ =	___
	___ +	___ +	___ +	___ =	___
3c Releases to water (including nonhazardous liquid wastes)					
______	___ +	___ +	___ +	___ =	___
______	___ +	___ +	___ +	___ =	___
3d Solid wastes					
______	___ +	___ +	___ +	___ =	___
	___ +	___ +	___ +	___ =	___

Using this or any other life-cycle worksheet has specific advantages and disadvantages when compared to conducting a complete LCA. The principle advantage is that it provides a more-detailed analysis of the process than the checklist, and it is easier to conduct than a complete LCA. On the other hand, it does not encompass the full environmental impacts of a process or activity life-cycle stage.

Here are some guidelines for applying Table 9. First, the reader should view this worksheet as an aid or general outline. The worksheet is intended to help managers gain a more complete understanding of the life-cycle environmental consequences associated with existing processes, potential P2 projects, and competing project alternatives. When completing the worksheet, do not worry if data are not available to complete all requested information. The information contained in even a few completed sections of the worksheet can still prove useful in evaluating and comparing the environmental performance of existing processes and potential projects.

However, be aware that completing only certain sections of the worksheet may provide misleading results. For example, completing sections on solid wastes and releases to air without entering data on releases to water may bias the analysis toward projects where the primary environmental consequences result from water pollution. Similarly, collecting and analyzing data on a limited number of life-cycle stages may bias the analysis toward projects whose primary environmental effects occur upstream or downstream from stages under analysis.

The information requested on the worksheet can be indicated either numerically or by text description. Descriptive information is sometimes the only information available. Specific instructions follow.

Line 1. Indicate the process steps to be reviewed. For example, a life-cycle analysis of a solvent degreaser tank system might examine the following three activities: acquisition of solvent, use of the tank, and disposal/recycling of waste materials.

Line 2a. For each of the process steps indicated in Line 1, identify the raw materials used. Examples of typical materials include chemicals, parts, and minerals. Do not forget to include associated packaging materials, such as cans, cardboard, plastic wrap, etc.

Line 2b. Indicate the energy involved with operating the process activity. We have included three common energy-source categories - electricity, natural gas, and fuel. Include other categories, if needed. If numerical data are available, it is possible to sum together all entries from the same energy source (i.e., electricity usage from each of the process steps examined).

Line 2c. Indicate the quantity of water consumed in each of the process steps being evaluated. Note that water could be coming from surface sources (e.g., pumped in from a nearby river), from a well, or from purchased city water.

Line 2d. Indicate other inputs, as needed. Some process steps that can generate additional inputs include pre-process cleaning, process cleaning, and maintenance supplies required in the upkeep of the process.

Line 3a. For each process step, indicate the products that result. Be aware that the products often become the inputs to the next step in the sequence.

Line 3b. Indicate numerically or by description the air releases associated with the process step. Examples of typical releases from an industrial process include particulates/dust and solvent vapors. Numerical records of air emissions can often be found on permitting applications or in engineering records. If numerical data is not available, provide a narrative list of emissions.

Line 3c. Indicate the wastewater discharges and liquid hazardous wastes associated with each process step.

Line 3d. Identify the solid waste generated from each process step. If possible, list the type and quantity of solid waste, and how it is managed (e.g., 10 pounds of paper products that are recycled, or five cubic yards of sludge that is landfilled).

FINAL COMMENTS

This chapter has provided an overview of the financial tools, including life-cycle assessment techniques, needed for evaluating P2 project opportunities. Please refer to the list of recommended reading materials at the end of this book for additional explanations of the financial calculations already discussed.

Some final comments about cost items and about placing value on future costs and benefits: Both of these can pose some difficulties in a project's financial analysis, because quite often true costs can be buried in overhead items, or masked by other operating costs and categories within an organization. It is important to try and better define these items to properly assess the financial benefits of the project.

Allocation of Cost Categories

Compared with traditional project-analysis processes, expanding the analysis to include broader cost inventories requires a more detailed data-tracking system. Many enterprises use tracking systems that group inventory categories together into facility-wide overhead accounts. These types of tracking methods make it very difficult to identify all of the discreet costs that will be affected by proposed project alternatives. P2 activities, in particular, are at a disadvantage because many of the savings that result from these projects (such as energy, sewage, water, permitting, and waste disposal) often occur in areas lumped into overhead accounts. To overcome this, staff performing project analyses must first identify the exact data needs for the project under review. Then, a comparison can be made to information available from traditional record keeping systems, to identify information gaps resulting from items being lumped together or reported on a facility-wide basis. To eliminate the data gaps, one of several approaches can be employed:

- For the simplest of cases where several inventory categories have been combined, a review of the input data developed by each department in a facility may reveal the data for the particular project in question. For example, the accounting department indicates on its books only the total quantity of valves used at the entire facility. However, a review of department-specific expenses would likely reveal a more-detailed account of valve use by location.
- For categories that are aggregated for the whole facility and not by specific project (such as water usage), engineering estimates or a facility walk-through may be used to generate an estimate allocation to specific projects.
- For aggregated categories that cannot be easily allocated on a project-specific basis by either of the above two methods, it may be useful to discuss the data needs both with the vendors that supplied the original equipment, to see if any baseline consumption data exist, and/or with auditing professionals, to identify what types of measurement devices or meters could be located at the specific project to meet the data needs.

Placing Value on Future Costs and Benefits

Developing reliable estimates of future costs and benefits can be a difficult task. Quantitatively estimating future costs for items such as property clean-up and environmental compliance for a facility's decommissioning and post-closure can be extremely difficult. A generalized approach to this problem is to group future costs into one of two categories: *recurring costs* and *contingent costs*.

Recurring costs include items that are currently incurring costs and are anticipated to continue incurring costs into the foreseeable future, based upon regulatory requirements. These include permits, monitoring, and compliance with regulatory requirements. The first step in estimating the future costs of these items is to determine how much the facility is currently paying. Then, estimate how much the cost can reasonably be expected to escalate in the future. For example, if monitoring costs are currently $10,000, and are expected to rise with inflation, a conservative estimate would be a 4 percent annual increase. Consequently, the monitoring costs for the following year could be estimated at $10,400, assuming that monitoring requirements do not become more stringent. Note that when using the P3AW (Table 6), it is not necessary to escalate these values, because nominal discount rates from Table 7 can be incorporated into the worksheet. In other words, Table 6 already takes inflation into account when calculating present values.

Contingent costs include catastrophic future liabilities, such as remediation and clean-up costs. Though current activities can lead to these future costs, quantitative estimates of these liabilities are difficult to obtain. Often, the only way to include these future liabilities in the budgeting process is to qualitatively describe estimated liabilities without attempting to define them using dollar amounts. If a P2 option is being considered, make sure to include a comparison that highlights the areas in which future liability would be reduced by implementing the P2 option. For example, this approach could be used to describe the future benefit of switching from lead-based to water-based paint. Most likely, the best option may be to fully describe the potential liability if the change is not made and, if possible, document the remediation cost that could result if a liability event occurred today.

Final Remarks on Pollution Prevention

Pollution prevention works best in the context of an EMS. In fact, P2 should become an integral component of the EMS as opposed to a stand-alone program being implemented by a group of engineers.

Indeed, it would be wrong to place a P2 program entirely in the hands of the technical staff because the real driving force is improving bottom-line financial performance. Reducing the four cost categories associated with pollution management is the ultimate objective of any P2 program. For this reason, the proper mix for a P2 audit team should be a financial planner, technology-specific specialists, and either a compliance officer or attorney (see Figure 5). By matching the skills, backgrounds and experiences of the team members with the cost categories, we can ensure that P2 recommendations will focus on as many opportunities as possible.

Ultimately, if P2 is to have a significant impact in your enterprise, it cannot remain focused on one portion of the plant. The EMS is the vehicle by which the examples and benefits of localized P2 activities can be rolled-out to other parts of a plant operation, and to satellite and affiliate operations of your enterprise.

All too often management views P2 as a long-term collection of activities, which can present incremental advantages and improvements. Indeed, P2 can start off that way, but when embraced in a macro-sense throughout the business, its impact can be much more dramatic. We maintain that if enterprises were practicing elements of pollution prevention 30 years ago, Superfund sites would likely not exist today. The third-party liabilities resulting from off-site property damages and class-action toxic tort cases of today and over the last 20 years, are a direct result of practices that not only ignored fate and transport characteristics of pollution, but simply did not focus on efficient manufacturing. Clearly we can argue that strict enforcement of environmental laws is what has brought industry to understanding that it must be responsible for how its operations interact with the environment and public safety, but from a business standpoint we must recognize that the cost of compliance has become excessive in some cases such that more effective approaches are needed in order to sustain operations. Industry has always had the incentives to reduce waste-full by-products. The thousands of tons of waste and spent solvents and off-spec chemical products stockpiled in corroding drums, saturated soils from careless spills and dumping, the percolation of these wastes into groundwaters that ultimately impacted on drinking water supplies in parts of our country, the hundreds of millions of dollars spent on fines, penalties, litigations due to off-site property damages from careless waste management, not to mention site remediation efforts that have gone on for decades at some facilities – all represent the reasons for investing in pollution prevention and cleaner production.

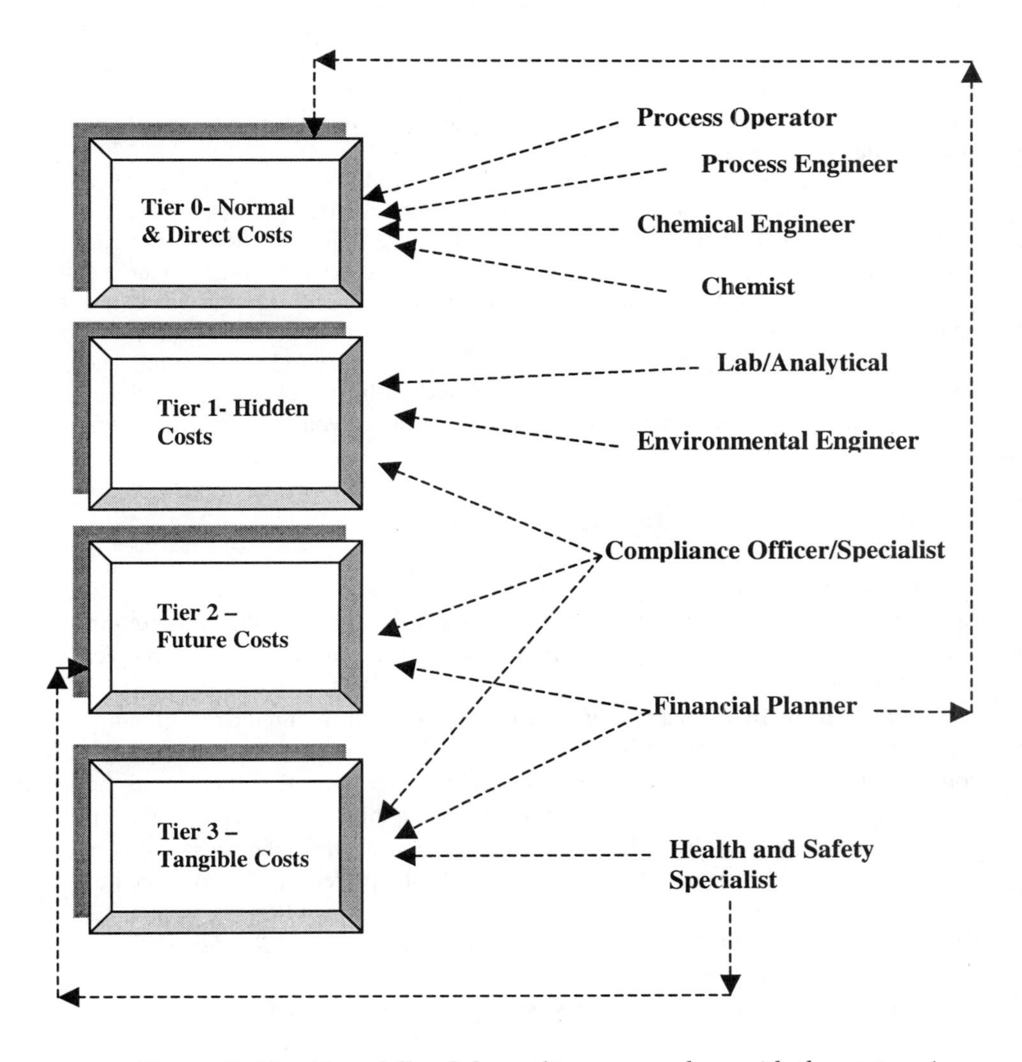

Figure 5. *Matching skills of the audit team members with the cost savings categories.*

And what about those parts of the world where many heavy industries operate within the framework of weak environmental enforcement? These enterprises not only face the same future liabilities that U.S. corporations did 30 years ago, but if they view pollution in its broadest sense – as waste and inefficiency, then their incentives exist. These enterprise simply have not quantified their losses and don't realize the money being lost through by-products flowing down sewer drains, flowing out of their stacks, and being washed away from stockpiles due to stormwaters. In addition, globalization of industry and business practices are making it exceedingly difficult for such operations to compete in world markets. Enterprises operating in transitioning economies are simply finding that their ability to penetrate markets in technologically advanced nations, where the general public recognizes the benefits of a *Green Stamp* of approval on goods and services are poor.

Bear in mind that rigorous applications of the procedures and practices outlined for both P2 and EMS are not what is important. Successful programs must be flexible and adapted to the specific needs of the enterprise. The objective should not be to restructure your entire enterprise, but rather to incorporate certain tools and to refocus priorities so that the economics of the operations benefit through improved environmental performance.

The following is a glossary of important financial terms to bear in mind. The appendix contains a list of P2 and EMS resources, which includes both printed references and Web sites we have visited and feel may assist you. And finally – if you have any questions, the authors are available to assist.

GLOSSARY OF IMPORTANT TERMS

Discount Rate: The interest rate (sometimes called the present-value factor) used to discount future cash flows to their present values. This represents the rate of return that could be earned by investing in a project with risks comparable to the project being considered.

Hurdle Rate: In internal rate-of-return calculations, the minimum rate of return that a project must generate to be considered worthy of investment. Although not common practice, in this book we have used a U.S. government discount rate as the hurdle rate. Projects that provide a rate of return below this rate will not be pursued.

Internal Rate of Return (IRR): The discount rate at which the net savings (or net present value) on a project equal zero. The IRR of a project can be compared to the hurdle rate to determine economic attractiveness. The general IRR rule is: *If IRR is greater than or equal to the hurdle rate, then accept project; if IRR is less than the hurdle rate, then reject project.*

Life-Cycle Assessment: A method to evaluate the environmental effects of a product or process throughout its entire life cycle, from raw-material acquisition to disposal. This includes identifying and quantifying energy and materials used and wastes released to the environment; assessing their environmental impact; and evaluating opportunities for improvement.

Life-Cycle Costing: A method in which all costs are identified with a product, process, or activity throughout its lifetime -- from raw-material acquisition to disposal -- regardless of whether these costs are borne by the organization making the investment, other organizations, or society as a whole.

Net Present Value (NPV): The present value of the future net revenues of an investment, minus the investment's current and future cost. An investment is profitable if the NPV of the net revenues it generates in the future exceeds its cost -- that is, if the NPV is positive.

Payback Period: The amount of time required for an investment to generate enough net revenues or savings to cover the initial capital outlay for the investment.

Total-Cost Assessment (TCA): A long-term comprehensive financial analysis of the full range of costs and savings of an investment that are or would be experienced directly by the organization making or contemplating the investment.

Appendix A
ADDITIONAL ISO 14001 AND POLLUTION PREVENTION RESOURCES

SECTION I: ADDITIONAL ISO 14001 RESOURCES

Author's note: Our aim here is to provide a sampling of recent literature that could be useful to readers of this book. The body of literature related to EMSs and ISO 14001 is very dynamic because experience worldwide with them is expanding rapidly. We've included nothing published before 1998 in the list below (the ISO 14001 standard was published in late 1996). Readers are encouraged to surf the Web for publications and other material not yet available when this book went to press.

BOOKS

1) Block, Marilyn R., and I. Robert Marash. *Integrating ISO 14001 into a Quality Management System.* Milwaukee: ASQ Quality Press, 2000.

Explains how to expand an existing quality-management system to encompass the requirements of ISO 14001. Provides a comparison and guide to integrating ISO 9001 and ISO 14001.

2) Block, Marilyn R. *Identifying Environmental Aspects and Impacts.* Milwaukee: ASQ Quality Press, 2000.

Not available for review at press time, but is said to use examples from six very different types of companies registered to ISO 14001, and to provide easy methods and worksheets for identifying environmental aspects and impacts consistent with the requirements of ISO 14001.

3) Cascio, Joseph, ed. *ISO 14000 Handbook.* Milwaukee: ASQ Quality Press, 1998.

Includes checklists, case studies, and a great deal of practical information and guidance for implementing ISO 14001. Goes beyond to discuss other standards in the ISO 14000 series. Includes contributions from professionals working on various ISO 14000 standards.

4) Edwards, A.J. *ISO 14001 Environmental Certification Step by Step.* Oxford, U.K.: Butterworth-Heinemann, 2001.

Written primarily for small and medium-sized enterprises, this book sets out an overall program for establishing an EMS, and provides step-by-step directions for implementation in conformance with ISO 14001.

5) Goetsch, David I., and Stanley B. Davis. *ISO 14000: Environmental Management.* Upper Saddle River, N.J.: Prentice Hall, 2000.

Publisher information says that this book is written as a practical teaching resource and how-to guide. Each chapter contains a list of key concepts, review questions, critical thinking problems, and discussion cases with related questions.

6) Hillary, Ruth, ed. *ISO 14001: Case Studies and Practical Experiences.* Sheffield, U.K.: The Network for Environmental Management, Greenleaf Publishing, 2000.

Contains case studies that provide experiences of companies and what certifiers look for when they visit firms. Highlights ISO 14001's strengths and weaknesses. Analyzes environmental and economic performance improvements under ISO 14001.

7) Kanholm, Jack *ISO 14001 in Our Company: Self-Study Course for Personnel.* Pasadena, Calif.: AQA Press, 1998.

A booklet that offers a general orientation to the ISO 14001 standard for self-study or group training. Could be useful for basic EMS training for employees.

8) Kanholm, Jack. *ISO 14001 Manual and Procedures: EMS Manual, 21 Procedures and Forms*, Pasadena, Calif.:AQA Press, 1999.

A CD-ROM containing a full set of formatted document templates and procedures, including a model EMS manual, that conform to the ISO 14001 standard.

9) Kanholm, Jack. *ISO 14001 Requirements: 61 Requirements Checklists and Compliance Guide*. Pasadena, Calif.: AQA Press, 1998.

Interprets and explains the ISO 14001 standard as 61 distinct requirements and discusses, with respect to each of these requirements, how to implement, required documentation, and what auditors will look for as evidence of compliance.

10) Kinsella, John, Annette Dennis McCully, and David Burngasser. *Handbook for Implementing an ISO 14001 Environmental Management System: A Practical Approach.* Bothell, Wash.: EMCON, 1999.

A guide for implementing an EMS that meets the requirements of the ISO 14001standard. Based on work with five companies, ranging in size from 30 employees to 42,000, enabling readers to learn about the challenges they encountered and how they solved them.

11) Perry, Pam, ed. *The Bottom Line: How to Build a Business Case for ISO 14001.* Boca Raton, Fla.: CRC Press-St. Lucie Press, 1999.

Based on the experiences and data of companies that have implemented ISO 14001. Describes how implementing the standard can benefit a company and how to build a case to convince top management to adopt ISO 14001.

12) Roberts, Hewitt, and Gary Robinson. *ISO 14001 EMS Implementation Handbook.* Oxford, U.K.: Butterworth-Heinemann, 1998.

A guide for implementing and maintaining an ISO 14001 EMS. Includes recommendations, checklists, templates, certification tips, case-study materials, and more.

13) Saponara, Anthony, and Randy A. Roig, Ph.D. *ISO 14001 Environmental Management Systems: A Complete Implementation Guide.* Vancouver: ERM-West, Inc. and STP Specialty Technical Publishers Inc., 1998.

A very extensive, two-volume ISO 14001 implementation manual containing dozens of checklists, prototype charts, templates, document examples, and other tools. Uses "Green Acres Hotel and Resort" as a case illustration for presenting implementation principles, techniques, and tools.

14) Shell , David J. *A Green Plan for Industry: 16 Steps to Environmental Excellence.* Rockville, Md.: Government Institutes, Inc., 1998.

Offers a 16-step process for developing a comprehensive environmental management plan. Written for environmental managers, this book also contains techniques and tools for implementing ISO 14001.

15) Schoffman, Alan, and Allan M. Tordini. *ISO 14001: A Practical Approach.* Oxford, U.K.: OUP, 2000.

Presents tools that industry professionals, environmental managers, or officers of companies have used to plan and implement an EMS that conforms to ISO 14001. Includes case studies of medium and small companies.

16) Woodside, Gayle, and Patrick Aurrichio. *ISO 14001 Auditing Manual.* New York: McGraw-Hill, 1999.

A guide to developing and carrying out an internal ISO 14001 audit program and managing the ISO 14001 registration (certification) audit process. Includes a set of questions that are typically asked during the registration audit process, as well as their answers.

17) Woodside, Gayle, Patrick Aurrichio, and Jeanne Yturri. *ISO 14001 Implementation Manual.* New York: McGraw-Hill, 1998.

A guide for implementing ISO 14001 that is, in part, a result of the authors' experience at IBM. Uses "Quality Seat Belts, Inc." as a case illustration for presenting the basics.

18) Zottola, Vincent, and Vincent Zottola Jr. *The ISO 14001 Implementation Tool Kit.* New York: Richard D. Irwin/McGraw-Hill, 1999.

Contains materials to help with implementation planning, developing EMS procedures and work instructions, corrective action, internal auditing, and more, including a gap-analysis checklist.

WEB SITES

1) *http://www.awm.net/iso/*
Advanced Waste Management Systems, Inc. Describes ISO 14000, registration process, ISO 14001 FAQs. Lists EMS-auditor training course providers, and ISO 14001 consulting firms and their Web sites.

2) *http://cei.sunderland.ac.uk/envrep/iso14000.htm*
Centre for Environmental Informatics (CEI). Provides information and advice on ISO 14001 and directories of environmental management registration bodies and consultants and trainers around the world. Contains a bi-weekly on-line newsletter, *ISO 14000: News and Views*. Also has editorials and case studies about EMS implementation.

3) *http://www.ceres.org*
Coalition for Environmentally Responsible Economies (CERES). Standardizes corporate environmental reporting and advocates corporate environmental responsibility. A forum for multi-stakeholder dialogue consisting of more than 50 investor, environmental, religious, labor, and social justice groups.

4) *http://www.epa.gov*
U.S. Environmental Protection Agency (EPA). Provides answers to FAQs, articles, and documentation of case studies and pilot projects concerning EMS and ISO 14001, including tips and lessons learned.

5) *http://www.epa.gov/owm/wm046200.htm*
Features *Environmental Management Systems: An Implementation Guide for Small and Medium-Sized Organizations*, intended to help small and medium-sized organizations implement an EMS. Contains a detailed description of an EMS and instructions on how to implement an EMS in an enterprise.

6) *http://es.epa.gov/partners/iso/iso.html*
The EPA's Standards Network. Describes the ISO and its environmental management standards.

7) *http://www.ecologia.org/iso14000/*
Ecologia is an international environmental nongovernmental organization (NGO). Describes ISO 14001 and how this and other EMS tools are used in practice, and hosts a discussion group on environmental and sustainable development implications of these tools. Provides a list of related links.

8) *http://www.frey.demon.co.uk*
Frey Environmental Associates. Describes general ISO 14001 implementation and benefits. Provides a list of related links.

9) *http://www.getf.org/projects/enviromgmt.cfm*
Global Environment and Technology Foundation. Contains information on municipal EMS pilot projects completed in partnership with the U.S. EPA.

10) *http://www.globalreporting.org*
Global Reporting Initiative (GRI). Strives to establish a generally accepted global framework for sustainability-reporting at the organization level. Provides draft June 2000 guidelines for preparing, communicating, and requesting information about corporate environmental, social, and economic performance.

11) *http://www.inem.org*
International Network for Environmental Management (INEM). Includes case studies, general information on ISO 14000, publications, interactive check-lists, downloadable how-to guides, related links, calendar of conferences and events.

12) *http://www.iso.ch*
International Organization for Standardization. Includes basic information on what ISO is, what ISO does, the members of ISO, ISO's organizational structure, and information on ISO's publications and long-range strategies.

13) *http://www.iso.ch/presse/presse12.html*
Clarifies the implications of ISO management systems, rather than product-performance standards, and what certification to ISO management-system standards really means.

14) *http://www.iso14000.com*
The ISO 14000 Information Center. Contains news releases, ISO 14000 supporting companies and organizations, a discussion group, overviews of the standard, EMS articles, resources, links and an aid to finding training on implementation, auditing, or promoting ISO 14001.

15) *http://www.iso14000.net*
GlobeNet. Includes news briefs, case studies, databases for EMS accrediting bodies, registrars, and registered companies. Contains a "Status of ISO 14000 Documents" link.

16) *http://www.mgmt14k.com*
MGMT Alliances, Inc. Contains the "ISO 14001 Pizza," which includes guidance for setting environmental policy, planning, implementation and operation, checking and corrective action, and management review.

17) *http://www.ncedr.org/guides/iso/htm*
National Center for Environmental Decisionmaking Research (NCEDR). Provides downloadable ISO 14001 guidance manual, to help organizations develop an EMS that is consistent with ISO 14001, achieve ISO 14001 registration, and improve environmental performance. Provides a list of related links.

18) *http://www.eli.org/isopilots.htm*
National Database on Environmental Management Systems. Contains information on an ISO 14001 Pilot Project study, which evaluates the performance of U.S. enterprises implementing ISO 14001-style EMSs. Provides links to ISO 14001 implementation, training, and resources sites.

19) *http://www.worldbank.org/nipr*
New Ideas in Pollution Regulation (NIPR). Targeted at individuals and organizations interested in public-policy issues relating to the cost-effective control of pollution. Maintained by the World Bank's Economics of Industrial Pollution Control research team. Contains ISO 14001-related articles and links.

20) *http://www.nsf-isr.org/html/body_iso_14000.html*
NSF International Strategic Registrations, Ltd. (NSF-ISR). Contains basic information on ISO 14000, the registration process, ISO 14001 auditors, ISO 14001 training services, EPA/NSF EMS pilot projects, EMS articles and publications, FAQs, and ISO 14000 links.

21) *http://www.pri.sae.org*
Performance Review Institute. Discusses the standard in general terms, trends in regulation, advantages of implementation, common problem areas found during an audit, and the relationship between ISO 9001/2 and ISO 14000.

22) *http://www.quality.org//lists/quest.list.txt*
QUEST (Quality in Environmental and Safety Training). ISO 14000 information e-mail discussion list. Provides Internet forum for research and information exchange on harmonization of standards like ISO 14001, ISO 9001/2, OSHA standards, etc.

23) *http://www.rabnet.com*
Registrar Accreditation Board (RAB). Provides information on RAB-approved ISO 9000 and ISO 14001 registrars, training course providers, individual auditors, how to select a registrar or auditor, directories of companies currently certified to ISO 14001, and directories of accredited ISO 14001 auditors. Discusses requirements for QMS and EMS certification.

24) *http://susdev.eurofound.ie*
European Foundation, sustainable development theme. Searching for "ISO 14001" under the "tools" heading produces a set of hyperlinks to other Web sites that together offer a wide array of practical tools related to ISO 14001 implementation.

25) *http://www.trst.com/iso-14001-pages.htm*
Transformation Strategies. Contains links to pages of smaller, less well-known companies that have registered to ISO 14001. Contains articles, benefits of implementation, books, case studies, keys to success, and steps to take.

SECTION II: ADDITIONAL POLLUTION PREVENTION RESOURCES

Author's note: We've listed the resources in this section to correspond to the chapters to which they relate.

The following references will help readers of Chapter 6 gain further information on pollution prevention case studies.

1) Ahmed, Kulsum. *Renewable Energy Technologies: A Review of the Status and Costs of Selected Technologies,* World Bank Technical Paper No. 240. Washington: World Bank Group, 1994.
2) Bernstein, Janis D. *Alternative Approaches to Pollution Control and Waste Management: Regulatory and Economic Instruments.* Washington: The World Bank, 1993.
3) Cheremisinoff, Nicholas P. *Handbook of Pollution Prevention Practices.* New York: Marcel Dekker Inc., 2001.
4) *Pollution Prevention and Abatement Handbook 1998 — Toward Cleaner Production.* Washington: The World Bank Group, 1999.
5) *Profile of the Petroleum Refining Industry,* EPA 310-R-95-013. Washington: U.S. Environmental Protection Agency, 1995.
6) *A Planner's Guide for Selecting Clean Coal Technologies for Power Plants,* Technical Paper No. 387. Washington: The World Bank, 1997.
7) *Guide to Accessing Pollution Prevention Information Electronically.* Boston: U.S. Environmental Protection Agency — Northeast Waste Management Official's Association, 1997.
8) *Profile of the Motor Vehicle Assembly Industry* EPA 310-R-95-009. Washington: U.S. Environmental Protection Agency, 1995.
9) *Prototype Study of Industry Motivation for Pollution Prevention,* EPA 100-R-96-001. Washington: U.S. Environmental Protection Agency, 1996.

Readers of Chapter 7 can use the following references for information, examples, and procedures useful in working with material and energy balances, and general process calculations.

10) Azbel, David, and Nicholas P. Cheremisinoff. *Fluid Mechanics and Unit Operations.* Ann Arbor, Mich.: Ann Arbor Science Publishers, 1983.
11) Chopey, Nicholas P., and Tyler G. Hicks, editors. *Handbook of Chemical Engineering Calculations.* New York: McGraw-Hill Book Co., 1984.

Readers of Chapter 7 can use the following references as background reading materials on equipment specifications criteria for unit operations and processes.

12) Cheremisinoff, Nicholas P. *Handbook of Chemical Processing Equipment*. Boston: Butterworth-Heinemann, 2000.

13) Cheremisinoff, Nicholas P. *Chemical Engineer's Condensed Encyclopedia of Process Equipment*. Houston: Gulf Publishing Co., 2000.

14) Peters, Max S., and Klaus D. Timmerhaus. *Plant Design and Economics for Chemical Engineers*. New York: McGraw-Hill Book Co., 1968.

15) Cheremisinoff, Nicholas P. *Liquid Filtration — 2nd Edition*. Boston: Butterworth-Heinemann, 1998.

16) Cheremisinoff, Nicholas P., and Paul N. Cheremisinoff. *Compressors and Fans*. Englewood Cliffs, N.J.: Prentice-Hall Inc., 1992.

17) Cheremisinoff, Nicholas P., and Paul N. Cheremisinoff. *Carbon Adsorption for Pollution Control*. Englewood Cliffs, N.J.: Prentice-Hall Inc., 1993.

Readers of Chapter 8 will benefit from the following references.

18) Shim, Joe K., and Joel G. Siegel. *Managerial Finance*. Schaum's Outline Series (067306-9).

This instructional text, which describes financial analysis, includes sections on the time value of money and on capital budgeting.

19) Shim, Joe K., and Joel G. Siegel. *Managerial Accounting*. Schaum's Outline Series (067303-0).

This instructional text, which describes management accounting, includes sections on cost concepts, terms and classifications, cost allocation, and capital budgeting.

20) Purcell, A.H. *Hazardous and Solid Waste Minimization,* ISBN/ISSN: 0865871361. Government Institutes Inc.

This document, which describes waste minimization and resource recovery, includes the whys and wherefores of waste minimization; considers the economics of waste-management decisions; and covers waste-minimization planning, auditing, and implementation.

21) *Waste Minimization Manual,* ISBN/ISSN: 0865877319. Government Institutes Inc.

This document discusses waste minimization, economic imperatives, legal and regulatory incentives, and how to conduct waste-minimization audits. It also contains waste-minimization case histories for Dow, DuPont, Chevron, Hewlett Packard, and the U.S. Navy.

22) Wittman, Marlene R. *Costing and Financial Analysis of Pollution Prevention Projects.* Massachusetts Office of Technical Assistance.

This text provides a curriculum that is intended to familiarize environmental professionals with basic business terms, and to increase professionals' awareness of the factors that influence an investment in P2 options.

Index to Tables

Subject Index

Truth and Hope

TRUTH *and* HOPE

The Fürst Franz Josef und Fürstin Gina Lectures Delivered at the International Academy of Philosophy in the Principality of Liechtenstein, 1998

PETER GEACH

University of Notre Dame Press
Notre Dame, Indiana

http://www.undpress.nd.edu

Manufactured in the United States of America

Library of Congress Cataloging-in-Publication Data
Geach, P. T. (Peter Thomas), 1916–
Truth and hope / Peter Geach.
p. cm.
Includes index.
ISBN 0-268-04215-2
1. Truth. 2. Hope.
3. Catholic Church and philosophy. I. Title.
BD171 .G43 2001
121—dc21
00-011588

∞ *This book is printed on acid-free paper.*

Contents

Preface

The greater part of this work was delivered as a series of lectures at the Internationale Akademie für Philosophie in Schaan, Principality of Liechtenstein. I have thoroughly revised and supplemented my texts of these lectures. I am deeply grateful to the Princely House whose munificence established the Academy and to the Rector and all my colleagues. *Ad multos annos!* Chapter 6 was delivered in view of a request for a paper from the boys of Shrewsbury School in England; its content turned out to have so many difficult sides to it that no boy ventured upon a quick response, though I gather it provoked discussion afterwards. The truth of prophecy clearly belongs with the other aspects of truth discussed in this volume, so I decided to include it.

1 Truth, Love, and Immortality

The book I published under this title was written to expound the philosophy of McTaggart; he had been the mentor of my youth, and in youth I began writing a commentary on his philosophy. I completed one draft in 1939; after many intermissions, and after several drastic rewritings, I finished the work just short of forty years from beginning it. I consciously adopted the style of defence counsel for McTaggart: counsel is not personally committed to his client's being right in fact or in law, but is concerned that the case for him should be fairly presented and harmful preconceptions be removed from the jurors' minds; counsel's personal convictions are not to be obtruded.

My present aim is quite different. I shall not fail to characterise some of McTaggart's positions as grave errors; but I wish to 'accentuate the positive' and bring out what I take to be the main truths I learned from McTaggart—and therefore did not unlearn when by the mercy of God I became a Catholic.

McTaggart's magnum opus, *The Nature of Existence,* begins with a careful laying of foundations: with a general ontology, concerned with the most abstract features of reality. People's impression of McTaggart as a bold, even wild, speculative philosopher is in a measure justified; but the ontology developed in the greater part of his first volume is on the contrary extremely sober, and a thorough knowledge of it saved me from many endemic errors of English philosophy. His world is a world of actuality: it consists of actual individuals, and he uses the traditional term 'substances'.

Characteristics have their being only by entering into facts about substances; there are no universals *ante res*. This holds even for uninstantiated characteristics; they have their being in negative facts; if a thing isn't red, then *red* is what the thing isn't. The actual world is not one world picked out of a magazine of possible worlds. Existence, for McTaggart, can neither have gradations nor be relativized; one conception may be nearer truth than another, but a thing cannot half-exist; talk of people existing in the world of a novel or in a belief-world is wanton self-deception if anyone takes it literally and seriously.

The characteristics of substances include relations; McTaggart, like Aquinas, rejects the view that relations have no extra-mental reality, and refutes the sophistical arguments of Bradley against their reality. An actual individual is identified by its characteristics (this is McTaggart's thesis of the Dissimilarity of the Diverse), but is not identical *with* its characteristics, either severally or all together. We might use the simile of a triangular area and its sides: the area is marked out by its sides, and two areas cannot share the same three sides, but the area is not any one of the sides nor yet all of them together. McTaggart utterly dismissed the idea of a 'bare particular' which the mind can discern *under* the characteristics it wears, as a lady is naked under her clothes; taught by him, I have always rejected various allied notions, like that of the 'pure ego' which has no mental states in its constitution *because* it is their 'owner'!

Facts are not 'loose and separate'; there are real relations between them, there are language-independent entailments ('intrinsic determinations') and incompatibilities. Causal laws, for McTaggart, state intrinsic determinations, which are opaque indeed to our understanding in many cases, but are none the less necessary. Only such causal laws can give us a firm grip on what we are talking about when we come out with subjunctive conditionals: we may assert that if an Australian archdeacon were beheaded at the North Pole he would die, but only because the beheading of a human body 'intrinsically determines' the death of that body. Failing an appeal to such laws, we need not suppose there is any fact of the matter about what would have happened if. . . . (This last teaching of McTaggart has certainly stayed by me: as a Catholic I was not likely to be a Molinist, especially as I was instructed by a Dominican.) We should notice that McTaggart carefully avoids the sloppy talk about an *A* invariably followed by a *B;* the causal law about beheading requires not barely that any beheading be followed by *a* death, but that the beheading of a human body be rapidly followed by the death of *that* body. The point ought to be obvious, that in general the causal law will take the form of an *A*'s being followed by a *B* in a specific relation *R* to *that A:* but how often elaborate writing about causal laws misses this point, and is reviewed by other philosophers who also miss the point!

The one article of McTaggart's general ontology that I now find not just wrong but barely intelligible is his unhesitating assumption that the terms 'part' and 'whole' are univocal; in the same sense of 'part', a liver is part of a body, a man is part of a club, and a thought is part of a mind. I am merely surprised at many contemporaries who build elaborate structures of 'mereology' upon this shaky foundation.

In general, however, I acknowledge a great debt to these ontological sections of *The Nature of Existence:* here I found a model of clear English style and sustained strict reasoning. It has done me lasting good that I took as a mentor a man of such clear and candid mind. From later parts of the work I imbibed serious errors; but when I shook these off, it was certainly with relief, but without resentment, and with abiding gratitude for truths never forgotten.

The tale of McTaggart's main errors, in which for five years or so I followed him, is quickly told: he disbelieved in God, in the reality of time, in the freedom of the will, and in the reality of matter. I shall here pass over the last error very quickly. McTaggart's arguments for the conclusion that what we call bodies are misperceived spiritual realities are subtle, complicated, and very much of his own devising. Their premises are in some instances very plausible, like the Dissimilarity of the Diverse (already mentioned), and the principle that every characteristic that occurs, occurs in a completely determinate form. (This last principle has of course played a part in the thought of many philosophers.) In my book on McTaggart I expound at length his way of reducing *ad absurdum* all attempts to conceive consistently the internal structure of matter, and his account of why the like trouble does not arise over conceiving purely spiritual beings.

The other main errors all hang together. Clearly, if there is no time, there can be no choice between alternative possible futures. Moreover, classical theism, Jewish, Christian, or Muslim, depends on the unbridgeable distinction between a changeable world and a God who is eternal and changeless, who created the world by his will and governs it by his providence; any such theism is unsustainable if we regard time as an illusion.

When I ceased to follow McTaggart, it was for me a single act to accept the reality of time and the unchanging eternity of God; I never for a moment thought of accepting a changeable God. I held on here to something I had learned from McTaggart: a changeable God, who can be affected causally by changes and activities of his creatures, is no God at all, and there can be no reason to believe in the existence of such a being. As McTaggart argued, it would be quite arbitrary to allow the question 'How come?' for changeable things in the world, but disallow it for a changeable God. If on the contrary God is supposed eternal and changeless, then to ask 'Who then made God?' is manifestly silly.

I learned this lesson, and much besides, from McTaggart's other great work, *Some Dogmas of Religion*. Upon retirement from the Indian Educational Service my father was appalled to discover that the guardian who had had charge of me had inculcated a narrow variety of Protestantism. Part of this was a belief in (what I have later called) an *absolutely* omnipotent God: a God capable of feats whose very description is on the face of it self-contradictory. When he judged me to have reached years of sufficient discretion, my father decided on the drastic remedy of making me read McTaggart's book. The remedy was entirely successful; I felt towards McTaggart thereafter as Lucretius did towards Epicurus; he had delivered me from a mentally crippling superstition, and I could now think like a free man. It was only some years later that I read *The Nature of Existence* and came to accept McTaggart's positive conclusions.

Mentally crippling as it is, belief in absolute omnipotence is not confined to the naive and unreflective. Notoriously, Descartes came to this belief, though it can have been no part of his Catholic education; and held that truths of arithmetic hold only by God's free decree, which could have been quite other. I have reason to believe that this sort of thinking is not extinct: possibly such thinkers would say that the feats they ascribe to their God as within his absolute power are not in truth logically impossible, but only appear so to the corrupt mind of the unconverted. Quite recently I read an article maintaining that since God created everything other than himself, he must have created his own attributes! But just as Wittgenstein said we could not describe how an illogical world would look, so we can say *nothing* about a God who 'transcends' logic. Worship of such a God is of a piece with, and may easily lead to, adulation of a 'daemonic' man of genius who is 'above' the restraints of honour and decency and humanity.

From this same work I learned valuable lessons about what may be called the ethics of belief and about criteria that may be applied to discredit alleged revelations. Later, we shall be considering cautions that are needed in ascribing human moral virtues to the Divine Nature. It is another matter where theologians ascribe to the Divine Nature attributes modelled on human vices. Such a God would be *capable de tout;* one could not even trust him, let alone worship him. For example, an early English Calvinist held that God consigns some of us to everlasting damnation for his mere good pleasure—*quia voluit.* At that rate we could not be sure that God will not damn the whole human race for his mere good pleasure; as McTaggart points out, allegedly revealed promises to the contrary could give no assurance, for the character of such a God might not exclude the will to fool us. Happily, there can be no reason to trust the alleged revelation of such a God; if he can lie, still more may those who claim to speak in his name.

As against this I have encountered the argument that a man in many ways vicious may have a quirk of conscience that restrains him from lying. Perhaps so; but if the man were in other ways very vicious, I'd not stake anything I very much minded losing upon his veracity. And it would be plain crazy to stake one's *all* upon the veracity of a being who must be very vicious indeed if he is the truthful source of some alleged revelation.

From McTaggart I also derived an extreme dislike for certain sorts of Christian apologetic. For example, some apologists exploit confusions about terms in which doctrines are formulated, whether by fostering existing confusions or by generating new ones. An unclear understanding of 'person', 'nature', or 'substance' is pleaded as a reason why the listener's or reader's mind cannot rationally doubt a dogma formulated in such terms: surely such a doubt is presumptuous! An even worse form of apologetic is an *argumentum ad horrendum:* it would be simply too dreadful if such-and-such a dogma were false! Sometimes people have used this argumentative move in mere unawareness that the person addressed might find some of *their* dogmas horrendous. In any event the move is no good; on any view, except the Christian Scientist's, some of the undoubted facts about the world just are grim.

Incidentally, McTaggart's use of the word 'dogma' is not idiosyncratic but quite traditional; for example, Aquinas in his commentary on the Book of Job remarks that Job's friends held the dogma that great misfortunes are always punishments for great sins, but that this dogma is false; the term should not be taken to connote even authoritative teaching, still less truth. McTaggart insisted that dogmas are necessary to a religion, as bones are to an animal. In his time non-dogmatic 'religion', what Matthew Arnold had called 'morality touched with emotion', still enjoyed a certain vogue; the idea was that this should replace the Christian religion, though still retaining its name; somehow a need was felt to keep as well some of the language of the old religion, stories from the Bible, and so on. McTaggart insisted that if the world really is not as Christians have believed it to be, we had better say so frankly and not deceive ourselves and others with pietistic language. Most recently, dogma-free 'Christianity' has taken a specially repellent form; statements are made about 'the believer' which are clearly false about the overwhelming majority of Christians, past or present, of whatever denomination: e.g. that 'the believer' does not pray for things to happen and has no expectations about the future to distinguish him or her from other people.

Dogma, for McTaggart, is essentially a metaphysical concern; and he explains metaphysics as the systematic study of the ultimate nature of reality. 'Ultimate' is here used to contrast with 'apparent'. It is of course mere common sense to distinguish between appearance and reality. Only highly sophisticated, and indeed sophistically argued, systems of philosophy will maintain, like Protagoras, that all appearances are correct and all opinions true; or, like Epicurus, that the Sun and Moon are just about as big as they look; or, like Berkeley, that there is no more to a body than the various ways it affects our senses. Reflection on the difference between appearance and reality is one way into metaphysics.

Another way into metaphysics is reflection upon time. Once in a Leeds pub I listened in respectful silence as two working men in dungarees drank their pints and argued the question whether or not it is right to say that tomorrow never comes; perhaps one of them had casually come out with this familiar saying; anyhow, the part of their conversation I heard began with an attempted refutation of the dictum:

> 'Of course tomorrow will come. Tomorrow is May 18th, isn't it? But May 18th will come; so tomorrow will come.'

The other working man's retort went like this:

> 'That can't be right. If it is right now to say something *will* come, then some time it will be right to say it *has* come. But it will never be right to say tomorrow *has* come; so it cannot be right to say tomorrow *will* come.'

I could easily spell out a few of the logical and semantical problems that arise about the validity of the two opposing arguments. Raising such problems is part of what makes the difference between merely producing theses and arguments of metaphysical interest, and that *systematic* study of them which constitutes metaphysics. Unsystematic interest in such matters is part of human nature; Gareth Matthews wrote two fascinating books about the metaphysically interesting talk of young children aged from 8 to 11. Lewis Carroll's *Alice* books are full of philosophical problems, and their abiding appeal testifies to their stimulating an intellectual appetite widespread in the human race; they have been translated into many languages—I possess a version adapted to the use of Australian aborigines.

Time in particular is an abiding source of perplexity. To many philosophers McTaggart is known only as having written about time. Even here, they know only his argument that time is illusory; they take no interest in his account of the reality underlying temporal appearance. Of his negative argument I shall say nothing here. As with Anselm's *Proslogion* argument and Zeno's argument against motion, refutations are numerous; and their authors disagree with one another very widely. Quite recently Michael Dummett in Oxford and Hugh Mellor in Cambridge have each accepted part of McTaggart's argument—though not the same part.

McTaggart was deeply convinced that the difference between past and future, which appears to us so important, is not a mere illusory appearance or a quirk of our minds due to our needs as living organisms, but points to a feature in the structure of reality. The British Idealists adopted a pose of superiority towards hope, which is so important in Christian spirituality. As on many other matters, McTaggart was wholly out of sympathy with them. He thought it could be shown to be a rational object of hope that we shall obtain an eternal life of love, which he even often called Heaven. In the first Epistle of St. John such an eternal life of love is paradoxically presented both as now actually possessed and as an object of hope, a life that transcends our conception. McTaggart, who was much impressed with this Epistle, claimed that his philosophy enabled him to resolve this paradox: all actual existence is for him timeless, including our apparently time-bound states of mind; what seem to us to be successive experiences are, in his language, fragmentary perceptions, which do not really succeed one another but coexist with a total experience that subsumes and transcends them. This total experience is the heavenly experience, which not merely is timeless, as McTaggart holds all experiences really are, but is self-consciously eternal; he might well have adopted the Boethian definition—the complete possession, all at once, of unending life. In Heaven the illusion of time no longer exists; the fragmentary perceptions that constitute our present life are contemplated, not indeed as really past, but as mere fragments, which represent themselves as transitory, of the heavenly perception; and the joys of perfect knowledge and love in Heaven make the sorrows and faults and follies of present experience appear infinitely insignificant. For this consummation we must wait; we cannot shake off the illusion of time by religious exercises. While we are under the veil of time, Heaven can only appear to us as future. Heaven is not really future; but if we speak strictly and philosophically, neither is tomorrow's breakfast. McTaggart's philosophy affords no grounds for a puritanical or Manichaean attitude towards such pleasures as our present experience affords; the beatific vision is good, he said, and so is a bottle of champagne.

Soaking myself in McTaggart, I imbibed a desire for Heaven and eternal life, which of course I had not to abandon on becoming a Catholic; and meanwhile I was preserved from giving my heart with total devotion to some less worthy end, as I saw many contemporaries doing. Even as regards the relation of time and eternity I had no need to find McTaggart wholly mistaken. God's life, the life of the Blessed Trinity, really is the sort of Boethian eternity that McTaggart ascribed to all persons; and we have the great and precious promise that, in a way we cannot now begin to understand, we shall transcend all the delusion and misery and wickedness of this life and become sharers in that eternal life.

With much hesitation, I even venture to say that there may be some truth in McTaggart's idea that any later state of a person's experience somehow incorporates any earlier state of the same person. He in fact thought that the appearance of a temporal order among a person's experiences is based on the real order of an 'inclusion series'; the experience of Heaven, which incorporates all present experiences, is thus in the same relation to today's experience as tomorrow's experience is; so for practical purposes Heaven is to be regarded as future and as the object of hope. For ordinary people a future state is more important, *ceteris paribus*, than an equally distant past state. Spinoza regards this as an illusion by which the free man is unaffected (*Ethics* IV.62), and some recent philosophical writers have argued for the same conclusion; McTaggart firmly refused to condemn ordinary people's attitude as irrational. Certain thinkers who emphatically asserted the reality of time and change, William James and Bergson, agreed nevertheless with McTaggart in regarding the series of a person's experiences as an inclusion series with the later stages regularly including the earlier.

However this may be, in some respects McTaggart saw more clearly and wrote more worthily about time and eternity than certain Christian writers, e.g. C. S. Lewis and Dorothy Sayers. I have to say that when I encounter the word 'eternity' or 'eternal' or 'eternally' in Lewis's writings it is a pretty well infallible sign that he will be talking nonsense. I refrain from citing what I find the worst examples of this; I confine my criticism here to a manner of speaking Lewis shares with other Christian apologists. These writers will have it that time is real from man's point of view only: God sees all history as a timeless whole. Under God I have to thank my old mentor McTaggart for making me see that this is just an absurdity. On the one hand, the way God sees things is the way they simply are; if what we men regard as a successive history God sees as a timeless whole, then our way of seeing things is just what McTaggart called it—misperception; but in that case traditional theism, with its contrast between a changing world and a changeless Creator, is destroyed. On the other hand, if there really is an aspect of time and change in things, then how can God be imperceptive of it? Discussing the thesis of Empedocles that like is known by like, Aristotle remarked that in that case the deity or principle of Love, having no strife in his own constitution, must be wholly ignorant of Strife and her works. But for Anaxagoras, Mind is apart from all things in the cosmos and unchanged by their changes; and thus it is that Mind has total control of all that is in the cosmos and total knowledge of all happenings, past, present, or future.

The British Idealists spoke depreciatively of 'finite selves', which they regarded as pretty low-grade manifestations of the Absolute. The State (they invariably used this reverential capital letter) was on a far higher level of manifestation. Some of Bradley's essays exalt the claims of the State in a way reminiscent of Nazi ideology: he stresses the State's right to cut off members regarded as harmful and even to decide that some tribe of man must be extirpated. For all this McTaggart had the strongest possible dislike. The State for him was a mere means, not an organism; worship of the State was a more degraded superstition than worshipping a bull or crocodile, which is at least a living unity.

For McTaggart, not the State mattered but persons. (As a pure matter of English style, I regret that he always wrote 'selves' rather than 'persons'.) Persons such as we know were for him the unique bearers of value, since for him there was no God; those whom the British Idealists devalued as mere 'finite selves' were for him eternal realities, 'primary parts' of the Universe, the heirs of beatitude. Their beatitude consisted in the clearest and most intimate mutual knowledge and the most intense love. For McTaggart, as for Aquinas, the beatific *vision* was the source of love. Love involves an intense consciousness of union with another; even in our present experience the consciousness of union may bring love in spite of intensely dislikable traits in the beloved which cause extreme pain to the lover. (Consider Catullus' *odi et amo* and Baudelaire's *C'est tout mon sang, ce poison noir!*) But in the heavenly state many undesirable emotions will be excluded by its self-conscious eternity and by the absence of any wish that things were otherwise; free from such obstructions, love will have free play and will be far more intense than any earthly love.

The difficulty I eventually came to feel about this was to conceive how non-divine persons could be beatified merely by contemplation, however close and intimate and accurate, of *one another*. Even on the part of pure angelic spirits such mutual beatification is barely conceivable. It seems like imagining that mirrors sufficiently well polished could generate light just by reflecting one another.

On love as we find it in our experience McTaggart wrote nobly. Love is for a person, not in respect of this or that characteristic, but just as this person; although love will be accompanied, except in perverse and tragic cases, by desire for the happiness and good of the beloved, fundamentally love does not consist in this but just in a desire for the life of the beloved and for union with the beloved. McTaggart's own life was a life filled with love: the love of a good son and brother and husband (the story of his courtship and marriage is quite romantic) and of a wise, trusty, unjudgmental friend to many people in many walks of life.

In my schooldays (I cannot assign responsibility more precisely) the view was somehow conveyed to me that God loves all men equally, and we too, so far as our frailty allows, ought to imitate God in this: this, I gathered, was what the Christian religion was all about. This love of any person just as *a* person is quite different from McTaggart's conception of love for a person as *this* person. In his talks on the radio C. S. Lewis propagated the former idea; surely here the medium was the message—the tone of the talks was cheapened by their being delivered for a vast audience in a popular idiom. Lewis really knew better, and my criticism here does not apply to his book *The Four Loves*. For me, taught by McTaggart, love was a sacred mystery, which the lover should contemplate in himself with 'self-reverence'; for what was presented to me as Christian love I felt no reverence at all. I was merely repelled by the idea that one should try to love everyone equally, enemies as much as friends; moreover, people's talk of this seemed to me manifest cant—they were clearly making no effort to love that way themselves. Happily, when I was being instructed in the Catholic Faith, the Dominican who taught me led me to study Aquinas on the matter: *De ordine caritatis, Summa Theologica* IIa IIae q. 26: as a Catholic catechumen I need not endeavour to love everybody else equally even within my own acquaintances, nor try to love declared enemies just as much as my friends. From the same source I learned to reject another doctrine of the like edifying sound: that I ought to try to love *better* people *more*.

McTaggart believed that the final beatific vision of the beloved was a consummation not to be lost by any folly or vice, because it is a relation in eternity. When love seems to occur in our fragmentary experiences, this cannot be illusion; he held indeed that we each greatly misperceive the content of our minds, but A's misperception could not misrepresent the union of A to B so that it appears closer than it really is. When the lover sees all perfection in his mistress:

> To you, to you, all song of praise is due,
> Only with you are miracles not wonders—

the world may mock, but for McTaggart it is the world, not the lover, that is sunk deep in illusion; this love is a glimpsing of the perfection that he will enjoy forever in Heaven.

Love of God, however, is in McTaggart's eyes a delusion, a pouring out of love on a mere reflection. He ought to have considered the love people have had for Jesus Christ; I know of no passage in his writings where he does so. I cannot account for this; certainly he was not the sort of man to be impressed by the arguments of bogus scholarship to show that Jesus is a mythical character. He was however not attracted by the depiction of Jesus in the synoptic Gospels; nor could he be induced to believe 'for the very works' sake'; even if the miracles could be proved to have happened, power over events, he remarked, need not go with insight into religious truth; Napoleon greatly excelled the average English clergyman in the first respect, but it would be rash to infer that he was also a man whose religious views contained more truth than the clergyman's.

A rather bad hymn represents God as one who could 'live and love alone' and indeed did so before the creation of the world. McTaggart cast doubt on the possibility of even conceiving a person who could exist without an Other, and on his view of love a solitary person could not love. Under his influence I viewed with aversion a Monarchian kind of theism. McTaggart can scarcely have encountered the Trinitarian conception, by which each Divine Person is an Other to the other Persons: each, in traditional language, is *alius,* some*one* else, in regard to the other two (though not *aliud,* some*thing* else). McTaggart, coming from a Victorian rationalist background, was sent away from one school for arguing against the Apostles' Creed; then at his boarding school he was plunged into a milieu of 'muscular Christianity', a mixture of Modernism, Pelagianism, and athleticism. It is very doubtful if he found better guides to Christian thought among the Fellows of his Cambridge college. I myself was to learn that Monarchian theism is not part of the Catholic religion; and now I could believe that in the beatific vision of the Blessed Trinity McTaggart's description of heaven is verified: 'we shall know nothing but our beloved, and those they love, and ourselves as loving them, and only in this shall we seek and find satisfaction'.

McTaggart's vision of love was sustained, not merely by confidence in the soundness of his arguments, but by mystical experiences; these made him *see* the whole universe as a world of love; they came to him unsought and quite unpredictably. For himself these experiences were utterly convincing, but he knew well that for other people they would not even count as evidence. Before he had this gleam to follow, already as a schoolboy, he had decided on philosophy as his vocation; he now conceived it as his task to validate by hard metaphysical arguments the truth he had grasped intuitively. If his vision was true, then patient metaphysical reasoning must establish its truth: just as a schoolboy with an algebra problem is confident that by carefully following the rules and methods he has learned he will reach the answer given at the end of the book. Although McTaggart's magnum opus, *The Nature of Existence,* lacks his final revision, he died in the conviction that in essentials his lifelong task had been accomplished.

At the end of the Trinity College MS. of *The Nature of Existence,* I read the words in faint pencil: 'Heart of my heart, have I done well?' McTaggart might answer himself in the words of Paracelsus, in Browning's poem of that title:

> Meanwhile, I have done well, though not all well.
> As yet, men cannot do without contempt;
> 'Tis for their good, and therefore fit awhile,
> That they reject the weak, and scorn the false,
> Rather than praise the strong and true, in me;
> But after, they will know me. If I stoop
> Into a dark tremendous sea of cloud,
> It is but for a time; I press God's lamp
> Close to my breast; its splendour, soon or late,
> Will pierce the gloom; I shall emerge one day.

What Is Man?

The City Council of Cambridge published some bizarre notices outside the Guildhall at the time of a Socialist majority. One notice proclaimed the City to be 'a nuclear free zone'; I trust the Russians took heed. Another, which is our present concern, affirmed that man is only one of a million species of animals on Earth; man could claim no preeminence; mankind ought to defend 'threatened species' instead of selfishly pursuing human aims. All this was typical of a movement in contemporary England against what is called 'speciesism', against a loyalty to the human race regarded as chauvinistic. Quite a number of philosophers are influenced by fear of 'speciesism'; I should account for this by supposing that they began by giving primacy to man out of mere instinct and tradition without any clear rationale. John Locke and Voltaire attacked the idea that human beings share a real essence; the attack has been enormously influential, and thus many have felt there is no longer rational support for the dogma of man's primacy. Obviously, Darwinian currents of thought have swept people along in the same direction.

The question 'What is man?' is pointless unless there is one general answer to be given. This would sometimes be in effect denied; people sometimes talk as if membership of the human species were like membership of a club and new members were admitted upon election by some committee of existing members. Do *we* then decide whom to call human? and who are *we?* Let us not forget Lenin's 'Who whom?' It has happened that one class of people should deny another class to be truly human: Nazis acted thus towards Jews and Gypsies, and while 'the peculiar institution' prevailed in 'the gallant South' some white people had a similar attitude towards black people. Rationalisations have not been lacking, based on spurious genetic theories or perverse reading of scriptural texts. But all this has just been elaborate self-deception; for example, whites in the South showed in many ways that they *knew* blacks were their fellow-men and not subhuman animals. They lived in constant fear of conspiracy and rebellion, to forestall which they had strict laws against teaching blacks to read; obviously they had no fear of conspiracy and rebellion by horses, nor were farmers forbidden to teach cattle to read.

I shall therefore take for granted that there is a common human nature. Assuming this, people have answered the question 'What is man?' variously. 'Man is a highly evolved primate.' 'Man is an immortal spirit somehow united to an elaborate mechanism.' 'Man is an immortal spirit charged with the control and taming of a beast'—thus Tennyson. 'Man is a computer working by a program that nobody put into it.' And finally there is the old answer: 'Man is a rational animal.' This I shall defend.

There is one old objection to this traditional definition: that men are rational in various degrees, and some of them only to a very slight degree, so that by this definition their humanity too would vary considerably in degree. This is an error of category, precisely in the Aristotelian sense of the word: it confuses 'rational' as a differentia of a species with 'rational' as expressing some actual capacity. 'Rational' as a differentia relates not to capacities of first level, capacities for certain specific performances, but rather (as has well been put by Anthony Kenny) to a capacity *to acquire capacities.* There is a wide variety of capacities that human beings can acquire and other animals cannot: what is innate in every man born is the capacity to acquire such capacities. At birth no human being can use a language, but normally human beings master some language or other. (Admittedly there has recently become popular the idea that babies *are* born knowing a certain language, an unspoken language: they then frame hypotheses in this language for translating the spoken language of their milieu. Speculations of this sort are developed in great detail; they are laughably called 'cognitive science'.)

It would take an immensely long description to cover even roughly the activities manifesting the rational capacities men can acquire. Most of these activities have some close connection with language, but they extend far further than linguistic performances. What I give here is a mere sampling. The composition of music, and the invention of forms of dance and ceremonial. The building of houses and other structures, whereby man becomes able to live in varied and very mutable climates and is not confined to a particular ecological niche. The mastery of fire and flames, as means of heating and lighting, which also serve for preparing foodstuffs and other materials for human needs. The playing according to rules of all manner of games, athletic and intellectual. The devising of methods and standards of measurement. The devising of tools and machines for all sorts of purposes under human handling: the forelimbs of beasts are specialised, e.g. for walking or clawing, man's hand alone is *organon organōn*, the tool adapted for making and using tools. The devising of instruments of observation that supplement human sense-organs. The devising of means of transport, by land and sea (and now in the air). The preservation of traditions about ancestors and about the past generally, so that the inventions and discoveries of one generation are not lost to later generations. The use of language, not just for messages about the immediate environment, but for telling stories and for speculations about the origin and nature of the world we live in. All manner of representative arts. And finally, activities aimed at getting into contact with some superior being or beings who can govern the course of man's world and grant his petitions.

Such are rational activities; we may see and hear all around us the huge difference they make to our environment. Many of these activities are to be found in any human culture, not just in the cultures called civilised. And on this planet man is *the* rational animal: no productions or activities of the other animals remotely compare with those of man. Denials of human uniqueness are mere sophistry, laughable or pitiable. One characteristic sophistical move consists in saying: Mention any human activity you regard as peculiar to man, and I will find you a parallel activity of another species. Obviously we need to consider the whole range of human activities, not some one activity; but waiving this objection, we find the parallelisms for single human activities are often extremely unconvincing. Is it not obvious, some ethologist asked, that a man's putting on the door of his office a plaque saying G. KATZENELLENBOGEN, MANAGING DIRECTOR, is just like an animal marking out its territory with spurts of urine? No, Sir, it is not obvious.

Nowadays we hear stories of apes learning language. I was once assured by a psychologist that an immature ape could utter words at the same level as a young human child; I told him he could tell that one to the Marines. He protested that the report came from workers at a prestigious university; I did not keep concealed my own conviction that all the same this was a case of scientific delusion or fraud; since the story is not now told, I conclude that I was right. Different stories are now current. For example, it is alleged that an ape was taught to arrange movable symbols in straight lines on a baize board, from which the ape's teacher could read off sentences. Some of my pupils at Leeds proved themselves less efficient at expressing thoughts in a linear logical symbolism; and they, unlike the ape, had already mastered a rich spoken language.

We may here remember the earlier story of Clever Hans, a horse that allegedly carried out complex mathematical operations in its head and conveyed the answers by tapping with its forefoot. I no more believe in the symbol-using ape than in the mathematically skilled Clever Hans. In both cases there was alleged symbolic behaviour that had no connection with the creature's life in its environment: calculations for Clever Hans, for the ape expressions of thoughts like 'Red is not a shape but a colour'. Normal human beings do develop to the stage of using language not relevantly connected with their immediate environment, if not for abstract thought, then at least for telling stories. But this development presupposes what Wittgenstein called *language games,* where language is interwoven with other human activities: such a foundation was clearly lacking for the symbolising ape, just as it was for Clever Hans.

The alleged teaching of American Sign Language (Ameslan) to apes is a rather different matter. I think the most that could here be claimed is that apes can associate *manual* signs with things or performances as dogs can associate *vocal* signs; it is not of significance here that the dog *receives* artificial signs and the ape *produces* them. Nothing here counts as the production or understanding of *language.* When an ape facing a refrigerator comes out with gestures that in Ameslan mean 'open' and 'eat', each sign has an association with the ape's own life in this environment; it had previously seen the refrigerator opened and been fed from its contents. It would be quite absurd to speak of the ape's having invented a sign to *mean* 'refrigerator', or inventing any word at all; only into a language with sentences could a *word for* 'refrigerator' be introduced.

Creativity, creation of complex signs from simple signs previously learned, is indeed essential to language-learning; wishful thinking on the part of the observers may have led to their ascribing such creativity to an ape in the refrigerator case. Every language has a vocabulary, and new sentences are learned whose several words belong to this vocabulary. This creativity does not lend itself to description in terms of stimulus and response and learned reactions and reinforcement and all that. In man it may show itself very early; a child of mine, not yet three years old, said to her mother, 'Don't be gone!': she had certainly never heard an adult say this.

Every language has a grammar as well as a vocabulary. Like other well-known truths, this has been denied. Philosophers are prone to believe that in some very alien language, say Eskimo, there are not even separate words, let alone vocabulary and grammar. By this story, one Eskimo will come out to another with a salad of phonemes which as a whole says 'I am just going over to the nearby village to borrow a kayak'. The story is a reckless falsehood: this could be checked by looking up 'kayak' in the nearest dictionary that gives etymologies; and any decent encyclopædia will have some information about the grammar of Eskimo languages. To human articulate speech there is no parallel in the 'language' of beasts or birds. Darwin indeed may have held that the unhappy inhabitant of Tierra del Fuego came out with strings of sounds halfway between human language and brutish noises; but Darwin was wrong if he thought so; some of these natives came to England and learned English.

Vitally important as language and language-games are, language-related activities are not the only manifestations of rational nature. There is an old story of some shipwrecked mariners landing on a Pacific island and presently observing a man hanging on a gibbet: 'Thank God!' they cried, 'we've come to a civilised country!' Rationality and a certain level of material culture were needed for the spinning of the rope and the carpentry of the gibbet; *a fortiori* they could be certain that they were not on an island where the highest animals were orangutans. Speaking here of rationality is of course not praise of the inhabitants; if a man is hanged, he or his judges must be much to blame; in the Law of Moses gibbeting a decaying corpse was a grave offence, what Rabbis called desecration of the Name, *hillul ha-Shem;* but rational nature in an agent is presupposed to our having a right to either praise or blame of that agent.

Men's lives are wonderfully various; there is no regular pattern of life for men, as for other kinds of animal. Pico della Mirandola appealed to this to show the uniqueness and primacy of man; writers of our time perversely use it to show that man has no end that suits his nature. But to the questions implied by the Cambridge City Council's notice: What has man got that other species have not? What gives man the right to dispose of other species? we may rightly reply: Only to man can such questions be addressed. Only to man can it even be said 'your species is just one among a myriad animal species': no member of another species on Earth could grasp the thought.

People who make a show of treating other species as rational beings will mention bees and dolphins and claim that these species too have language. But why then do they not attempt linguistic communication with bees or dolphins? Of course we cannot dance as bees do, nor will human throats shape themselves to make dolphin sounds; but surely we could make artificial radio-controlled bees and electronically imitate dolphin sounds. Such attempts at communication are never made, and surely the reason why is that the projects are foreseeably futile. People who claim to recognise dolphins as fellow rational beings all the same perform vivisection experiments on them; either this shows the insincerity of the claim, or else we have to do with wretches who are willing to torture rational beings to gratify curiosity and would do the like for human beings but for fear of the criminal law. Either way they have no right to call upon the rest of us to trust them as being 'all, all honourable men'. Science can be built only on a presumption of good faith, and here we cannot admit such a presumption.

Man alone is a rational animal here below, and all members of the human species are rational animals in the sense I have been explaining. Human disabilities which block the human capacity to acquire capacities do not prove the absence of this higher-order capacity. In many cases that would once have appeared hopeless we have learned how to remove or circumvent the impediment; we have learned how to communicate with children blind and deaf at once or severely spastic; many disabilities can be prevented or cured by medical means, like supplying a missing chemical substance or devising a diet free from something poisonous to an individual. Where no such beneficent devices yet exist, we need not recognise a distinction in principle. One thing certain is that remedies would not have been found, and no more will be found, if the policy were simply to kill defective infants.

No animals now living on Earth are halfway between human beings and brutes. Wittgenstein once suggested that we need not regard feeble-minded people as leading a defective form of life; rather, their form of life is just simpler than ours. A great philosopher is likely to say something pretty silly some time; of course there could not be a variety of humankind all living together with this 'simpler form of life'. Even in the most favourable physical environment this feeble-minded community could not survive as a flock of sheep might, for they would perish from accidents they could neither ward off nor remedy; feeble-minded individuals survive only by the kindness of their healthy fellows.

Such a 'simpler' community life might be imagined to have been an intermediate evolutionary stage in the coming to be of *Homo sapiens*. Before discussing the evolution of man, I want to point out how inconsistent it is to combine the treatment of man as simply one form that the evolutionary process has thrown up with noisy protests about 'threatened species' which human activities may extirpate. For one species to destroy another is just part of the process of nature; in England the grey squirrel is exterminating the native red squirrel, and the 'Norway' or 'Hanover' rat is exterminating the black rat. Nobody seriously means to restrain the destroyers, even by force; persuasion is obviously out of the question. If the human species similarly 'threatens' other species, then on the inter-specifically neutral view favoured by Cambridge City Council this is just one more of the conflicts that are part of the evolutionary process. If *I* am asked to act against the 'threat' because *my* species originates it, then I can no longer be asked for inter-specific neutrality, ignoring which of the myriad species *I* belong to.

Again, any living species necessarily 'disturbs the environment' (or 'the ecological balance') just by continuing to exist where it might not; if this is a sin for man, then men sin by existing: *el delito mayor del hombre es haber nacido*. Some people in England seem to accept Calderon's verse as a motto; one such (grossly disturbing the environment by driving a car!) displayed a label saying PEOPLE ARE POLLUTION. I have heard of a society, Doctors against Population, with a declared aim of reducing our population in Great Britain to some 30 million (I have my own idea about where the cuts might *begin*). In the same spirit some people deplore the absence of wolves and bears from the island. But the ecology will be greatly disturbed by roads and houses, and even with a vegetarian regimen, agriculture must bring about the disturbance of many animals by planting and the destruction of many more to protect the crops. A human way of life that minimised interference with the living world would be that of naked food-gathering non-hunting savages; life would be poor, nasty, brutish, and short. When I hear a university professor telling about the equal consideration and protection due to all animal species, I am tempted to ask him why he does not abandon his Chair and his house and his clothes and his shoes and take to the woods. Solicitude for other species may make sense if man is God's viceroy on Earth, a steward who must give account of his stewardship and must not waste or spoil his Master's goods; it makes no sense at all if man is just one more species struggling for survival among others.

Aware of man's loneliness as a rational animal, people sometimes dream of extraterrestrial contacts. But we may be reasonably certain that the rest of the Solar System is devoid of rational life; and the nearest star is four light years away. People try sending out radio messages, or capsules containing details of men's and women's physical appearance and human scientific knowledge; this is a good deal less rational than Shelley's writing his 'wisdom' on pieces of paper and enclosing them in bottles that he cast into the Atlantic: outer space is wider than the Atlantic. Thus do people worry about possible extraterrestrial rational animals, a worry that afflicts even some Christians. It really is not our business: 'What is that to thee? Follow thou me.'

As against this, I have seen calculations about the proportion of stars in the skies that will have evolved life on some planet: of this great multitude, it is argued, a small fraction, but still a large absolute number, will have had life evolve to the point of harbouring rational animals. (One author, I remember, calculated the number of worlds on which, if God ever were incarnate, there would have to be simultaneous incarnations: a number so great that even Divine Wisdom could not keep track of all the incarnations without getting into a muddle!)

We really need not go into this. We are well used to the coming-to-be of rational animals; but we ought to reflect more than we do upon the wonder of it. I shall argue that the coming into existence of any one rational animal is a miracle, something inexplicable by the processes and laws of subhuman nature: not just long ago in Eden, but every time there is a new baby. In old language, the human rational soul is not developed but *created.* If so it is quite vain to speculate about rational animals elsewhere; their origin, like men's, would be miraculous and not calculable in advance; we could know of it only if God chose to tell us; and he has not so chosen.

An evolutionist who has any sense will say: Admittedly man now stands on a high eminence above other animals, but this achievement is not a miracle but a very long series of small steps. It would just be the notorious *sorites* fallacy to deny man's eminence *now,* and the evolutionist really need not go in for such nonsense as comparing a notice on a door with the demarcation of territory with urine. But the series of small steps may appear to be a concept that rules out the idea of miracle: no one step, surely, could make the difference between being and not being in the image of God.

I shall not here discuss the origin of life; and I do not in the least wish to rule out either the origination of life from non-living materials, or the development of new species. If experiments produced rudimentary living things from a broth of inanimate chemicals, intended to mimic conditions in a primeval sea, some fools would court this as a devastating blow to Christianity. (Of course in one way life would even so not come to be in a lifeless environment: let us not forget the flesh-and-blood experimenter responsible for the whole setup, who may now forget his own presence as one forgets the spectacles on one's nose! However, I waive this point.) It is gross ignorance of the history of science to forget how long it took people to accept that in the present conditions on Earth, life is generated only from previous life; when people widely believed the contrary, nobody found this perilous to faith. It was by the scrupulously careful experiments of men like Pasteur and Tyndall in the nineteenth century that people came to accept that even lowly organisms are not generated just from lifeless chemicals.

St. Thomas Aquinas readily accepted that the solar radiation could generate life from the non-living even in our existing circumstances; still less would he, or should anyone, be upset if irradiation or electric discharges produced life from the non-living under laboratory conditions.

Aquinas would accept with equanimity the idea that natural processes may generate new species not originally created; he cites the production of mules, specifically different both from horses and from donkeys.[1] (The barrenness of mules clearly has no metaphysical importance.) Scholastic writers can be found (cited for example in James's *Principles of Psychology*) who argue *a priori* against the natural origination of new species, but their arguments are sophistical and silly. Whether natural selection in conjunction with casual variations is a *sufficient* explanation of the huge actual variety of living things may very well be doubted; the doubt is not to be dismissed by referring to the *actual occurrence* of the factors whose *sufficiency* as explanations is alleged; still less by pretences that to raise this doubt amounts to doubting the evolution of species altogether. Happily we need not go into these discussions. The question is whether the coming of men into existence is something radically new; in the Middle Ages the great leap is not from non-life to life or from one animal form to another but from subrational to rational existence; and this is the view I shall defend.

Let us remember that the problem arises not only for the origin of the human species but also for the origin of the human individual. An ovum and a spermatozoon are individual beings of a human rather than animal sort, if human beings produce them; but they are not two men or women nor even together one man or woman. They are on the same level as any other human cells that live outside a human body. I doubt if even the zygote formed by the union of two human gametes is a human being from the very first stage of its development. If it should prove possible to divide a zygote *in vitro* at an early stage of development, implant

1. *Summa Theologica* Ia q. 73 art. 1 ad 3 um.

one half in a woman's womb, and then implant the other half in another womb, this sinister technology would mean that one zygote is not necessarily programmed genetically to develop into one human baby; if both implantations succeeded, one zygote would have yielded two human beings at the mere whim of a physiologist. The like is plausibly supposed to happen when identical twins are born; the famous Siamese twins, Chang and Eng, had before birth a shared placenta and umbilical cord; so they were one organism, like a tree with double trunks, but clearly two human persons. Very rarely there is born a human chimera, whose body cells do not all bear the same genetic constitution; this is supposed to occur because two zygotes have fused to develop into one baby. Less rarely, by cell-division a zygote may become a mere lump, which doctors call a mole (Latin *mola*); moles sometimes threaten the life of the women who carry them. Nobody would scruple to remove a mole even for the mere health and convenience of a patient; a mole has no trace of human structures, it merely grows.

Hatred of abortion and embryo experimentation makes people favour the view that a zygote is from the first a human being with full rights. But let us not confuse moral and ontological questions. Doctors and nurses are indeed trained in double talk; what is in the womb is called a baby or a product of conception according as live birth or abortion is planned; but of course it is as clear that a woman's womb contains a little boy or girl as that a cat's womb contains little cats. However, this does not settle the moral question. The philosopher F. H. Bradley claimed for that great organism the State the very 'right to choose' as regards its citizens that is now claimed on behalf of a woman as regards the child she is carrying: namely, the right to cut off if the interests of the greater organism demand this. A person who favours abortion need not fuddle his mind by denying the humanity of the unborn; he may be openly and clear-headedly nasty like Bradley. Conversely, there may be reasons for respecting even the earliest stages of human development without prejudging the issue when full human life begins; though surely a doctor need not respect the life of a mole developed in the womb.

In any case, human nature is inexplicable as a development from subhuman animal nature, because of the way the human mind works. Evolution could at best explain human reason only as a development from what Hume calls 'the reason of animals'. 'The reason of animals' serves to let an animal survive in its environment and propagate its kind; obviously this may go with gross misapprehension even of the environment. Let us remember Bertrand Russell's chicken: a hen forms the useful habit of coming to be fed at such-and-such a time and place each day, but one day the farmer wrings her neck; meanwhile, however, she has propagated her kind.

Man alone has a world; an animal has only an environment. Natural selection could at best explain 'the reason of animals' as serving their adaptation to an environment that suits some species; human reason can adapt a variety of environments for human survival. Living in a world, men devise stories of how it all began. All tribes of men have traditions of their ancestors. Now for thoughts about the long ago and the far away there is not the least reason to expect natural selection and the struggle for existence to have led to men's observing the right standards. Human persistence in wondering about the ultramicroscopically minute structure of things and devising apparatuses to investigate this is even of negative value, so far as survival goes; but this does not show our theories are false, out of kilter with reality.

I have heard that towards the end of his life Darwin wrote down some theological musings which he was aware were not much good—but then asked himself whether in any case a brain developed in the struggle for existence carried on by apish ancestors could be expected to cope with theological problems. By the same token, a brain so evolved could not be expected to think adequately about the gradual evolution of primates over hundreds of thousands of years; for this thinking must depend on projection into the remote past of regularities presently observed, and the reliability of such projections seems to have little to do with the survival value of human inductive habits.

Theories of how man came to be, whether Darwinian or anti-Darwinian, of course have nothing to do with our criteria for judging deductive validity. An argument that purports to be conclusive must be judged simply on its merits, not by its provenance or the motives of its proponent or anything like that; we need not even concern ourselves as to whether it proceeded from a human mind or from a man-made computer. (Some popular writings of C. S. Lewis tend to obscure this point.) But inductive reasoning notoriously is not demonstrative; here the question is not whether a conclusion strictly follows from premises, but whether our acquired styles of projection accord with the nature of things. Logic cannot guarantee that they will accord, even in the long run; anyhow, as Lord Keynes said, in the long run we are all dead.

There are many tricky problems about inductive procedures into which I cannot enter here. The question how we can know it is *really* reasonable to go by inductively 'reasonable' procedures can be brought into focus by one simple case. Consider the form of reasoning '90 percent of the As are B, *x* is an *A*, therefore *x* is a *B*'. Given true premises and finite classes, it is logically guaranteed that 90 percent of such arguments have true conclusions. But that does not mean that 90 percent of the conclusions *actually drawn* from true premises by human beings will have been true when this schema is followed; men might turn out to have been very unlucky in their use of it.

It is another thing if men are the living images of a Divine Providence which controls not only the humanly observed course of events but also the movements of men's thoughts. God is not out to ensnare his rational creatures; to quote a dictum ascribed to Einstein, '*raffiniert kann der Herrgott sein, bös ist er nicht*'; God may be subtle but he isn't mean, he will not lay traps in our path. The ancient Hebrews regarded our reliance on the course of nature as founded on a covenant, e.g. 'the covenant of day and night'. To Christians this way of speaking may be unfamiliar but should not be alien. We can rely on the order of nature only because we rely on the promises of God that are manifest in his works; failing such trust in God we ought to say, in the words of the musical show *Helzapoppin,* 'anything may happen and it probably will'.

Many writers have tried to eliminate the problem of induction by telling us that we simply cannot help the sort of inductive thinking we go in for, that we could no more try to live by an alternative style of thinking than we could stop breathing by an effort of will. On the contrary, it is easy to imagine other life-styles, and quite credible that there have been such. Let us consider Prescott's account of the Aztecs' beliefs and practices. Every fifty-two years, Prescott tells us, the Aztecs awaited a crisis in the order of the world: regularities that had hitherto prevailed might break down. When the last nightfall of the era came, they extinguished all the fires and locked up all the pregnant women (who might bring forth monsters in the new era). In the middle of the night the high priest offered a human sacrifice and on the victim's breast tried to kindle a fire by friction in the traditional way. When the old method had worked, messengers ran with the new fire throughout the land: with huge relief Aztecs decided that the utterly unprecedented was not to be expected for the next fifty-two years.

One might easily suppose that this was a mere ceremony comparable to the Catholic ceremony of the new fire on the Easter vigil. But not so: it was a serious belief, as the Aztec reaction to the Spanish invasion showed. A new era had begun shortly before the invasion: although the new fire had been successfully kindled, people were alarmed by sinister portents. Then the Spaniards came with all sorts of unprecedented things: horses, iron armour and iron swords, guns and gunpowder. . . . The old prophecy whereby the Aztecs had long expected the unexpected in a new era had been fulfilled, their resistance crumbled. When their courage revived, it was too late.

Of course the Conquistadores and the Aztecs had very different expectations as to the course of the world precisely because they were of different religions. In our culture even atheists hold a belief in the order of nature which is historically a tradition inherited by Christians from the Hebrew prophets: the belief was in the blood and bones of the Conquistadores. Success against the Aztecs does not of itself show that the Conquistadores were right: if we say the Aztecs' worldview led them to disaster, we may reflect that our trust in the laws of physics has enabled us to construct weapons that might bring upon us a vastly greater disaster than conquest by the Spaniards was for the Aztecs—which would not show that our physics was wrong. It is not such contingent fears that should make us frame our view of natural regularities: without trust in the unspoken but manifested promises of God, there is no ground on which to make any plans. And man alone can know these promises: promises not only for future well-being but for security in building theoretical knowledge. Man can indeed fall away from such knowledge into gross superstitions; but only man can be even superstitious.

The anointing oil of man's royalty all too many are in our time willing to wipe off, putting themselves on a level with the beasts that perish or with machines that men's wits have devised and men's hands have made. Man is just a highly developed beast, or a computer whose program nobody wrote. But we ought to laugh at such follies: particularly at a man surrounded by complicated and expensive apparatus made to his own design, whose aim is to show how the coming to be of himself and all his apparatus could be explained by a description of events that at no point brings in any such concept as aim or design. The use of words to deny freedom, the design of apparatus to disprove freedom, is specifically human and is itself only a perverse use of freedom.

I have concentrated on 'rational' in the phrase 'rational animal'; we scarcely need reminding that we are animals. Though man's thoughts range through all space and time, he is corruptible and weak and mortal. From a natural point of view there appears no remedy for this; not that the human mind's surviving the body appears not to be consistently conceivable, but that Nature affords little reason to hope for this. Anyhow it could not be rationally demonstrated that a separated soul will exist *endlessly* if it survives at all. God is not, so to say, stuck with human souls once they have come to be; their life depends on his will, and his will could end it 'like a tale that is told'. What God will do only revelation can tell us.

To the mystery of man, so nobly fashioned in mind and will, so vile in many of his ways: so continually driven to think and plan beyond the limits of an animal life, so fettered within these limits to all appearances: the Christian revelation supplies the key. This can give endless scope, depth beyond depth of light, to the demands of man's enquiring intellect; and aspirations that might seem vain dreams, futile longings, especially as regards our dead, are now taken up into the glorious hope of the resurrection of the dead and the life of the age to come.

3 Consistency

Do we need to be consistent? Some have thought not. Walt Whitman wrote: 'Do I contradict myself? Very well then, I contradict myself. I am large, I contain multitudes.' And I have heard of a Japanese physicist who was quizzed by his Western colleagues: when he was with them, he talked about the constitution of the Sun in an ordinary scientific way; then what about the belief that the Sun is a goddess and the remote ancestress of the Emperor of Japan? 'That's all nonsense' he said: 'but when I am in Japan—I believe it!' Perhaps it would not be fair to take this remark literally; after all, some Westerners appear able to divide their life and thought into watertight compartments. Certainly a man may be conscientious to the point of devotion in one sphere and very far from virtue in another sphere, like Harry Graham's organist:

> Alas, in his domestic life
> Our hero sadly fails;
> He starves his children, beats his wife . . .
> But dash it all, what can you say?
> You haven't heard the beggar play!

This raises the question of the unity of virtues and the unity of vices. Some vices are clearly incompatible, so nobody can seriously think that if a man has some one vice he has every vice. It is however more plausible to hold that if a man has some one grievous vice, all his apparent virtues are bogus, and this doctrine of the unity of vices may be put contrapositively as a doctrine of the unity of virtues: that a man can have any one virtue only if he has all the virtues of a virtuous man.

If true, these doctrines would imply a pretty grim view of the world; but then, many facts about the world simply are grim. To be sure, ordinary people's judgments are not in conformity with these doctrines. It would generally be accepted that a sexually immoral man may as a statesman be upright, incorruptible, wise, and patriotic: though such immorality from time to time spectacularly brings some English statesman to utter ruin, other men whose lives are fairly well known to be no better models of chastity somehow get away with it. Actors and actresses devoted to their profession often meet with similar indulgence. So judges the world; but then the world is no competent judge.

Any seriously vicious habit damages a man's practical reason; if all judgments of a seriously impaired practical reason had to be corrupt, the doctrine of the unity of virtues or of vices would stand. But we should not conclude that it does stand; men are not so logical as to pass regularly from one false judgment to another logically implied by it. In some theological disputes each side has accused the other of holding a doctrine that logically implies some heretical falsehood; the matter in dispute may sometimes have been such that one side or the other must have been right in discerning heretical implications in the other side's view. It does not follow that in that case one side or the other must have been a party of heretics: to justify so grave a charge, the heretical conclusion must have been actually drawn and pertinaciously upheld, but given human illogicality it may never have been drawn. Aquinas would certainly concede this point about views with heretical implications; but the same point may be made, as regards a practical judgment impaired by a particular vice, against his view of the unity of virtues; such impairment of practical judgment *need not* bring with it a general vitiation. A devil no doubt directs all his will and action to consistently evil ends; but this is not true of most men, even of men notably bad in some ways.

I maintain the further thesis that minor vices may save men from major ones. Tyrants could do much more harm in the world if all their servants were flawlessly efficient, untiringly industrious, and financially incorruptible. Kant would pretty certainly have held that efficiency, industry, and financial probity in the service of a tyrant made the servants more detestable; this would at least make them more potent for mischief, even if it did not make them worse men. But I think an incompetent, lazy, and venal servant of a tyrant may be overall a less bad man than if he had been without these faults; the evil of pursuing bad ends is much mitigated by slackness in the pursuit. Hume in his *Dialogues Concerning Natural Religion* has his character Philo complain that if men had 'a greater propensity to industry and labour' then there would result 'the most beneficial consequences, without any allay of ill'; the complaint is ill considered. Laziness is a fault Providence uses to mitigate our vices and lessen the mischief they do.

Inconsistency of theoretical judgment also has its uses, in saving men from going the full length in drawing conclusions from false premises. What then is the advantage or need of consistency, or again the bane of inconsistency?

In his book *Introduction to Logical Theory,* Strawson regards our drawing attention to inconsistency in discourse as a 'charge' that does not 'refer to anything outside the statements that the man makes'. This is as wrong as can be. In the first place, the man may consider whether or not a set of propositions are inconsistent without turning any one of them into a statement. Strawson goes on to compare inconsistency to writing something down and then erasing or cancelling it; he is clearly thinking of a head-on collision between two statements; but it is extremely easy to construct a set of inconsistent propositions with more than two members such that no pair of them are inconsistent, e.g. by conjoining the premises of a valid syllogism with the contradictory of the conclusion. Somebody of confused mind may easily commit himself to such an inconsistent trio, and when challenged oscillate between maintaining A and B, maintaining B and C, and maintaining C and A. For this sort of inconsistency the simile of a man who writes something down and straightaway cancels it is altogether inept.

Strawson's idea of inconsistency as a merely internal fault of a body of discourse is diametrically opposed to the correct view of the matter. In fiction, indeed, inconsistency is a merely internal fault, and does not matter so long as it does not offend the reader. This holds precisely because the indicative sentences in a work of fiction do not latch onto reality: the author and the reader merely make believe that they do so. When discourse is meant to latch onto reality, then inconsistency matters: not because falling into inconsistency means perpetrating a specially bad sort of error, logical falsehood; but because inconsistent discourse inevitably has some non-logical fault. Like it or not, an inconsistent history will somewhere be factually false, an inconsistent set of orders or instructions cannot all be carried out, an inconsistent moral code will at some juncture be prescribing morally objectionable conduct, and so on.

There is a highly relevant story about the Waterloo campaign. Marshal Grouchy, whose army was at some distance from the field of Waterloo, had two orders from Napoleon: to intercept an approaching Prussian force, and to march to Napoleon's aid as soon as he heard the cannonade begin. In the circumstances both orders could not be obeyed. When they heard the cannonade, Grouchy's staff officers urged him to hasten to the field of Waterloo; he overruled them and gave battle to the approaching Prussians. He won in this engagement, and afterwards was able to bring his force back to Paris comparatively unscathed; but his decision was fatal to Napoleon, who had hoped till the end of the day for Grouchy's arrival. Of course Napoleon's orders to Grouchy had not been logically inconsistent, but their being inconsistent in the actual circumstances was quite bad enough.

Since the penalty of inconsistency works out in a concrete case as the penalty of a false factual belief, an unexecuted order, or the like, why worry specially about inconsistency? Is not inconsistency unavoidable for human minds, just as error in general is? And then why not adopt, as regards different areas of one's life and thought, a watertight compartments policy?

Indeed, *humanum est errare*. But that does not dispense us from the task of forming good mental habits that will lessen our liability to error. And acquiescing consciously in a detected inconsistency is like cherishing a viper in one's bosom: it is to ensure that somewhere one's thinking will be at fault in a non-logical manner. We are tempted to such acquiescence by that grave intellectual vice which Quine has labelled the desire to *have been* right; we may wish for the advantage of accepting some newly discovered truth without the discomfort of frankly confessing and recanting old errors.

A peculiar defence of inconsistency is to be found in some popularising books about the foundations of physics. When each of two theories, on the face of it incompatible, covers part of the facts but neither covers the whole ground, a pretended solution is to say the two theories are *complementary;* neither is to be discarded as false, but care must be taken not to assert or use them conjointly! Some show of logical work is then made in order that this solution shall accord with a revised logical code; and then this convenient notion is sometimes extended to appease other conflicts, e.g. between a belief in universal determinism and a belief in human free will. All this is just self-deception. If neither of two theories is satisfactory on its own account, let us not fancy we can mend matters by adopting both and calling them 'complementary'. As for proposals to bend logic, logic must remain rigid if it is to serve as a lever to overthrow unsatisfactory theories; otherwise refutation of a theory by contrary facts could always be staved off by enfeebling the logic that shows the contrariety.

Logic can never be constrained to withdraw a thesis by reason of a rival thesis established in some other discipline; for in a sense logic has no theses, being merely concerned with what follows from what. Logic is like a constitutional queen of the sciences: a queen who can never initiate legislation, but unlike the British monarch can put in a veto—on the score of inconsistency or fallacious reasoning. Of course in saying this I do not claim infallibility for logicians. A logician may be somehow misunderstanding the sentences into which he reads inconsistency or fallacy. And error may creep, indeed has crept, into accepted logical teaching; I myself have long been endeavouring to show this as regards the traditional doctrine of distribution. To resume my constitutional analogy, logic is like the British House of Lords in its role as a court: this court is not a court of first instance and does not by legal fiction claim infallibility, on the contrary, it can revoke its past decisions; but its jurisdiction cannot be declined. Nobody is in a position to say: 'The results of my research upset, or at least cast doubt upon, the logic you use to criticise me.' I myself once encountered this response to an argument I had stated: 'That's a very logical argument, but if you knew as much of the history of logic as I do, you wouldn't trust logical arguments.' But the only remedy for bad logic is good logic.

If the doctrine of double truth ever rears its head, it must be combated, now as in the time of Aquinas. Truth cannot clash with truth. If the doctrines of the Faith are true, they can conflict neither with one another nor with truths in some other domain. An argument that purports to show such a conflict must thus be regarded as no proof but a fallacy; and any fallacy can in principle be exposed by producing some unexceptionable counterexample, an argument of the same form in which the premises are uncontroversially true and the conclusion is uncontroversially false. And the logic needed will just be ordinary logic, universally accessible and acceptable, not logic accessible only to 'baptised reason', whatever that may be.

What I have just said has been assailed both by unbelievers and by Catholics. An unbeliever has spoken of 'Wittgensteinian fideism' and accused me of making the Faith into a norm of logical validity. This is sheer misunderstanding. Anybody who holds a body of doctrine, whether theological, geological, or biological, if he is confident of its truth, is entitled to doubt the validity of a pretended logical refutation of what he holds; but so far this is not a refutation of the alleged refutation. For that, as I have said, one needs exposure of a fallacy by straightforward logic.

From the opposite side, my approach to logic has sometimes offended pious ears. When I remarked that a certain argument against the doctrine of the Trinity was invalid, a Catholic sarcastically replied, 'Of course you think you understand the Trinity!' Of course I did not and do not claim to understand the inner life of God; but the *doctrine* of the Trinity, as Newman pointed out, can be stated in a small number of theses, each of which can be grasped and held with real assent. Proving that this is a consistent set of theses is quite another thing, and I made no claims to have such a proof. All that I claimed was ability to refute a particular argument against Trinitarian doctrine, and such a claim need not be held presumptuous.

A *general* proof of consistency, which would at once enable one to dismiss all such attacks, is certainly not going to be available to mortal man. If the propositions in which the doctrine of the Trinity is formulated could be proved to be a consistent set, then *a fortiori* each one would have been proved possibly true. But as regards the inner life and mutual relations of the Divine Persons, there is no difference between possible and necessary truth. I have been told that some scholastic writers investigated the consequences of false suppositions such as 'Suppose the Son had proceeded from the Holy Ghost and not only from the Father': such investigations are mere folly, as though one enquired what factors 97 would have if it were not a prime number. Since in this realm possible and necessary truth coincide, proving the possible truth of Trinitarian theses would mean proving their truth, which is certainly not possible to mortal man.

An adversary might now ask: 'Why should I even consider whether the doctrines you hold might be true, when you have given me no reason to think they are even coherent?' 'Coherent' is something of a vogue word in recent philosophical writing: writers who use the contrary term 'incoherent' seem very often to feel themselves dispensed from making it clear whether they mean 'inconsistent' or 'nonsensical'. Now nobody can fairly be asked to *prove* that he is not talking nonsense. A form of words that at first sounds intelligible but is really nonsensical does not admit of logical manipulations, for these will themselves be nonsensical. Logic cannot handle nonsense. (To be sure, there is a bogus art called 'significance logic' that purports to track the frontier between sense and nonsense as falling *within* the territory of logic, like that between truth and falsehood!) Given a piece of latent nonsense, there is an art or skill of reducing it to patent nonsense; but this is not a kind of logical derivation, and there is no such procedure as proving that some piece of discourse is *not* nonsense.

We should be utterly clear in our heads about the difference between a nonsensical pseudo sentence like:

> 'Tom and Bill respectively kicked and punched each other.'

and a sentence containing a buried inconsistency, like:

> 'In this university there are two teachers each of whom teaches everybody else here and is not taught by anybody else here.'

Reductio ad absurdum works by deriving a patent inconsistency from a set of premises, which shows that one or other of the set is false; this valuable method of proof would be a ridiculous procedure if patent inconsistency were not to be distinguished from unconstruable nonsense.

In modern logic a great deal of work has been done on consistency proofs; it is very demanding work. Since a theory may be consistent without being true, consistency might be expected to admit of easier proof than truth does. But the reverse is the case. If we have true premises and proceed by valid steps of reasoning, the conclusion needs must be true; but given a theory of some complexity, it may be formidably difficult to prove its consistency while leaving the question of its truth in abeyance. So it is not reasonable for an enquirer to demand that the 'coherence', i.e. freedom from contradiction, of a theory shall be shown to him before he will even consider whether it may not be true.

Many philosophers are ill informed about consistency proofs; and so we get books like Richard Swinburne's *The Co-*

herence of Theism. It is only too plain that he has no real grasp of the distinction between nonsense and self-contradiction, and is quite unaware of the nature and difficulty of consistency proofs. And when he comes to the 'coherence' of ascribing to God this or that attribute, the results are painfully comic. He relies heavily on imagination, although Leibniz had rightly called this an *indice trompeur;* we can *imagine* things that on reflection are self-contradictory. (One of Escher's engravings shows a stairway running round the four sides of a tower, on which by continual ascent one gets back to the starting point.) Swinburne 'proves' that it is 'coherent' to call God omnipresent because he can imagine a process by which *he himself* could progress towards omnipresence! My imagination suggests a simpler way, the White Knight's way:

> To feed oneself on batter
> And so go on from day to day
> Getting a little fatter.

If I were fat enough, my mass would attract into itself all the mass of the universe, and then I'd be omnipresent.

I return to the question of consistency in life and character. There is a deep inconsistency in a man who retains his faith but is well aware of falling into damnable sins. But the suggestion that he 'might as well' abandon his faith also, since by itself faith will not save him, is merely a suasion of the devil. As Newman remarked, God might have willed that loss of charity by grievous sin should carry with it loss of faith, but in his mercy God has not so willed. Faith without charity is like a tree cut down to its roots: if the root still lives, a new shoot may yet spring from it. The conquistador Pizarro, after a career stained with many crimes, was dying utterly unprepared, desperately resisting a sudden murderous attack. Pierced by many wounds, he fell to the ground; he dipped his finger in a puddle of blood and traced a cross on the flagstone, and then with a last effort he kissed the cross, called out the name of Jesus, and died. Let us hope he found the mercy he sought.

On the other hand, a man's state is very bad if he has lost the virtue of hope, by presumption or by despair; both are ways of committing a 'sin against the Holy Ghost', incurring a disease of the soul comparable to terminal bodily illness; though no disease of body or soul is beyond the healing power of God.

Again, any virtue may be corrupted by a failure of courage, in a world when there are dangers to face and afflictions to endure: if a man is prepared to intermit the practice of some virtue when that becomes too dangerous or too toilsome, then he does not really possess that virtue any more. It would be no absurdity to say of Harry Graham's organist: 'He is really devoted to his art and conscientious about his performances, although he is a very bad husband and father'; but it would be absurd to say of some judge: 'He is really an upright judge, even though he bends the law sometimes under threats from the Mafia.'

Charity, the love of God, is what we are all made for, and without charity we are nothing. From this great truth two contrapositive conclusions have been drawn: that seemingly good actions are really bad, because done without charity; and that because some act is good of its kind, a man who did it must have been in charity at the time. Neither conclusion follows from this premise alone; we need the extra premise that apart from works of charity people are totally depraved, and all they plan or say or do, however fair seeming, is evil continually. Aquinas sensibly remarks that even people who lack grace and charity are up to performances good of their kind, for example cultivating the soil and defending their country; total depravity does not exist outside Hell, where the will is wholly concentrated on an evil end. Such activities good of their kind are of themselves unavailing towards salvation and the vision of God; but they are not on that account wicked activities.

I ought to say a few words about the relation between charity and unselfishness. Ronald Knox in a sermon described unselfishness as *the* characteristic Christian virtue, mentioning however, without its giving him pause, that the word has no real equivalent in Latin or Greek! But if an invalid 'selfishly' develops talents that are unlikely to benefit her fellows, she is cultivating a garden God gave her to cultivate. On the other hand, for the sake of the vilest ends men have endured hideous sufferings and in the end given their lives; this, as St. Paul might say, is a sacrifice possible to a man who has not charity and in God's eyes is *nothing;* the goods he sacrifices only show the hellish intensity of his evil will.

Charity is called the form of the virtues. It is not true that apparent virtues are only *splendida vitia,* showy vices, without charity; but real unity of mind and heart, 'to will one thing' as Kierkegaard puts it, comes only with charity. *Ama et fac quod vis* is a sound maxim only if it is taken to mean, not that if you (in some sense) love you may do as you please, but that if indeed you love you will manifestly act so that men see your good works and glorify your Father in heaven. It is not for me to say whether a man's long pursuit of truth or justice or compassion will in the end turn out to have been love of God even when he was in profound ignorance as regards explicit judgments about God. That will all be sorted out on Judgment Day.

Truth, Truthfulness, and Trust

Arguing with my daughter Mary Gormally, a man was casting doubt on the very existence of any fixed standard of justice: after all, one cannot measure justice! Mary immediately retorted, 'Without justice there is no measurement.' A little thought shows how deep this retort penetrates. For measurement there must be true weights and measures; the results of measurement must be truly reported and recorded; and we must trust our fellows that there are such true reports. Otherwise the practice of measurement is futile. So there can be no measurement without a species of justice.

Ethics and epistemology are here closely intertwined. Many philosophers have developed, or perhaps merely sketched, a do-it-yourself theory of knowledge; but the toolkit that would be needed is lacking. 'No man is an island': everybody is epistemically 'part of the main' (mainland). I imagine the nearest approach to a do-it-yourself physicist was Henry Cavendish: he did a great deal of solitary research, and how far he had developed the theory of electricity was not known until his posthumous papers were edited by James Clerk Maxwell long afterwards. Those days he could not purchase or bespeak a galvanometer; instead he devised a method of using his own sensations of electric shock to measure currents—with surprisingly accurate results. I remember mentioning Cavendish during a discussion in Cambridge of a paper by Karl Popper; I asked Popper whether by his well-known method of 'demarcation' the electrical researches of Cavendish would count as scientific; he unhesitatingly said they would not—I imagine his reason was that Cavendish did not lay open his research to critical examination by his peers. But if we do not accept a Popperian 'demarcation' of scientific work, we can easily show that Cavendish too was 'part of the main'. He was in a way his own galvanometer, but certainly not his own barometer, or thermometer, or balance; for these 'philosophical instruments' he had to rely on the integrity of some maker, and for his chemicals, on the honesty of some supplier. So his measurements too relied on other men's justice.

Natural science of all sorts flourishes only because deliberately fraudulent claims about experimental results are rare and are generally known to be rare. Precisely because they have this high code of honour and expect its observance by others, eminent men of science have made arrant asses of themselves about psychical research; in those regions fraud is all too common and men not trained to look out for it have been sadly deceived. Here the real experts are not natural scientists but conjurers, who never pretend that the effects they produce for entertainment are anything but illusions and are disgusted with those who use such illusions in a pretence of supernormal powers.

Trust in our fellows' testimony of course plays a far larger part in our lives than its role in scientific work. Philosophers often harbour a vague idea that the reliability of testimony can, or in principle could, be established by induction from 'experience'. That word is used with a hopeless oscillation between two senses. Hume says that 'experience' tells him to expect to see a visitor when he hears a certain noise (the creaking of hinges); and also that 'experience' tells him what happens to a letter between its being posted and its reaching his hand. But of course no one man's experience covers what happens to a letter in all that long time; most of the time anyhow it is in a bag unobserved by anybody! Very often, as in this example, 'experience' signifies the result of pooling many individuals' experiences; and that can be effected only by individuals trusting one another's testimony. Similar considerations apply to John Stuart Mill's assertion that we know watches are designed to tell the time only because we know from 'experience' that they come from watch factories; for visits to watch factories are not within the scope of many people's actual memories.

Of course we all know that men may be deceived, and what is more may lie. When one of us considers some extraordinary piece of testimony, it will be proper to apply general knowledge of circumstances in which men are deceived and of circumstances in which men are tempted to lie. But one who thinks of 'our' general knowledge may easily forget that *his or her* knowledge of human propensity to make mistakes or to lie is accessible only through others' testimony; an individual's *strictly personal* knowledge of occasions when *he or she* has blundered or lied or has been victim of someone else's blunder or lie would be a miserably inadequate basis for formulating general principles whereby to sift out unreliable testimony. Just as we mostly eat and drink without testing for poison, so acceptance of testimony has to be the rule, scepticism the exceptional case; and nobody can live on purges and emetics.

We may compare our reliance on testimony with our reliance on memory. Given the things any individual may seem to remember, a history of events based on individual memory would tell of an extraordinary world, in which objects vanished out of locked drawers and the like; indeed one and the same person's ostensible memories uttered at different times may form a logically inconsistent corpus. But this does not justify a general scepticism about memory; if anybody did succumb to such general scepticism about memory, no inductive inference could remedy the situation (though I have read a philosopher's attempt to validate memory by induction!)

Suppose a man seeks to remedy a faulty memory by writing memoranda. How is he then to know that the MS. of the memoranda has not changed over time? Surely he could then reply only that he is confident that his memory affords no precedent for such change in a MS. So this remedy for defective memory itself works only by reliance on memory in another area.

On the other hand, any individual's memory is subject to checks: what other people say, public records, and the like. The clearest ostensible memory by a lady of having proposed marriage on February 29, 1900, could not prevail in 1910 against the public record that there was no such date in 1900. It is clearly not the case that the intuitive character of memory trumps non-intuitive information from testimony and records.

Testimony is indispensable as a foundation for the house of knowledge. Only by accepting testimony can we establish criteria for weeding out unreliable testimony. The Royal Society has used a motto *Nullius in verba*. This is a fragmentary quotation of a Latin hexameter:

Nullius addictus iurare in verba magistri.

As originally meant, this is a warning against swearing to the truth of some matter on the say-so of some dominant personality, in the way Pythagoreans are supposed to have settled disputes by appeal to what HE (Pythagoras) said. But I fear the tag has often been cited in the much less respectable sense: Take nobody's word for it. The book from which I learned chemistry at school solemnly warned the young experimenter to rely on his own observation rather than on what a book tells him he ought to observe. The author of the book was a teacher at my school; I am confident that if a boy had excitedly told him, 'Sir, please Sir! When I did that experiment I found the products of the reaction had less mass than the reacting substances had before the reaction', he would have wearily replied, 'Rubbish, boy, you must have let some of the gas escape' (or something of the sort); and the boy would do well to accept his master's authority about how the experiment ought to have resulted, rather than rely on his own observation.

Reliance on authority is a kindred topic to reliance on testimony. Those who claim authority cannot all be trusted, for they contradict one another. The question which authority to trust is difficult and inescapable. But we must steeply, most steeply, rebut the sophists who would argue 'In accepting an authority you are relying on your Private Judgment that the authority is reliable: so Private Judgment trumps authority.' Inevitably my judgment is my judgment, my very own judgment, thus my Private Judgment; but this is a mere tautology, from which nothing interesting can follow.

Similarly, when I decide to follow one authority rather than another, I am not in effect setting up myself as a superior authority. It would be quite difficult for me to give good reasons for trusting one lawyer or doctor rather than another; but such trust on my part need not be merely blind, nor on the other hand am I claiming to know more law than my lawyer and more medicine than my doctor.

If I cannot remember what a Greek word means, I consult the *Lexicon* of Liddell and Scott. A U.S. citizen called Knoch, laudably desirous of finding out what the Greek New Testament actually means, decided that this lexicon was untrustworthy: Liddell and Scott were dignitaries of the Church of England, so they might have strong motives for distorting their account of the Greek language to make the New Testament express their own theology! So Mr. Knoch decided that the dictionary, and even the grammar, of the Greek language needed to be remodelled by someone free from Anglican preconceptions. I have not had enough curiosity to procure Mr. Knoch's Greek grammar and dictionary, though I know they exist in print; but I was sent a copy of the New Testament in a Knochian translation. Its quality may be judged by the fact that the adverb *eutheōs*, 'immediately', is always rendered by the ejaculation 'Good God!' because *eu* means 'good' and *theos* means 'God'. By the study of Knoch's New Testament an academic philosopher whose career and publications had hitherto been quite conventional was led to abandon his post in order to evangelise the world for a new Christian sect whose canon is fixed by Knoch. I suspect that his decision resulted from extreme cultural deprivation; many U.S. citizens grow up unable to read any language but English, and have not a clue towards finding out what it takes to learn any other language properly.

If I myself tried to spell out my reasons for trusting Liddell and Scott on the meaning of Greek words rather than Knoch, I am sure they would be wretchedly inadequate. This does not make me feel a shiver of doubt in the matter. Moreover, when I decide to follow Liddell and Scott rather than Knoch, I am not claiming to be a greater authority on the Greek language than any one of the three—though I do claim to know better than Mr. Knoch.

There is a good old maxim: *addiscentem oportet credere*, a learner must trust. Obviously Liddell and Scott learned Greek by trusting their teachers; the distrustful Mr. Knoch learnt his Greek all wrong. And the teachers of Liddell and Scott will have learned Greek by trusting *their* teachers; and so on back, till we reach someone learning ancient Greek from native speakers. I am afraid, however, that tradition raises problems too difficult to engage with here.

The need for trust on a learner's part was sharply brought home to me with my attempts to teach logic to a Mr. Lal. I was using Quine's *Methods of Logic* as a textbook, and Mr. Lal was supposed to do the exercises. After the first one or two chapters, Mr. Lal ceased to do the exercises and instead produced weekly essays rebutting the logical doctrine Quine set forth in successive chapters. The merits of his criticisms of Quine may be judged by the fact that he maintained that only 'truth-functional dogmatism' stood in the way of realising that in English 'and' and 'if' mean roughly the same. I was unable to persuade him that he ought to try to work through the exercises instead of looking for fallacies in Quine. Naturally he learned nothing and failed in the examination.

Logic is indeed a very special case. The teacher's task is not to get the pupil to accept the validity of formulas or methods of proof on authority, but rather to lead him to a point from which he can judge validity for himself: in Plato's words, the learning of logic should be 'like a flame leaping from mind to mind'. The trouble with Mr. Lal was that he did not trust his teacher and his textbook enough even to believe that there is here a right way of looking at things generally available to learners given their patience and good will. A measure of industry is needed too; logic can no more be learned without doing exercises than music can. Unfortunately in Mr. Lal's case patience and good will and industry were lacking.

As a basis for an inductive generalisation we need a mutual trust among observers that cannot itself be inductively justified. Ever so many philosophers have discussed the rationale of passing from 'all hitherto observed swans are white' to 'all swans are white'; they are interested in the fact that though before the discovery of black swans in Australia it was reasonable to perform this inference, the conclusion did turn out false. I believe Wittgenstein was the first philosopher to ask how on earth we get the premise: once this question is raised, it should be clear that the premise is attainable by anyone only by reliance on other people's reports and records about the sighting of swans.

Both in theory and in practice men have to rely on the truthfulness of fellow-men: life could not be carried on unless lying and fraud were the exception rather than the rule. I now turn to consider what is the malice of lying, i.e. what makes lying bad, and how grave in various cases this malice is.

As I have said, epistemically no man is an island. The language a man speaks is an institution, just as money is; and it is a *per se* effect of lying that it damages the institution, just as in accordance with the law of Copernicus or Gresham, bad money drives out good. The damage done may be grave or only slight. Jocose lies of the April Fool type do negligible damage; and serious harm is rarely done by lies aimed at baffling a man's curiosity about contingent matters that are not his business. Such lies are even less culpable if the enquirer who is deceived would use true information for some seriously wicked purpose. Even in this case, however, it is better to mislead the enquirer without lying. A Dutch woman whom the Nazis suspected of hiding Jews told the official who questioned her that there were Jews under the table: there were in fact Jews under the table—under the floorboards. The Nazi official took her remark as a mere piece of cheek and enquired no further.

One sort of lie that is utterly inexcusable is a lie that of its nature is likely to damage the general fabric of human knowledge. And false teaching about religion, whether by teaching false doctrine or by attempting to serve the cause of a true religion with falsified history or faked miracles, is the worst lying of all. The version of Job on which Aquinas wrote his commentary contained the dictum 'God has no need of our lies'; since God is truth, says Aquinas, an attempt to serve him by such lies is in fact not honour but gross insult.

An attack on the currency of language need not consist in making false statements; the very vocabulary may be corrupted. Translators of the Old Testament nowadays often sedulously avoid ascribing *mercy* to God: this is calculated to darken men's appreciation of the quality of mercy. For example, the text in Jonah 'Those who observe lying vanities forsake their own mercy' is mistranslated with 'true loyalty' instead of the word 'mercy'; but the Hebrew word never means that; the plain sense is that God alone can pardon our sins against him, and pardon can never come by following false gods (called 'lying vanities' in the usual unecumenical style of prophetic vituperation). Another word regularly mistranslated is the word *emeth,* 'truth'. On these and related matters it is profitable to read James Barr's book *The Semantics of Biblical Language*.[2]

Too many people know Greek for the word for 'truth' in the New Testament to be thus boldly mistranslated in modern English versions without the translators' being discredited. It is easier for learned men to get away with mistranslation of Hebrew and misstatements about the Hebrew mentality. The ancient Hebrews, we are told, must not be credited with the Hellenic concept of truth, telling things the way they are: *emeth* did not mean this, but rather had the sense of steadfastness or reliability, particularly on the part of God towards his chosen people. That ancient Hebrews lacked the 'Hellenic' concept strikes me as a racist assump-

2. Oxford, 1961.

tion; it could easily be used for anti-Jewish propaganda. 'That wretched people', a Nazi might say, 'lack the very idea of truth: the word mistranslated 'truth' actually signifies in their Scriptures merely the steadfastness of the tribal god in looking after his favourites!'

On the merely linguistic point of what *emeth* means in Hebrew, it is enough to cite James Barr. Barr remarks, for example, that, overwhelmed by actual acquaintance with King Solomon's wealth and wisdom, the Queen of Sheba is reported as exclaiming, 'so what I heard of you in my far distant kingdom was truth (*emeth*)': her informants had told it to her the way it was.

The duty of A to give true information to B is correlative, one might suppose, to B's right to be truly informed. But A's possessing information which can be available to B only by A's telling B does not in general give B even a *prima facie* right to the information, which would impose on A a *prima facie* obligation to give it to B. Even if A has previously lied to B about the matter, A has not in general a duty of restitution to B which he must discharge by now telling the truth about the matter. Such a duty of restitution would be decidedly onerous, and an attempt to fulfil it would often be embarrassing and even mischief-making. (I have heard that a certain religious sect actually imposed this duty on neophytes, with predictably awkward results.)

The case is altered if A is B's teacher or is in some other such relation in which B specially trusts A; then if A has knowingly given B false information, B has been *harmed* and A must undo the damage if possible. A may similarly incur such an obligation if he said in good faith what he later finds to have been untrue. But in very many cases of information about contingent matter, B's mind and his general intents are very little harmed, if at all, by his being misinformed. In that case A has no clear duty of telling B the actual truth of the matter, and may have sound prudential reasons for letting things rest.

The behaviour of Bertrand Russell is interesting to consider at this point. As regards episodes in his own life, Russell had no scrupulous care for truth; his autobiography is notoriously mendacious. I myself once heard him claim at a party that in order to write his *History of Western Philosophy* he had read the complete works of Aquinas: 'some of them were quite interesting'. Given the time he took composing that work, he must have been lying; I fear it will have added to his pleasure in this lie that his hearers included several people who, as he knew, would be well aware that he was lying, but whom he expected to be too polite or too embarrassed to expose him. But logic was sacred for Russell; about logic he absolutely would not lie.

A young friend of Russell, Philip Jourdain, constructed a 'proof' of the Axiom of Choice from the axioms of *Principia Mathematica*. This would have been a great achievement if the 'proof' had been valid; but alas, it was invalid. Jourdain, as Russell well knew, was afflicted with a slow and inevitably fatal degenerative disease of the nervous system; but even to comfort a dying man Russell would not lie about logic. Like many a squarer of the circle, Jourdain fancied that adverse expert judgment on his work could not be honest, rather must proceed from envy at a man's succeeding when so many had failed to reach the result. In this state of mind Jourdain died; his widow wrote a bitterly reproachful letter to Russell.

By telling Jourdain the truth, Russell foreseeably did Jourdain's mind and soul no good at all. Jourdain was no better off intellectually, since he did not accept Russell's demonstration of his error; and when Russell disappointed him, his futile rage against Russell's supposed dishonesty in criticism must have got him into a worse spiritual condition than he would have been in if Russell had lyingly congratulated him on a splendid achievement.

I do not know whether Russell at this time still held the ethical views of G. E. Moore; he himself wrote a mini-treatise on ethics expounding Moore's ideas. If he still followed Moore in ethical theory at the time of corresponding with Jourdain, then his actions belied the theory and showed him in a better light than if he had acted in accordance with it. Russell plainly did not weigh up the respective 'intrinsic values' of Jourdain's states of mind and soul that would result from his telling the truth or lying. The lie would not have harmed anybody except Jourdain and his poor wife; I believe Jourdain's 'proof' came to be published in a learned journal; in that case its error will have quickly become known to mathematicians and not have been further disseminated. All the same, Russell would not comfort Jourdain with lies.

Aquinas does not measure the *malitia,* the badness, of lying by any such evaluation of consequences. For Aquinas, Russell's autobiographical taradiddles will mostly count as merely venial sins; whereas the lie Russell would not tell to Jourdain would count as a mortal sin. For in judging an act, Aquinas holds, we must consider whether it has a *per se* tendency to grave evil; and deliberate misinformation about a matter of science has just this character.

For Aquinas the *malitia* of lying consists in speaking contrary to one's own mind, *locutio contra mentem*. A man's mind is of course not a rule or measure of truth, but unless he is exceptionally corrupt he at least desires to orient his mind to Truth. And Truth is God, even though in aiming at truth one may not explicitly realise this. So *locutio contra mentem* of its nature makes it harder to orient oneself to God. Even a habit of telling little lies about contingent matters is very harmful; a soul that should be a mirror of Truth is covered with hundreds of tiny scratches. And one fine day the habitual teller of little lies will in a sudden emergency come out with a big, whopping, utterly inexcusable lie, because he sees no other way out of his predicament. On the other hand, the brave Dutch woman of my story did not lie even with the concealed Jews' lives at stake; doubtless she feared God and had an ingrained habit of avoiding lies; so when an emergency arose she saw a way of escape without lying.

My readers may well have gathered that I do not object to equivocation as a way of avoiding actual lies. Many people will regard my attitude as Pharisaic or Jesuitical hypocrisy: surely an equivocal utterance may be just as misleading as a lie?

It may indeed; but there is no general obligation to avoid misleading people. Just consider how very onerous a duty it would be if one had to avoid all words and deeds that might, or foreseeably would, mislead some people! There is no such obligation. Equivocal utterances may produce the very same false impressions as direct lies; but they do not, like lies, debase the currency of language. And as Catholic martyrs of Elizabeth's and James I's time pointed out, to the fury of their persecutors, by the Gospel record Jesus Christ said, 'Destroy this temple and I will raise it up in three days', meaning his own Body by 'this temple'; this was as strained an equivocation as many allowed by casuists.

There are indeed casuists who defend outright lying as preferable to equivocation: surely a plain lie shows a character more straightforward, *plus simple et plus loyal,* than that of an equivocator? I think this attitude shows an insufficient realisation of how odious lying is to the God of Truth; an excellent formulation of replies to specious defences of lying is to be found in the Catechism of the Council of Trent. Doctors of the Sorbonne went even further than those casuists: in controversy with Descartes, they deprecated the key role played in his philosophy by appeals to the Divine veracity, and appealed to the Scriptures as evidence that God himself sometimes lies! Of course it would wholly discredit revelation if it were supposed to proceed from a deity who may lie when he sees fit. As against this, someone has argued that a man's credit is not *wholly* destroyed by our knowledge that he may now and then lie. But trust in our fellow-men in spite of occasional lies affords no precedent for trust in a possibly mendacious deity 'whose dwelling is not with men'.

People often have an odd attitude about lying to children, as if such lies really did not count. Since a child's very learning of language is possible only if, and insofar as, its parents give it true rather than false information, it is hard to see the rationale of this. Lies to children are often not confined to trivial matters of fact. A child's information about its own birth and parentage is necessarily hearsay. Even if some concealment is felt to be necessary, lying is a huge wrong; excused as being 'for the child's own good', it proves all too often to be extremely harmful. My daughter Barbara, who taught and practised psychiatric nursing in the United

States, had several patients who suffered severe psychical damage from discovering that such false identities had been imposed upon them; therapy was not always successful.

Another aspect of adult conduct towards children as regards lying is the way that the concept of lying is stretched to cover actions and omissions that ought not to count as lying. 'Lying by hiding', for example: there is no obligation to pass on just to any enquirer information one possesses, particularly about one's own doings, and parental authority does not simply annul a child's right to privacy. 'Acting a lie' is an even more dubious concept. Only misuse of a conventional means of communication can be a lie; in very many cases, however misled spectators may be, no lying is involved: e.g. if a priest in time of persecution dresses like a courtier or a workman, or an escaping prisoner carries a plank past a building site.

It has often been a way of bullying a child to begin by inculcating a sense of the disgracefulness of lying and then to affect to regard as lies instances of 'lying by hiding' or 'acting a lie'. A frightened child's escape routes must be closed by scaring it off all such so-called lies, as well as actual lies; Kipling and others tell grim autobiographical stories about this. Still stranger ways of educating a child against lying have been the grossly fabulous stories about the horrid fate of young liars.

'Lying by silence' is not an empty or ill-formed concept, but great care is needed in applying it. There are indeed cases where a failure to speak up may have the purport of an actual statement, and thus of a true or false statement. Silence when a chairman is about to say 'Carried unanimously' is in effect a vote for the motion; but in other cases silence in face of an assertion amounts to dissent. Describing the circumstances in which someone asserts something by remaining silent would be quite difficult. In a huge number of cases, remaining silent does not amount to any assertion at all.

These considerations have a bearing on the general ethical topic of acts and omissions. It is often said that the difference between acts and omissions is morally irrelevant;

people are persuaded to accept this thesis by talk about some particular cases, in which a particular omission is indeed morally on all fours with a particular action (e.g. failing to switch a life-saving machine on, as compared with switching it off). Once the general thesis is accepted it leads moral philosophers to extravagant conclusions: e.g. that failure to send a cheque to Oxfam for some impoverished community is morally on all fours with sending a can of poisoned meat to that community. In that case all of us would be 'in blood stept in so far' that we might well develop a cauterised conscience about killing. But the matter of truth-telling and lying supplies clear counterexamples to this pretentiously high-minded thesis about acts and omissions: omission to come out with the truth about some matter within one's knowledge simply is not in general tantamount to lying about the matter. Nobody has a general duty to pass on all the information he possesses to everybody he has to do with.

I now turn to the topic of making and keeping promises. One might think that an insincere promise is simply a lie, to be condemned like other lies. However, this has been disputed: philosophers, as the saying goes, will argue the hind legs off a donkey (because a donkey has *only four* legs). It has been argued, with apparent seriousness, that an insincere promise cannot be a lie. It is no lie if the promisor makes it in the form 'I promise so-and-so'; for then he is, as he says, promising so-and-so. But suppose he does not use the word 'I promise' or some equivalent, but just says 'I will'? 'I will' certainly constitutes a promise in certain circumstances; but then, though the words are different, the speech-act would be the same as if the form 'I promise' had been used; since with the 'I promise' form the speech-act could not be a lie, neither is it a lie in the 'I will' form.

We need not spend much time on the 'I promise' form. When a promise is made in this form, there is a different relation between 'I promise' said by N. N. and 'He promises' said *about* N. N. from the one that holds between 'I shave' said *by* N. N. and 'He shaves' said *about* N. N. But that does not mean that N. N. did not make any statement when he used the 'I promise' form; he was making a statement, not as to what he promised, but about his own future actions. So let us switch to considering promises in the 'I will' form.

Suppose a rascal goes around promising farmers (in the 'I will' form) that he will kill their rats, and collects payment in advance. By a quirk of English law, he could not be successfully prosecuted for obtaining money by false pretences: the false pretences have to be false statements as to past or present facts, not as to future contingencies like human actions. But this aspect of the law need not stop any sensible man from saying that the pretended rat-catcher was a shameless liar, seeing that he had no skill to catch rats nor any intention to try.

The 'I promise' form neither is needed to constitute a promise (some very solemn promises are made in the form 'I will'), nor at all accounts for the obligation to keep promises. Using a grammarian's term, we may call 'I promise' a back-formation; just as 'swashbuckler' meant not a man who goes in for swashbuckling, but a man who swashes his buckler (shield) with his sword. In circumstances where others might say of N. N. 'He promises', N. N. might say 'I promise' in making a promise instead of the plain 'I will'; 'I promise' is then not a statement *that* he promises, but stands in for an 'I will' statement, truthful or not, of what N. N. is about to do, and it is this statement that would in the circumstances be a promise. You cannot, like the legendary Oxford don, escape the grave obligation of promises by only saying 'I will' to express your intention! Life is not so easy as that; not thus can a man live without promising anything.

People arguing as to whether any actions are invariably wrong often assume that one who says some kinds of action are invariably wrong will certainly count breach of promise as one such kind. This is amazingly thoughtless: obviously someone who thinks there are absolutely forbidden kinds of action will not think that N. N. is on the contrary obliged to perform such an action because he has so promised. I suspect that ideas of the binding force even of atrocious promises may historically derive from a misreading of the story in Judges about Jephtha's daughter. But a deed's being narrated in the Bible nowise implies that the narrator approved of it. Jephtha's vow to sacrifice the first animal that was his which he encountered on coming home was already discernibly wrong in advance of the event, because he might well meet with an animal not apt for sacrifice: he acted foolishly and wickedly in taking it that this vow could apply to his virgin daughter. (Here I follow St. Jerome and Aquinas.)

What then is wrong in deciding not to keep a promise sincerely expressing someone's intention at the time it was made? I have encountered the idea that the promise is then made to *have been* a lie. This account is not easy to expound or defend. Perhaps it should not be dismissed out of hand as involving the absurdity of a changing past: when most of an uttered sentence is in the past, a man can clearly alter its sense by the way he chooses to end it; and Existentialist writers have manipulated the idea that a man's later actions may alter the significance of earlier actions. But the line of argument might prove too much: viz. that a sincere expression of intention which was *not* a promise might come to have been a lie through a change of intention. So I shall look for a rationale for the binding force of promises that involves less questionable notions.

A promise, if sincere, is an expression of intention. But so long as a man does not change his intentions frequently and frivolously, there need be nothing blameworthy in his changing them: all the less if he announces changes of mind. Again, if I now were to plan ahead on the basis of others' expressed intentions, these need not amount to promises. In the good old days of British railways there was regularly published *Bradshaw's Guide*, which purported to give timetables for all passenger train services in Great Britain. This did not constitute in law a contract made by railway companies with passengers; nobody who bought a ticket on the faith of *Bradshaw* could legally succeed in an action against the publishers or the railway company if a train did not run as announced and he incurred financial loss in consequence. Of course it was in the clear interest of the railway companies to endeavour to run trains as announced in *Bradshaw*, and in the publishers' interest to report the railway companies' plans as carefully as possible. So reliable planning *need* not depend on more than sincere intention. What then is the role of promises? How do they differ from sincere expressions of intention?

I believe we get a clue from the familiar complaint 'You let me down' when a promise has been broken. It is as though the promisee had built something on foundations supplied by the promisor and the foundation turned out rotten. It is obviously of great advantage in human society that when a man has once expressed his intentions to his neighbour, he should in certain circumstances be *held to* fulfilment of those intentions, even if on his own account he would now choose to act otherwise. (Of course the circumstances in which N. N. is held to what he now says he wills ought to be reasonably ascertainable by N. N.; we cannot regard as acceptable some convention by which N. N. might find himself entrapped into marriage by using some form of words although he had no intention of making marriage-promises! This situation is alleged to have been possible under Scots law.)

Given this account of what makes promises binding, we can understand what are the circumstances in which promises no longer bind. First, a promise no longer binds if the promisee says to the promisor, 'I don't hold you to that'; he need not feel grateful to a promisor who will not let himself off but on the contrary 'encumbers him with aid' that had been promised. Again, the binding force of promises depends on the benefit, or avoidance of damage, that fulfilment of promises signifies to the promisee; so if circumstances changed and thus the promisee would be damaged or even notably inconvenienced by the promise's being kept, then the promise would no longer bind. Other examples may be cited, e.g. if keeping a promise to A would result in damage to B. It would be wrong to pass on the other side of the road and ignore a stranger who lies bleeding; it would be monstrous in that case for a host to protest, 'You let me down!' because by playing the good Samaritan his guest was late for dinner.

Promise-breaking differs considerably from lying as regards being excusable or justifiable. Lying is always something to feel some shame and sorrow for. Even someone who lies in an emergency for his neighbour's necessity ought to reflect that if he had not accustomed himself to telling little lies under no pressure at all, he would very likely have had it given to him, as it was given to the brave Dutch woman, to devise a way of escape without lying. To be sure, the guilt of lying in such a case is not very grave. But in all sorts of ways the obligation to keep a promise may lapse: the promise may then be broken without scruples or remorse. It would be wrong to pass by on the other side when a stranger lies bleeding sooner than break a dinner engagement, as I just now said.

So much for human promises; I now turn to God's promises. For God to make a promise to a man or a people is not something new occurring in God's mind; an intimation of God's will to man is something that occurs in the human mind. I have already spoken of God's naturally known promises or covenants to man, on which we must rely for our trust in the order of nature; the story of Noah and the rainbow is a naive representation of such a promise that as sure as there are rainbows after rain there will never be a deluge to wipe out the human race. (At a time when theories of the rainbow were familiar on the continent of Europe, many good Protestants in England had a picture of 'The First Rainbow' hanging on the wall; the Scotch poet Campbell wrote a poem protesting against the profane scientific account of rainbows!) I argued that our confidence in the order of nature makes sense only if we, unlike the Aztecs, believe this is a matter of God's promises. Let us be clear that this is not a matter of belief in unchanging natural laws. The Bible speaks of the covenant of day and night; but there is clearly no natural law guaranteeing this as a regular alternation on a particular body in the heavens.

Why can we be certain of God's promises? The thought that God who is truth cannot lie is not a sufficient basis for such trust; for a man, I have argued, can break his promises without necessarily making himself a liar. Indeed, some Catholic authors think God has already reneged on his promises to Israel; they were after all made only conditionally upon good behaviour, and Israel did not observe the condition. To be sure, Jews reading the Prophets would be excusable for not seeing the Divine promises as thus conditional; the Psalms represent the Sun and Moon as witnessing in the skies to God's faithful word, and Jeremiah compares the promises to Israel to the covenant of day and night; as regards forfeiture for misconduct, when Ezekiel repeats the message 'I the LORD have promised it and I will do it', he expressly adds that the promise hangs not upon the good conduct of Israel but upon God's will that his Holy Name shall be honoured, 'which you have profaned among the Gentiles'.

If despite all this we could believe God to have reneged on his promise to the Jews, we could hardly be confident that there is not a similar escape clause ('given good behaviour') in his promises to the Church. Indeed, some theologians clearly show that they take God's promises to the Church no more seriously than his promises to Israel. We often find expressions of fear that God may allow not just his Church but the whole human race to perish, by war or pollution say. This is a sin against faith and hope. I possess a book quoting with approval a theologian who held that 'defensive' measures to protect 'spiritual values' might turn out to involve as a foreseen (but not, God forbid!, intended) effect the total destruction of the human race; this would be a mere *malum physicum*, incommensurably less important than the 'spiritual values' being defended; demurring to this view was characterised as 'eudaemonistic pacifism'! Attitudes like this amount to saying to Almighty God, 'Sorry, we may have to break your commandment against murder of the innocents because we do not trust your promises'; have such writers forgotten the doom denounced in the Apocalypse upon 'the fearful and unbelieving'?

God who is Truth cannot make deceitful promises; but can God ever break promises? We may dismiss the argument (which a writer falsely ascribed to me) that God cannot do wrong and all promise-breaking is wrong; the second premise is false. But the reasons why a man may decide, rightly or wrongly, to break a promise do not apply to God. God is eternal and cannot change his mind. Unforeseen circumstances may make a man no longer wish to keep his word, or he may not see all the implications of a promise; God knows all things and cannot be taken by surprise. A man may make separate promises to two people which turn out incompatible in the event; God cannot be entrapped.

Moreover, God is almighty; he could not try to do something and fail. In a sermon for children, Newman taught them that 'the Father Almighty' stands in the creed to remind us that God's good will towards us cannot outrun his power to fulfil it. When the writer of the Epistle to the Hebrews uses the word 'impossible' in relation to God, he is referring precisely to the necessity of God's keeping his word once given, because he can neither undo the past reception of his promise by men nor find himself powerless to fulfil the promise. God is not 'a man that he should lie', nor 'a son of man that he should repent'.

On Truth as an End

Many people who are far from the Christian religion or any other religious tradition have had a deep devotion to the pursuit of truth. The last three words were used as a title for a book by Willard Quine, whose long academic life the title epitomises. Many writers of the high Victorian rationalist school, among them W. K. Clifford and Sir Frederick Pollock, expressed themselves very strongly about the supreme value of truth. In his *Study of Spinoza* Sir Frederick Pollock wrote, 'Seek the truth, fear not and spare not . . . and the truth itself shall be your reward, a reward beyond length of days or any reckoning of men'.

It is not surprising that men are found to value truth even apart from being enjoined to do so in revelation; men are made for the truth and are not totally depraved; their pursuit of truth attests to the root of good in human nature. But it is all too easy for the natural desire for the truth to be diverted or corrupted. Many philosophers are lured by one or other false light, such as relativism or pragmatism; and from the philosophy schools the corruption is spread around in 'intellectual' chatter. Recently Stephen Stich wrote a whole book to prove that we need not worry if our beliefs largely lack truth; truth, he holds, is a very dubious merit of beliefs as compared with other features they have. And Jane Heal wrote an article maintaining that truth has no intrinsic value, but just a general instrumental value for all sorts of purposes: thus, a burglar will want a *true* plan of where the gas, water, and electricity conduits run under the bank he means to tunnel into. This illustrates the way truth is devalued. Philosophers, Heal says, need not regard themselves as servants of some Goddess of Truth: whom they do serve is another question, I might retort.

Christians who take the Bible seriously might be expected to take truth as a goal seriously; but one would not think so if reliance were placed on some who claim the Christian name. Acceptance of dogmas as true is of course only part of the Christian way; but just as a man would collapse without a skeleton, religion without dogma must become a shapeless jelly of emotion. At a meeting I once attended, there was a general consensus that Christian faith is not belief *that* so-and-so but belief *in* a Person: there was no creed in the early Church and Christian believers nowadays would be better off without any creeds. At last I was moved to protest that there is a creed often cited in the New Testament: 'Jesus is the Christ, the Son of God'. Indeed, there is a report of assent to the creed as a preliminary to baptism (Acts 8.37). The creed is short and simple, but just as much of a proposition and just as controversial as any longer creed used by Christians. Thomas Hobbes in *Leviathan* expresses the opinion that this supplies a necessary and sufficient foundation of Christian faith, on which all Christians can agree, and which will stand where superstructures men have reared upon it are consumed in the fire of judgment. Nowadays men are found who *reject* Hobbes's minimum creed altogether and yet claim the Christian name: men who regard the title 'Christ' or 'Messiah' as expressing an outdated dream of Hebrew nationalists, that their King should be one to rule all the world in righteousness; men who deprecate the idea of an only Son of God, and wish to wean us from employing it.

Again, some Christians seem to believe that religious authority may be claimed, and the claim not be self-stultifying, even if over an interval of time contradictory dogmas are taught. For very long it was taught both in the Jewish and in the Christian community that idolatrous cults are seeking after 'lies' or 'lying vanities'; Protestants taught this as strongly as Catholics. Nowadays there are Christians who regard such cults as ways of salvation for the tribes of men among whom they are traditional; this new teaching plainly contradicts the older teaching. Some allege ecclesial authority for the new teaching; I really need not go into that. As I once wrote in Hobbesian style: Bishops come and bishops go; and one Pope passeth, another cometh; ay, Heaven and Earth shall pass; but from the Law of Contradiction not one tittle shall ever pass; for it is the eternal Law of God.

To understand how truth can and should be man's end, we must approach by way of Pilate's question 'What is truth?'; and we must begin by ruling out false answers. At the outset it is possible to lay down a criterion immediately fatal to many false theories. There is no difference between believing (or supposing) something and believing (or supposing) that that very thing is true; any theory which would imply that there is a difference may be dismissed without much ado. For example, any variety of a pragmatist theory, whereby a belief's being true would consist in something like its being advantageous to hold it, must fail because 'It is true that the Earth is round' is tantamount to 'The Earth is round', and therefore cannot be tantamount to anything like 'It is advantageous to believe that the Earth is round'.

Again, 'true' cannot be regarded as a sign of the speaker's assenting to or confirming or corroborating what is put forward; for 'true' can occur without change of sense in a clause within an asserted proposition which is not itself meant to be asserted. A performative expression loses its force when it occurs in an 'if' clause: one who says 'If I assert that so-and-so' is not thereby asserting that so-and-so, and the like holds for 'If I confirm' or 'If I corroborate'. But on the contrary, 'If it is true that the Earth is flat . . .' *is* tantamount to 'If the Earth is flat', just as the two sentences would be equivalent if asserted.

We have to take more seriously the view that a belief or supposition is made true by its correspondence to a fact: a theory vigorously defended by Bertrand Russell and McTaggart. The theory of truth in Wittgenstein's *Tractatus* might also be called a correspondence theory: a proposition, whether in thought or in some sense-perceptible medium, is regarded as a fact, and truth would be the correspondence of this fact to the fact it purports to convey. But the question then arises whether this correspondence can *itself* be stated in a sentence that would satisfy the severe restraints of Wittgenstein's theory about what can be significantly said. Perhaps it would be necessary to invoke here the *Tractatus* doctrine of what *shows*, or comes out, in language but cannot be *stated* in a logically correct way. This doctrine has more to be said for it than appears at first glance; but it would take us too far to develop an exposition of its grounds.

A serious difficulty for the correspondence theory is the matter of identity for facts. As Quine has said, no entity without identity; but we have no clear and firm criterion for when two true beliefs or utterances correspond to the same fact.

A more decisive objection, to my mind, is that on the correspondence theory there would not be a single fact making true simultaneously a belief that so-and-so and a belief that this very belief is true. Suppose the former belief is true by correspondence to the fact F, then the latter belief will be true, not by correspondence to F, but by correspondence to a different fact, namely the fact of the former belief's correspondence to the fact F. This cannot be right.

An account of truth given by Aquinas (Ia q. 16 art. 3) may be confused with the correspondence theory, but to my mind is a different theory, not open to these objections. The key notion in Aquinas's theory is that a form may occur with *natural* or *intentional* existence: green, let us say, exists with natural existence in a green field, and with intentional existence in a mind thinking of green. This notion needs careful exposition, and I have tried to expound it;[3] I think it is strongly defensible. On the basis of this notion, Aquinas explains the truth of a judgment (say) that A is round as an agreement in form, conformity, between the judging mind and the object A. Roundness exists in A with natural exis-

3. In my essays 'Form and Existence' in *God and the Soul* (Routledge, 1969) and 'Aquinas' in *Three Philosophers* (Blackwell, 1961).

tence, in the judging mind with intentional existence, and in actualising an occurrence of this form, the judging mind is *ipso facto* aware of doing so, which would explain the identity between the *making* of this judgment and *the judgment that* this judgment is in conformity with the object A judged to be round. This escapes the difficulty about two different judgments, differently corresponding to facts, that I raised against the correspondence theory. And I do not find in Aquinas, as I do in some of his followers, what Elizabeth Anscombe has called the Fallacy of being Guided by the Truth: people concentrating on examples of judgments or propositions that are (or are supposed to be) true, and forgetting that the account given would break down for false judgments or propositions. On Aquinas's account, the occurrence of the form *roundness* brought about by a judging mind would still *present itself as* conformity with A, which is being judged to be round, even if A is in fact *not* round.

However, in the passage I cite Aquinas applies this theory only to judgments of very simple structure. Obviously judgments come in indefinitely many and indefinitely complex forms. It is not even clear to me how Aquinas's account would apply to negative judgments. For Aquinas the same form is actualised, brought into activity, whether a man judges that A *is* round or that A *is not* round: *eadem est scientia oppositorum*. But if I truly judge that A is not round, this truth cannot consist in a conformity in respect of roundness between A and my mind, for roundness then does *not* exist in A. I do not know of anybody who has developed Aquinas's theory beyond his own simple sort of examples; so I find myself at a halt as regards Aquinas's account of truth.

I think a clue to the labyrinth is given us by considering what logicians call *duality*. For any ordinary language there is conceivable an alternative language; if we call the alternative to English 'Unglish', a sentence of Unglish would look or sound just like an English sentence, but would have to be *translated* into English by *the contradictory opposite* of that equiform English sentence. Such an alternative language could be actually used, e.g. as a military code: allegedly

St. Joan of Arc used such a code, marking with a cross the messages that were to be read in the reverse sense. For translating between English and Unglish we need not confine ourselves to the rule for translating whole sentences; we could easily construct a dictionary pairing each English word with an Unglish word, and conversely; and using the dictionary we should end up with a pair of sentences equiform to a pair of contradictories in English (or, equally, in Unglish). English 'It is raining' and 'It is not raining' would respectively be translated into Unglish as 'It is not raining' and 'It is raining', which would be contradictories in Unglish as they are in English; the Unglish for 'not' is accordingly 'not'. (Commenting on the brief account of this matter in Wittgenstein's *Tractatus,* Max Black and Michael Dummett both make the same extraordinary mistake: they say what would amount, in terms of English and Unglish, to saying that 'not' in English corresponds to the *absence* of 'not' from Unglish sentences! Accordingly they find what Wittgenstein says 'wrong' or 'incoherent'.)

In modern logical jargon a pair of expressions related as expressions paired in an English-Unglish dictionary would be are said to be *dual* to one another. An expression may be self-dual: we have just seen that 'not' is unchanged by English-Unglish translation, and negation is indeed self-dual. But self-dual expressions do not all belong to the same category. Names are always self-dual, for if we switch from English to Unglish we shall still need to mention by name the same objects. Predicables on the other hand, and relation terms, are dual not each to itself but each to its contradictory. Quantifiers have duals neither as names have nor as predicables have: substituting for each expression its dual, we obtain from 'Tibbles/fears/some dog' the contradictory sentence 'Tibbles/does not fear/any dog'—'no dog' and 'some dog' would each give a wrong result.

What then happens with sentences? A free-standing sentence is dual to its logical contradictory. I hold that a coherent duality theory can be worked out only if we also take sentences that occur as integral parts of longer ones to be dual to their contradictories. To give a simple example, we obtain the correct dual to the sentence:

(A) Tibbles fears some dog and Bonzo does not fear any cat

if we replace each contained sentence by its dual and take 'or' as dual to 'and':

(B) Tibbles does not fear any dog or Bonzo fears some cat.

The names 'Tibbles' and 'Bonzo' are self-dual; 'some dog' and 'any dog', 'some cat' and 'any cat' are pairs of duals; the relative terms 'fears' and 'does not fear' are mutually dual. This gives us the duals of the two clauses joined by 'and' in (A); when we join these by 'or', which is dual to 'and', we get (B), which is the contradictory of (A) and is dual to (A). Sentences that occur in an indirect-speech construction raise some complications; but I think it can be worked out that for duality applied to the whole sentence these too must be replaced by their negations.

Plato in the *Sophist* gives us the fundamental insight that the unit of discourse in a *logos* is composed of signs differing in category: two names or two predicables (*rhēmata*) give us nothing intelligible, but rather gibberish like 'Tibbles Bonzo' or 'walks runs'. Aristotle in *De Interpretatione* focuses this more sharply: for truth or falsehood we need that a *logos* shall be *apophantikos*. This is the same point as Frege makes in his *Begriffsschrift*, when he says that the content of an expression must be *beurteilbar*, content of a possible judgment (*Urteil*). 'Priam's wooden house' and 'Priam had a wooden house' are each examples of *logos*, coherent discourse, but only the second is capable of truth or falsehood. And for a *logos* of this kind the principle holds that, whether freestanding or in context, it is dual to its negation. This sharply distinguishes a *logos apophantikos* from one that is not *apophantikos*. There are well-known puzzles about how negation attaches to a sentence containing such a *logos* as

'Priam's wooden house'; are the sentences 'Priam's wooden house burned' and 'Priam's wooden house did not burn' a pair of contradictories? Suppose Priam never had a wooden house! But no such puzzles arise over embedding the *logos apophantikos* 'Priam had a wooden house' in a longer sentence. And however we work out dualities in detail for sentences containing 'Priam's wooden house', we clearly get nowhere if we try to replace this phrase by something taken to be its negation; whereas dualities always work out correctly if we make an embedded sentence dual to its negation, and replace duals by duals in the surrounding framework that forms the completion of a longer sentence.

Frege unfortunately came round to assimilating an embedded sentence to a *logos* like 'Priam's wooden house'. In his mature work he held that an embedded sentence relates to an object, as 'Priam's wooden house' purports to relate to a house. In non-intentional contexts, Frege's *gewöhnliche Rede,* a sentence would relate to one of two special objects, the True and the False: everyone capable of forming judgments implicitly recognises these objects. (The word for 'recognise' is here *anerkennen;* it is used, e.g., for one government's diplomatically recognising another; quite a different sense would be given by *wiedererkennen,* used, e.g., for recognising an old acquaintance.) In intentional contexts, on the other hand, a sentence embedded would relate to a thought (*Gedanke*), regarded as something different thinkers may share.

Criticism of Frege's mature view is often superficial, put across by people whose logical insight is incomparably dimmer than his. Anyone who holds, for example, that a sentence is neither true nor false unless it is used to make an assertion is unworthy even to read Frege, let alone criticise him. But it was a serious error on Frege's part to abandon his previous insight that sometimes the content of a *logos* is distinctively *beurteilbar,* and that such a *logos* is not a complex designation.

The semantic role of an indicative sentence is utterly different from that of a name or again of a complex designation. Metaphorically, we may say that sentences *point,* and may be reversed, so that they point the wrong way if originally they pointed the right way and vice versa. On the Rome road where I see a signpost TO ROME, it makes all the difference which way it points; turn the signpost round, and the value of the information it gives is reversed. If the signpost has two arms pointing opposite ways and each says TO ROME, it is useless, and remains useless if it is turned right round. As readers of the *Tractatus* should easily see, such a signpost corresponds to the sentences Wittgenstein called *sinnlos*, tautologies and contradictions; they lack orientation.

Now to what are sentences oriented? Are all true sentences oriented to a single reality, or are different ones oriented to different realities?

First of all: is there just one way for a sentence to be true, or several different ways? Do true statements fall apart into logically insulated domains, which cannot be joined by any inferential bridge? I cannot here go into the complicated argument in favour of an insulated-domains theory of truth; I have argued elsewhere that they are sophistical, often in a pretty obvious way.[4] One direct argument for the contrary view is that the logic of truth-functions, though it does not cover as much of the field of logic as its creator thought, is clearly valid so far as it goes; and in it no distinction is made between different ways of being true. Again, any truth may be used in the figure of speech called asseveration: *so-and-so* as sure as *such-and-such*. 'He's guilty, as sure as I'm standing here; he'll be found guilty, as sure as eggs are eggs; and then he'll hang, as sure as God made little apples.' For this rhetorical device to work, the hearer must accept each of the clauses following 'as sure as' as true beyond cavil; it does not matter that one clause is observational in content, a second logical, and the third theological.

If all true sentences are oriented alike, does it follow that they all have the same sense? Not at all, as a further development of my parable about roads to Rome ought to make clear. Imagine a planet on which there is just one land mass

4. See my essay 'Kinds of Statement' in the book of essays *Intention and Intentionality*, presented to Elizabeth Anscombe, Harvester, Brighton, 1979.

and all the roads on this island lead to a city called Rome. Signposts on the roads would be reliable or misleading according as they pointed towards Rome or away from Rome. Two signposts both pointing towards Rome (or both away from Rome) would correspond to two sentences both true (or both false) but differing in sense. We may avoid the logical Manichaeanism of Frege's two objects, the True and the False, by holding that judgment and sentences *purport* to be oriented to just one object, the True, though they may be wrongly oriented.

People may still suspect that a sentence's being oriented towards the True can be distinguished only verbally from Frege's making true sentences into complex designations of the True. But insertion of a false sentence in the framework of a larger sentence in place of a true sentence is utterly different from inserting an empty complex designation in place of a non-empty one. Of course when I speak of orientation and pointing I am using metaphor in the hope that readers may catch on; but the difference between *being oriented towards* and *designating* is implicitly acknowledged by anyone who can recognise that characteristic of some *logos* for which Aristotle used the epithet '*apophantikos*', and Frege in *Begriffsschrift* spoke of a content that is *beurteilbar;* and in recognising this feature people implicitly recognise that to which true discourse is oriented. As Frege, to my mind wrongly, said that in judging everybody recognises (*anerkennt*) *two* entities, the True and the False, I say that in all judging there is an endeavour to orient the mind to *one* Thing, Truth.

I was once asked to consider the possibility of a race oriented to Falsehood as we seek to be oriented to Truth. I replied by quoting W. S. Gilbert's *Bab Ballads* (from 'My Dream'):

> But worst of all these social twirls
> The girls are boys—the boys are girls! . . .
> To one who to tradition clings,
> This seems an awkward state of things;
> But if to think it out you try,
> It doesn't really signify.

All that the supposed different style of thought and language amounts to is that the word 'Falsehood' could be arbitrarily used in place of the word 'Truth'.

My thoughts about the unique Thing that is Truth, to which all truths thought and uttered are oriented, however lowly the subject-matter, have of course been guided by Anselm's *De Veritate*. Is this then a proof of the existence of God? Not yet. I have not done anything to show that Truth is an actuality, *ist wirklich* in Frege's sense: nothing to show that Truth is living or that Truth *causes* true saying and thinking. So far as we have gone, it may be quite absurd to ascribe to Truth *will*, and efficient voluntary causality; Truth, like numbers, might lie outside the realm of activity and passivity; to think of Truth in causal terms might be on a level with the gross superstition of thinking 7 brings good luck and 13 bad luck.

However, if on other grounds we can show that there is a God, then we can hardly conceive of his needing to orient his thought to a Truth distinct from himself. That would be like the myth in Plato's *Timaeus* of a workman following a pattern. Rather, God must be conceived as constituting all truth: necessary truth by his nature, contingent truth by his operative or permissive will. God will then know all truth in this way, not by contemplating as distinct from himself something that is the measure of how far his thoughts are true. Only so could God be worthy of our total worship: not so the godling of some modern theologians, who is so far from constituting truth that his thoughts, his 'beliefs', do not invariably attain it! (Talk of God's beliefs is of course a danger signal; one who so speaks of God can hardly command serious attention. We may ascribe knowledge to God, but not belief. Belief is a disposition to think and talk and act in various ways; God, who is eternal, can have no dispositions, let alone variable ones. But knowledge is a capacity; the question often posed by philosophers ever since Plato's *Theaetetus* of what needs to be added to justified true belief to make it into knowledge is an unanswerable one, for *no* disposition can be made into a capacity by its satisfying some criterion. God's knowledge of things in the world is indistinguishable from his power and will: *scientia Dei causa rerum;* as we shall see, the best analogy is with our knowledge and control of our own voluntary acts.)

Recognising the identity of God and Truth has the most profound importance for our spiritual and practical life. Even before such recognition we could reject out of hand any religion whose god or gods were supposed capable of deceit. We can indeed trust men up to a point even though we know they sometimes lie; but how could the rationale of such trust be extended to gods whose dwelling is not with men? But if God is Truth and we can recognise a message as his, then this will command our absolute trust:

> Truth himself speaks truly, or there's nothing true.

I have encountered the objection that it is just bad grammar to say Truth speaks truly or speaks at all. One might as well call it bad grammar to say 'Her Majesty opened the bazaar'. I have said why I think we must acknowledge a unique Thing as the focus of all true thinking and saying. It is suitable to call this 'Truth'; a fig for grammar!

For ourselves, we can firmly impress on our minds the evil of self-deception and the harmfulness of a habit of petty lies: the first distorts the mirror, the second blurs it with a multitude of little scratches. Above all we must not lie in the supposed interests of religion; that is a gross insult to the God of Truth on the pretext of doing him service. 'God has no need of our lies'.[5]

5. For a splendid development of this thought cf. Aquinas's commentary on Job.

6 Prophecy

There are many remarkable stories of fulfilled prophecy. Given human mendacity one ought perhaps to regard any remarkable story with initial scepticism. Dr. Johnson has been thought eccentric for so regarding the first intelligence of the Lisbon earthquake; but we may well regard his reaction otherwise if we recall that more than a century later, when means of long-distance communication were incomparably better, a wholly false story reached Europe that there had been a general massacre by the Chinese of all the diplomatic missions in Peking; and a commemorative service in London was called off only just in time. On the other hand, many remarkable stories initially disbelieved have eventually turned out true.

Stories of fulfilled prophecy have been often regarded as calling for a special degree of scepticism; it is not easy to see why. People will offhandedly say that such stories, if true, would involve miracles and that there are well-known reasons for doubting stories of miracles. Happily I need go no further along this track. There is obviously nothing miraculous about somebody's claiming prophetic powers and uttering startling prophecies: this happens all the time. Men have an appetite to know the future more rapidly and reliably than from scientific forecasts; the demand is met by a supply of prophecies. Of the remarkable prophecies that are made, just a few are fulfilled; it is then natural that these should be remembered and the false ones mostly forgotten. It is quite irrational to correct the record that a prophecy was made and then was fulfilled by saying that the story of the prophecy's having been delivered *must* have been composed after the event.

In the reign of King Zedekiah Jerusalem was besieged by the Babylonians. Two rival prophets appeared, Jeremiah and Hananiah. Hananiah (called in the *Vulgate* Ananias!) prophesied that the yoke of Babylon would be broken and the siege be raised; Jeremiah on the contrary prophesied that the city would fall and the people be carried away into captivity, though they would eventually return. Jeremiah was severely punished for spreading alarm and despondency, as a similar prophet of evil would have been punished in England in the last German war. Jeremiah's was the prophecy that came true; and there is no need to suppose that his prophecy was not actually delivered but fudged up after the event. An uncommitted sceptic could simply say: 'Both prophets went in for a bit of wishful thinking about eventual good fortune for the Jews, but Jeremiah's gloomier prophecy was better fulfilled than Hananiah's, so it was better remembered. One or other of the two prophets was bound to be pretty well right. If the event had gone the other way, the Jews would have preserved a Book of Hananiah instead of a Book of Jeremiah.'

There are many cases in which a prophecy has been very exactly fulfilled and there is absolutely no question of error as to the giving of the prophecy or the date when it was given. I culled one very striking example from Lord Byron's notes to his tragedy *Marino Faliero*. The poet Alamanni, who died in 1560, wrote thus about Venice: 'If you do not change your thoughts, your freedom, even now fleeting away, will not count one century after its thousandth year.' The first Doge of Venice was installed in 697: 'a thousand years and one century' brings us to 1797; in 1796 the Republic of Venice fell, never to rise again, to the French revolutionary forces.

Savonarola uttered many prophecies, some fulfilled, others unfulfilled. One of his prophecies was that the wickedness of successive Popes and their courts would be punished by a sack of Rome at the hands of the ungodly. Savonarola died in 1498. In the year 1527 his prophecy was very amply fulfilled; the mercenaries of Charles V sacked Rome with an orgy of murder, rapes, and sacrilege. Many people who had heard Savonarola's prophecy lived to see it fulfilled. It is an extraordinary fact that in Philip Hughes's *History of the Church* (vol. 3, p. 494) this prophecy of the sack of Rome is given as an example of false prophecy. I can only conjecture that Hughes suffered a momentary lapse of care through living in an academic milieu oddly hostile to stories of fulfilled prophecy.

Christ by the record did foretell the fall and desolation of Jerusalem; there are of course those who infer that he never said anything of the sort and the story was composed only after the event. This inference appears irrational. Without any inspiration a patriotic Jew might expect that Jerusalem would eventually be sacked by the Romans, just as a patriotic Pole in the 1930s may have written a poem lamenting over the foreseen desolation of Warsaw. Rome's patience, like Hitler's, was eventually going to be exhausted; and then given a determination to dominate on one side and a long tradition of heroic resistance on the other side, it was easy to foresee the sequel in both cases—a capital city laid in ruins with a huge loss of life. It is even more frivolous to doubt the account of Christ's foretelling his own crucifixion. In history many a leader has foreseen that he would die by public execution; and since the Sanhedrin under Roman rule had lost the right of capital punishment, it would become necessary, if Christ's execution was once determined upon, to prevail on the violent but weak Pilate to inflict the Roman death penalty for treason.

The same records tell us that Christ foretold not only his crucifixion but his rising again on the third day. Here again scepticism of the record is unwarranted. If Christ actually had the power to lay his life down and take it up again, clearly he could foretell his resurrection. If on the other hand he had no such power, all the same he may have claimed it and have led his disciples to believe his claim. Many charismatic leaders have induced their followers to believe that they would never die, and belief has persisted after they did die: this happened with the Druzes' leader

Hakim and with the twentieth-century black cult figure who called himself 'Father Divine'. To a sceptic with no axe to grind it must appear a more likely story that Christ's disciples likewise believed in his survival in face of the grim facts because he had foretold it, than that they invented and wrote up the story *both* of the prophecy *and* of the resurrection when *neither* had actually occurred.

Sometimes a story gets doubted for fear lest if accepted it would give men a case of fulfilled prophecy, even when there is in fact no prophecy reported at all. In both Kings and Chronicles there is a record of a prayer delivered by King Solomon for the dedication of the Temple. There is really no reason why such a prayer on such an august occasion should not have been carefully composed, committed to writing, and carefully preserved. We are told, however, that it must have been composed after the Babylonian exile: otherwise we'll have to believe in a prophecy of that exile long before, and that is quite out of the question! Well, let us see what the prayer actually says. There is no mention of Babylon; Solomon's prayer does not even purport to prophecy that the people of Israel *will* be carried away into captivity for their sins. It is all a matter of *if:* Solomon prays that *if* captivity should in the future come upon the people for their sins, and then in the land of exile they repent, *then* may the LORD have mercy upon them and restore them to the Promised Land. The text contains a series of prayers introduced with an *if: if* people give sworn testimony in the sanctuary against one another; *if* the Israelites go forth to war; *if* there is a drought; *if* there is 'pestilence, blasting, mildew, locust, or caterpillar'; *if* a Gentile comes to the Temple with a petition. None of these petitions has the form of a prophecy; why then should another petition in the style be so taken? Given the geopolitical situation of Israel in his time, with powerful aggressive empires on either side, Solomon must surely have seen invasion and captivity as likely contingencies, no less than drought or pestilence or insect plagues; if he did not, his traditional reputation for wisdom has been sadly exaggerated.

Here then we have a record that does not even purport to be a record of a prophecy later fulfilled but which is rejected as true history through a blind fear of fulfilled prophecy's being accepted as an actual event. There are many other examples of bad arguments for the dating of documents.

Suppose I am told that a Psalm ascribed to King David cannot have been written by him because Hebrew of this style did not exist in King David's day. How much, I ask, do we actually know about the history of the Hebrew language? On the face of it we know incomparably less than we do of the history of the English language, including Lallans [Lowland Scottish]. Now in my hymnbook I find a hymn ascribed to William Dunbar (1460–1520), yet its language is clearly not Dunbar's Lallans but modern English with a few bogus archaisms like 'soulés' for 'souls' and 'comen' for the participle 'come'. Ought a reader then to reject the authorship of Dunbar? If he does he is dead wrong. Considering this case, we cannot exclude the possibility that over centuries of liturgical use a Psalm may have undergone the same modernisation that leads from Dunbar's text to the hymn as now printed.

Inferences from alleged allusions to events or personages of some epoch are likely to be equally shaky. Consider the following lines from Pope's *Dunciad*:

> The Goddess then o'er his anointed head
> With mystic words the sacred opium shed,
> And lo! her bird, a monster of a fowl,
> Something betwixt a Heidegger and owl,
> Perch'd on his crown . . .

The Goddess Dullness is here crowning her favourite, the Poet Laureate of the time. Pope adds this bogus learned note on 'Heidegger':

> A strange bird from Switzerland, and not (as some have supposed) the name of an eminent person who was a man of parts.

Enemies of the philosopher Martin Heidegger might well regard him as a paradigm of dullness, and he spent the last years of his life in Switzerland. So scholars of the year 3000 might conclude that such enemies corrupted the text of Pope in order to have a bash at Heidegger; but the inference would be wrong. (In fact, the man Pope here satirises was a Calvinist Swiss theologian; Pope greatly disliked Calvinism.)

Again, a passage in *Alice in Wonderland* might easily be taken as an allusion to the Dreyfus affair. When the Knave is on trial for stealing tarts, a document introduced as evidence in the case is a copy of verses. The Knave objects that he didn't write the verses:

> 'Are they in the prisoner's handwriting?' asked another of the jurymen.
>
> 'No, they're not,' said the White Rabbit, 'and that's the queerest thing about it.' (The jury all looked puzzled.)
>
> 'He must have imitated somebody else's hand,' said the King. (The jury all brightened up again.)

Now in the Dreyfus trial, the famous *bordereau*, which was supposed to relate to a betrayal by Dreyfus of military secrets to the Germans, was not in Dreyfus's handwriting, but expert witnesses were called to show it had in fact been written by Dreyfus imitating somebody else's handwriting. 'That proves his guilt,' the Queen of Hearts might say. The allusion might appear obvious; but in fact *Alice* was published in 1865 and the Dreyfus affair began in 1894. However, let us imagine that by A.D. 3000 men's knowledge of European and English history in the nineteenth century was in much the same state as our present knowledge of the history of Republican Rome: then scholars of that time might well argue that Lewis Carroll must have been alluding (for adult readers' understanding) to the Dreyfus affair. But like the inference about Pope's mention of Heidegger, their conclusion would be wrong.

What makes people uncritically accept weak arguments for redating of documents is this strange reluctance to accept fulfilled prophecy as a fact. There are even people around who insinuate, or even boldly state, that the prophets of ancient Israel were not concerned with predicting the future fate of Israel and the coming of Messiah: they were rather concerned just with social evils and their remedies. I remember, though I cannot attribute, the epigram 'Prophets were not *fore*tellers but *forth*tellers.' I shall not waste space on citing texts to refute this: the most casual reading of the Old Testament shows it is a total falsehood.

Men seek for prophets who can impart to them direct knowledge of the future, as opposed to prognostication by some scientific technique: is this indeed a possibility? Are the fulfilled unreasoned predictions no better than lucky guesses? As Wittgenstein pointed out, there is one way in which any normal human adult has knowledge of the future: each knows what he or she is immediately going to do. Human knowledge and human power here coincide.

Like other claims to knowledge, this claim to knowledge is open to philosophical attack. 'How can you know the way you will move your hand, when you know so little about the operation of your brain and nerves and muscles? Anyhow, how do you know you will not be struck with paralysis next moment?' The like sceptical ploy could be used to disallow claims to know anything by memory or by sense-observation. But whereas in both of these cognitive areas errors are familiar and of daily occurrence, it is excessively rare for somebody to announce a movement he is going to make and die or suffer a stroke next moment; so if we allow that despite errors much genuine knowledge is available by memory and by sense-perception, *a fortiori* we must allow that in general each of us has knowledge of immediately future actions.

In this little sphere a man is lord by knowledge and power; and it is in this way, as Aquinas says at the beginning of the *Summa Theologica* Ia IIae, that each bears the image of God who is Lord of all the world. For God's knowledge of the world is not a reflection of it, however perfect. Anaxagoras already said about *Nous*, Mind, that being unmixed with the things in the world and unaltered by the processes of the world, *Nous* has power over all things and knowledge of all that happens, past, present, or future. God's mind does not conform to the world he has knowledge of: God has perfect knowledge because everything in the world happens by his effective or permissive will.

Why this qualification 'effective or permissive'? A Victorian rationalist, Sir Leslie Stephen, wrote with derisive intent about there being a Great First Cause and ever so many little first causes. But this mockery in fact echoes what Doctors of the Church have said. St. Anselm chose to discuss the very first sin, the sin of Lucifer, because this is a pure case: Lucifer is a pure spirit, so no complex considerations arise about the relations of intellect and will to sense-based emotion; and no other sinner need be considered either. The work may be compared to a physicist's working out the thermodynamics of a monatomic gas. After a long careful discussion he answers the question *why* Lucifer chose to sin by saying: 'He willed because he willed. For the will had no other cause to push it or attract it; it was so to say its own cause and its own effect'.

The world is a huge checkerboard, not where 'Destiny with men for pieces plays', but where one Grand Master sees and controls the whole board and many limited players also make moves. Each of the human players sees only a little of the board and has limited powers of moving; some of them will wish to cooperate with the Grand Master's plans, others will try to thwart him; but in the end the master strategy of the Grand Master must prevail. He can take into account all possible moves of the other players; he cannot be thwarted or surprised or forced to improvise.

How then does prophecy fit into the picture of things? The Grand Master who has everything under control can safely announce, if he wishes, that he will achieve such-and-such a position on the board: 'on that square I will promote my pawn to Queen, which will mean a decisive defeat of my Archenemy in the end game'. Such is the representation of divine prophecy in my parable. How then are we to distinguish divine prophecies from lucky guesses? By their occurring as parts of a corpus of revelation, or being uttered by minds that are governed by such a revelation—given that the revelation itself is true. How a corpus of revelation may be recognised as true it would take me too long to discuss.

Obviously in this presentation I am not claiming to *prove* that my parable represents the real way things are; but I hope that in the earlier part of this chapter I have done something to remove initial prejudices and refute fallacious ways of argument that would keep people from seeing my position as even possibly reasonable.

It would not be candid to conceal what I see as the greatest difficulty against my view. By the scriptural record, among events prophesied are specific sins by individuals, e.g. Judas's betrayal, and Peter's threefold denial, of his Master. If these were divine prophecies, God must have foreseen these sinful acts; if his knowledge of events in this world is practical like one's knowledge of intended movements, not contemplative, then how is it not true that God wilfully brought about these sinful acts? Bad enough: and if, as in these instances, the sins involve deception, the God who is Truth is involved in deception, and then where are we?

We who are sinners cannot hope for clear understanding of sin. At a party everybody may be a bit drunk and some very drunk indeed; those who are less drunk may see that some who are more drunk are behaving in a dreadfully foolish and destructive fashion, but will not fully grasp what is being done wrong and how to check or remedy the mischief. They certainly will not then understand better by drinking more liquor themselves; here the maxim 'Like is known by like' clearly does not apply!

We cannot really understand sin; and I want to bar the way to a familiar solution, the so-called Free Will Defence. This would work only if the exercise of free will made sin inevitable. But free choices need not be between good and bad, right and wrong; any one of us often has a free choice between two goods, where it would not be wrong to choose either. So the Free Will Defence utterly fails.

Perhaps we can see in part why sin in general should be allowed: certain excellencies of character could not exist without sin as their foil, e.g. the courage of martyrs. But this does not answer the question why specific sins of individuals have been not only permitted but even foreseen and prophetically foretold, by God whose knowledge is not contemplative but practical. Even here, however dim our insight, we are perhaps allowed a little light. Caiaphas is described as saying 'It is expedient that one man should die and the whole people not perish'. As proceeding from the will of Caiaphas, this utterance was a piece of abandoned wickedness; as proceeding from God's will, the Evangelist tells us, the same utterance was an inspired prophecy of mankind's being saved from utter ruin by the death of one Man. Any event has multiple descriptions; even for a single person the same event may be willed under one description and not willed under another; still more may the same event be differently willed by a human will and by God's will. We may hope that 'where all is made clear', as King Alfred said, we shall see how the wickedest human acts have aspects under which they are positively willed by God.

This view of the future, and of God's knowledge and control and partial revelation of the future, is sharply opposed to a view widespread in our day: that there is simply *the* future, unique and fully determinate, that 'changing the future' is a self-contradictory concept: 'It was going to happen but it didn't happen' is on this view a violently improper use of language and can in no circumstances be literally true.

This view is again and again upheld by using a panoply of physicists' terms. We get a lot of talk about a plurality of frames of reference, and two-dimensional diagrams are drawn to represent a four-dimensional continuum, and these contain projections of light-cones and world-lines of observers. The decisive move is made quite unobtrusively: some future contingent event, let us say N. N.'s death, is marked as a *single* cross somewhere in the diagram; then, appealing to relativity theory, the author tells us that this single event is a past event for one observer but a future event for another. . . . Once that move is conceded, the game is clearly lost for those who, like me, wish to maintain the inescapable alternativeness of the future. All the diagrams, all the terminology of physical theory, just play the role of conjuror's patter, even if the author is deceiving himself as well as his readers.

The view of a future in which there is no alternativeness is so manifestly contrary to what is constantly presupposed in our practical thinking that people rarely uphold it without falling into plain inconsistency. For example, one and the same man will both maintain the definiteness of the future when engaged in metaphysics, and on switching over to ethics urge us to act so as to bring about the best alternative open to us. That is the way the inconsistency appears in secularist writers; but we find a similar inconsistency in Thomist writers like Maritain and Garrigou-Lagrange, who affirm *both* that there may be two contingent future propositions of which neither is 'determinately' true (although the disjunction of the two definitely is true), *and* that God does know ('determinately', I suppose) which is in fact true.

Of course some futures are certain; but that does not mean that as regards these there is no alternativeness. A prediction, say of a death, cannot specify in detail all features of the event; so far as the prediction goes, then, the event may happen one way or another. There is no reason to believe that 'in principle' this indeterminacy could be removed. The Grand Master's knowledge of which I spoke would not exclude variable lines of play by human players, but would be adequate to cope with any moves they made.

If the determinacy of the future is taken seriously, and the inconsistency I have just discussed is avoided, there is no reason to confine knowledge of this fixed future to God alone. This is of course psychologically easy: *rhetorically* 'Nobody knows' and 'God only knows' are almost equivalent. But people have also fancied that in rare circumstances human beings too may glimpse the fixed future. John Buchan wrote an excellent novel on this theme, *The Gap in the Curtain.* Some of the seers wish to avoid the future they glimpse, others to exploit their knowledge; nobody profits by this stolen glimpse. The event always happens as foreseen, whatever the seers do. Such fiction is harmless so long as it is not taken seriously; otherwise that way madness lies.

In this life we have real choices. Churchill used the fine phrase 'the hinge of fate'. In reality the hinge of fate is human free will. On that hinge a door may turn, to open and then perhaps to shut for ever; and it may be a door to Heaven or to Hell.

The Goodness of God

When talking of goodness, a Christian will think in the first place of God's goodness, which is the source of all goodness. And in a sense this is the simplest way to begin. St. Thomas plans the discussion of goodness in general and of God's goodness in the context of discussing God's simplicity; this is not the easiest part of the *Summa Theologica* to read and digest, in fact somebody remarked, 'If this is God's simplicity, what can his complexity be like?' The general aim in this part of the *Summa* is to show how God's nature excludes certain kinds of inner complexity that are found in creatures: this is the very reason why the discussion of God's goodness is simpler than the discussion of a creature's goodness.

To decide that a man is really good requires all sorts of complicated evaluations. What then do we mean by the goodness of God?

> Why do you call me good? Nobody is good but God.

On the face of it Christ's words are perplexing: it is not true that the epithet 'good' applies only to God. It is all right to speak of a good man: or indeed of a good cabbage or a good argument. For each thing we call good there is some feature needed to make it good of its kind; for different kinds of things the feature that makes one of a kind good is quite different. What is it, then, about God that constitutes his goodness? And why is nothing else good in the way God is good?

A man or a cabbage or an argument is good by measuring up to some standard. God is not good like that; God is good just by *being there*. Nothing else and nobody else is good that way. Even the human nature of Christ, which was all that the rich young man could discern in him, is not good that way; rather, Christ's human goodness involves his possession of virtues and graces beyond the mere possession of a human nature.

People often think of *being God* as some sort of role or position. This comes out in Martin Elginbrod's epitaph:

> Here lie I, Martin Elginbrod:
> Have mercy on my soul, Lord God,
> As I would do, were I Lord God,
> And ye were Martin Elginbrod.

And similarly La Hire, one of St. Joan's commanders, is said to have prayed before a battle: 'Lord God, deal with La Hire today as I would if I were Lord God and thou wert La Hire!' But expressions that make no better sense than these are to be found not as desperate cries but in what has the literary form of rigorous reasoning. Many times I have read an article that began like this: 'To avoid confusion over the use of the term 'God', we must distinguish between the role or status of being God and the being who in actuality is God: just as we distinguish between the office of being President of the United States and the man who now is President.' Whenever I have read this far I stop reading the article: I know it can contain nothing but sophistry and illusion. A president may exercise his office well or ill; but we must not think of the divine role as being played well or ill, not even as being played magnificently well. Creating and ruling the world is not a role that God plays, nor does God's goodness consist in playing this role supremely well; God, to repeat, is good just by being there. Creation was not a task set before God, so that he could be judged by his performance of the task; it is extremely comical when one of the participants in Hume's *Dialogue* adopts the attitude of a guest complaining of the design and running of an hotel. God had no need, and was under no obligation, to create. Least of all should we fancy that God was obliged to create the best of all possible worlds; 'creating the best of all possible worlds' is a non-task, like 'constructing the most sinuous possible curve'. However sinuous a curve is, one yet more sinuous is mathematically possible; however wise and loving and happy all the inhabitants of a world were, they could be wiser and more loving and happier, and there could be more of them. I learned this argument long ago from McTaggart. It is sad that, so long after he presented it, we find atheists proving that we do not live in the best of all possible worlds, as a means of refuting belief in God, and theists desperately trying to show that with all its evils this world is perhaps the best possible world.

We get a wholly false image of God's goodness if we think of it as a morally good disposition which regularly chooses good courses of action and rejects bad ones. People have sometimes conceived of a deity who *could* make bad choices but in fact avoids them! Praise of God in such a style deserves Aristotle's reproof: *phortikos ho epainos*, the praise is vulgar.

It should be immediately apparent that it is absurd to think of God as chaste or temperate or courageous. Virtues that regulate our human passions could be ascribed to God only if we also supposed God to have a range of emotional affects. Some theologians would indeed suppose this, appealing to scriptural expressions about God's anger and repentance and so on: though they would hardly appeal to Scripture so as to ascribe to God eyes and ears, nose and mouth, hands and feet! To show this error about God's emotions it is enough to remark that such a deity would be changeable in response to changes in the world: such a being is certainly not the true God. The question 'Who made God?' can be dismissed if God is changeless and eternal; but, as McTaggart pointed out, of two changeable beings it is quite irrational to allow the question 'How come?' for one and disallow it for the other.

One particular virtue of human beings that regulates emotional responses calls for separate discussion, not because the topic is of itself specially important but because nowadays people seem to worry excessively about whether

God's actions are governed by right feelings in this area. I am referring to the virtue of kindness to animals. Man is an animal, and sympathetic reaction to the feelings of other animals is part of our natural make-up; like other emotional reactions, this needs regulation by right reason. The casuistry of this is more difficult than many animal-lovers imagine: is one to sympathise with a seal, or with the codfish on which seals greedily feed, or with the orca (alias killer whale) which greedily devours seals? (Orcas in captivity are alleged to be friendly to man; and in a Cambridge newspaper I once saw an advertisement 'Adopt your own orca'!)

However men ought to feel or behave towards other animals, it is clearly foolish to look for the virtue of kindness to animals in the Divine Nature: for this is not, like human nature, a species of animality. To an uncaptive mind the living world manifests innumerable interlocking teleologies: innumerable phenomena of that world can be describe by the schema '*This* is so in order that *that* should be so.' This is not just a way of looking at the particular go of things once we have worked it out; it is heuristically valuable. Harvey, who thought of the circulation of the blood teleologically, gained much clearer understanding of it than Descartes, who on principle excluded teleological explanation from natural science. Quite recently J. Z. Young discussed the actual function of the pineal body in human beings by presuming at the outset that 'evolution' would not have let this organ survive if it were functionless. (As the poet said, some call it evolution and others call it God.)

The applicability of the teleological scheme in the living world does not mean that God sets himself ends and devises means to attain these ends: the ends in nature are not God's own ends. As Aquinas put it, *vult hoc esse propter hoc, sed non propter hoc vult hoc* (Ia q. 19 art. 5): God wills that there shall be a world in which teleological descriptions are found to work, but he needs no means to achieve ends, since as Xenophanes said, 'he does all things easily by the thought of his mind.'

Now this elaborately interlocked system of teleologically describable structures and processes does not manifest either deliberately contrived pain or devices to minimise pain. It is rather as if the Wisdom which produced the living world were indifferent to the pains incidental to the lives of animals. A bee's sting is as manifest an example of teleology as a hypodermic syringe; many poisons that plants generate serve no function in the economy of the plant itself, which goes on flourishing without the poison if grafted on the root of an allied non-poisonous species, but protect the plant against animals that would devour it. Predatory animals have structures plainly set towards killing and eating other animals; and the way a parasite reproduces often manifests most elaborate teleology. Is God there to be indicted for cruelty or callousness because things are so? Certainly not: God was violating no obligation towards animals; 'for thy pleasure they are and were created'. 'No sparrow falls without the Father' ought not to make us fancy a pang in a fatherly heart at a sparrow's fall; it merely means that the most trivial incident is within the order of Divine Providence. If the sparrow falls into the claws of my cat Gingerbold, that will be how God has been provident for him; like the lions in the Psalm, Gingerbold seeks his meat from God.

We may indeed in a sense ascribe to God virtues relating to actions: God is just and merciful and truthful and faithful to his promises. God's actions are changes that occur in creatures as he wills, not changes in himself; for example, God's promising something means that a man comes to know God's will, not that God's will changes.

As regards God's justice, we must realise that what are traditionally called *bona fortunae*, the goods of fortune, are in this life distributed neither equally nor in accordance with men's merits. There are indeed, as Hobbes observed, natural rewards that attend upon observance of the laws of nature and natural penalties that attend upon disobedience; but in this life such rewards and penalties are often overshadowed by good and evil fortune distributed solely according to the rules of a fair lottery. While there was a Prussian Kaiser of Germany, men in his cavalry regiments sometimes died from the kick of a horse: the deaths of cavalrymen in this way were distributed on a pure-chance pattern, as if each year every cavalryman drew a lot from an urn containing so many death-tickets. Innumerable such examples could be given. God is called by Thackeray the Ordainer of the lottery. By study of the theory of probability we can learn the terms of the lottery. Such chance events all fall within the order of Providence: the lot falls out of the lap, but the issue thereof is from the LORD. 'Justice as fairness' is ascribable to God in respect of the fairness of the lottery.

Statistics may thus teach us something of God's ways; but we ought not to use statistics in Sir Francis Galton's way to estimate the utility of prayer. It is natural to ask God for things as we ask our fellow-creatures for things; it would be ridiculous to consider whether asking fellow-creatures for things can be justified by statistics of how often petitions are granted. Nor is the granting by God of an individual petition likely to upset the fairness of the lottery; any single result is compatible with any statistical distribution.

Mercy is widely misunderstood nowadays, even as regards human relationships; Portia's great speech in *The Merchant of Venice* must often be read or heard with bleak incomprehension. Relaxing the penalty of a general law in a 'deserving case' is not mercy at all but mere justice: legislators cannot codify all the cases that may arise, and a judge may justly consider facts that the legislator could not provide for in advance. Mercy is mercy to the undeserving; and just this has become unintelligible to many people, including moral philosophers: the course of reasoning appears to be 'If it would be justifiable to punish someone, then if he were let off this would be *unjust*—in particular, unfair to others who are punished for the same offence.' George III, none too clear a thinker, is recorded to have refused a pardon to Johnson's friend the Rev. Dr. Dodd on the score that he would thus turn out to have *murdered* other criminals whom he did not pardon. In our day proposals for an amnesty for some class of offences have been resisted (and alas dropped) because there was a cry 'Not fair!'

Lewis's devils in *The Screwtape Letters* might well have counted it a battle won by the Philological Arm that the very word 'mercy' is erased from some modern versions of the Bible, replaced by 'steadfast love' or some such words:

though I notice that Jewish prayerbooks with Hebrew text and en face English translation still say 'His mercy endureth for ever', and the Jews ought to know. I have even seen the great text about doing justice and loving mercy perversely rendered with the adverb 'tenderly' replacing the noun 'mercy'!

I shall not here dilate upon this corrupt style of thinking. I merely point out that it is contrary to the constant teaching we find in Old and New Testament alike. By that teaching, God shows mercy to the utterly undeserving: pardon to the penitent, deferment of punishment even to the impenitent; that mercy is our only hope, and we forfeit mercy by being unmerciful to those who wrong us, and by seeking after false gods—'those who observe lying vanities forsake their own mercy' (Jonah 2.8).

As regards neither justice nor mercy does it make sense to wonder if God treats equals equally. All that anybody has comes as God's free gift: there is not even a conceptually prior status of equality with respect to which God might be supposed to treat equals equally or unequally.

Some readers may have been storing up the protest: 'Surely God's goodness consists in Love'. Granted: but we need to ask what is the object of God's love. A rather bad hymn might give us the idea that God could 'live and love alone'. On the other hand, some have thought that God needed to *create* other persons to love and be loved by. I am convinced that both views are wrong. McTaggart showed the difficulty of conceiving that an eternal person should even live without an Other; but this difficulty need not arise if there are several Divine Persons each of whom is an Other to each of the others. This is indeed standard theological language about the Blessed Trinity: each Person is Other to each, *alius* (someone else) though not *aliud* (something else). I can here go no further than to point out that Monarchian theology is not free from difficulties and certainly cannot claim to be on the face of it more reasonable than Trinitarian theology.

Spinoza said that he who loves God cannot desire that God should love him in return. This remark is often quoted with little understanding. To grasp what Spinoza means we have to look at his argument (*Ethics* V.19). Spinoza points out that though a man who loves God may have enhanced life and joy by thinking about God, God who is eternal and changeless cannot get such enhanced life and joy by coming to think about some man who loves God; nor is it possible, Spinoza adds, that a true lover of God should wish it were otherwise, since so to wish is in effect to wish that God were not God. Though the way of expressing this would be different, Spinoza's doctrine here is to be found also in Augustine and Anselm and Aquinas.

As I remarked earlier, it is often put about as Christian teaching that God loves all men equally and we ought to try to do the same so far as our frailty allows. But the authentic Christian doctrine is quite different: God the Father loves his only-begotten Son, the man Christ Jesus, more than he loves any other man, indeed more than all creation. Nor need we doubt, though it is scarcely our business to look for a rank-order of Saints, that the Father loves one Saint more than another; and Christ as man clearly had a special love for his Mother and for him called 'the Disciple whom Jesus loved', and did not love them merely on an equality with everybody else. And Aquinas lucidly explains that in our human loves there should be an order (*ordo caritatis*, IIa IIae q. 26). It is *not* virtuous to endeavour to love all our fellows equally, in particular to love enemies as much as friends.

God may be said to love better people more, because it is just his love that gave them existence and *makes* them better. Here again there is no model for us to try to follow: especially as we are such poor judges of who is really better, we ought not to set ourselves to love better people more.

In the *Summa Theologica* the article on God's goodness is placed by Aquinas in the middle of the articles where he is discussing *de Deo quo modo non sit*, the way God *isn't*. Thus far I have been largely concerned with refuting what I take to be errors: it is now time, as the song says, to eliminate the negative and latch on to the affirmative. But how shall I do that? As Thomas Hobbes said, the reward promised is one that 'we shall no sooner know than enjoy'.

I have latterly seen it argued that to all but a small class of intellectuals the idea of attaining beatitude by the vision of God's goodness is not specially attractive. But that vision is just what we are all made for, if by vice and folly we do not forfeit it: it is what Warren or Wrenno the weaver sought for as much as any Doctor of the Church:

> As for Wrenno the weaver, after he was turned off the rope broke with the weight of his body, and he fell down to the ground; and after a short space he came perfectly to his senses and began to pray very devoutly . . . and with that he ran to the ladder and went up it as fast as he could. 'How now?' says the Sheriff 'what does the man mean, that he is in such haste?' 'Oh!' says the good man: 'if you had seen that which I have just now seen, you would be as eager to die as I now am'. And so the executioner, putting a stronger rope about his neck, turned the ladder, and quickly sent him to 'see the good things of the LORD in the land of the living', of which before he had had a glimpse. (Bishop Challoner's *Memoirs of Missionary Priests*, London 1924, p. 343)

Index of Proper Names